SECOND ANNUAL REPORT

UPON THE

NATURAL HISTORY AND GEOLOGY

OF THE

STATE OF MAINE.

1862.

INTRODUCTION.

To the Senate and House of Representatives:

In accordance with the provisions of the resolve authorizing the continuance of the Scientific Survey, we have the honor to present herewith our Annual Report, describing the results of our explorations during the past season.

No change has been made in the organization of the corps of the survey; but we have been unable to employ all who were in the field the previous year. Mr. Houghton has not been in the service at all, and Mr. Packard only a few weeks; the rest have been hard at work in the field, some of us from the 10th of May to the 6th of October. Several persons of eminent ability have applied for the office of assistant to the survey, but we uniformly refused their services, and desire to accept no others until we have the means to employ aid in addition to those now on the list. Those who so kindly tendered us their services without charge shall be remembered first when we may have the means of remuneration in our hands.

On the 25th of March we conferred with His Excellency the Governor and the Secretary of the Board of Agriculture in reference to the plan of operations for the season. The instructions then received we have endeavored to carry out in the way now to be described. As before, we formed two parties, each of us either accompanying or having charge of one of them. The Naturalist, Dr. Holmes, spent a part of May, the whole of June and a part of July in exploring the distribution of the Lower Helderberg limestone, or the "marble layer" of Maine, in Aroostook county. About the first of July Mr. Fuller commenced to explore the marine zoology of the coast, under the direction of the survey, accompanied by Mr. Packard for a few weeks. He commenced his work near Eastport, and worked his way along gradually to Portland, having amassed a great number of facts and specimens.

He was at work dredging until the middle of September. Dr. Holmes explored the Natural History of the Kennebec and Androscoggin regions during the latter part of the season.

The Geologist, Mr. Hitchcock, was accompanied by Mr. Geo. L. Goodale (now Assayer to the State) during the whole season. About the middle of May they left Moosehead Lake in birch canoes, accompanied by Mr. O. White, of the scientific department of Amherst College, and explored the west branch of the Penobscot, the upper portion of the river St. John, returning by way of Churchill, Chamberlain and Chesuncook Lakes to Greenville. Mr. White became so much interested in the work of exploration, that at his own expense he explored the geology of Alleguash and Cancemgomoe Lakes, and presented us with a report upon them.

The Geological party spent the last of June and the whole of July in exploring the country near Penobscot Bay, and measuring a section from Eden to the Canada line in Somerset county. Then they spent more than a month in examining the country watered by the Schoodic Lakes and the St. Croix river, besides an important exploration of the iron ore and fossiliferous limestone of northern Aroostook county. The last work of the season was the exploration of the large lakes in Franklin and Oxford counties.

We propose to divide this report into three parts. Part I will embrace the observations that have been made in Natural History during the past year, in all the departments of Zoology and Botany. Part II will treat of whatever may have been learned respecting the rocks and fossils of the State since the publication of the Preliminary Report; while Part III will be devoted to the chemical portion of the general report. We shall not address your Honorable Assembly with each special report, but with this introduction present the several fragments in the most natural order possible. We have found it very difficult to devise an unexceptionable mode of presenting our materials, owing to their fragmentary character.

We regard the notices of iron ores in the State, and the description of the Botanical Map, as the most important practical results of the survey described in this report. The one may have an intimate connection with the welfare of our General Government, while the latter will be of value to those who propose settling in the best agricultural districts; and when the details are properly investigated, it will afford to the residents of all parts of the State the knowledge of the best fertilizers for each district.

We regard ourselves fortunate in being able to present communications from J. W. Dawson, LL. D., Principal of McGill College, Montreal, and from E. Billings, F. C. S., Palæantologist, of the Canada Survey, respecting new species of fossil plants and animals discovered in the State. These gentlemen kindly volunteered their aid in this department, and the former one visited the localities in person, thereby with his experienced eye gaining more information in a few days respecting this ancient flora of Maine than we could have done in as many weeks. We hope also to be able to present a valuable letter about the microscopic remains of animals and plants found in the "polishing powder" of our peat-bogs from Prof. L. W. Bailey, of the University at Frederickton, N. B., who has inherited in this department the skill of his lamented father, the late Prof. J. W. Bailey, of West Point.

We have been much gratified with the favorable reception our Preliminary Report has found, both within and without the State, and in foreign countries, as well as in the British Provinces on our own continent. You are earning for yourselves a reputation for wisdom, skill and liberality across the water. In a review of the report of last year in the *Daily Edinburgh Review* of July 7, 1862, your example is commended to the chiefs of the Geological Survey of Great Britain; not that science is not cultivated there, but that too little effort is there made to popularize scientific details and to present their practical bearings to the people. Of the fruits of the survey this writer says: "We can only say that they are well fitted to stimulate the chiefs of surveys on this side of the Atlantic." Also: "We cannot, too, but commend the wisdom of the Legislature's 'Resolve.' Great activity and enterprise are being thrown into agricultural and mining operations. Science might direct these energies. Geology might point out the connection between the characteristic rocks of a district and its soil," etc. "Zoology and Botany might also do much if their researches were set in popular and directly practical aspects. Maine has made the attempt, and the success is most marked. It will do a great deal more for the social comfort and morality of its people than the notorious 'Liquor Law' of that State ever could."

The estimate in which the report is held in the British Provinces may be seen in the following extract from the *Canadian Naturalist* for June, 1862: "The survey of Maine was commenced last year by Mr. Hitchcock, and his report shows a most praiseworthy diligence and an excellent combination of effort with others working in neighboring fields, along with great capacity for such work. The observations made and the fossils collected enable us for the first time to form just ideas of the parallelism of large portions of the rocks of Maine with those of New York, Canada, Nova Scotia, etc."

The *American Journal of Science and Art*, the leading scientific journal of our country, says in its issue for May, 1862: "It is worthy of remark that in these times of civil strife Maine has the courage to inaugurate a new scientific survey, while some other States are suspending work on surveys only partly completed. This is the more to the honor of Maine, inasmuch as in case of a foreign war she would be the first to suffer the liability of an invasion."

We have collected a large number of specimens illustrating the Geology, Zoology and Botany of the State, which we have accumulated at the rooms of the Portland Society of Natural History, preparatory to culling out suites of specimens for the State House and the higher literary institutions of the State. We are under great obligations to this society for the free use of their rooms, and especially because they are aiding wonderfully in promoting a scientific knowledge of the State by their publications and the opening of their collections to the use of the public. These publications assist the State materially, for they are filled with scientific details of great importance, which need not be repeated in our reports, and are accessible to all who wish to inspect them. The value of this society is well known to you, as in times past you have granted it aid. The society is now virtually paying back to the State all that it has received.

We must renew our thanks to the citizens of all parts of the State where we have travelled the past season, for their hospitality and earnest efforts to assist us in our labors. Last year we thought we had never seen people so obliging and generous as those who assisted us, but this year they have been still more courteous. In addition to those enumerated last year, we are under special obligations to the following gentlemen, as well as to many others whom we have not time to mention: Hon. Ira Fish of Patten, Joseph Pollard and Eben Trafton of Masardis, Daniel Stickney of Presque Isle, Hiram Stevens, William A. Sampson, J. W. Haines, Edward Fowler, Cyrus Estes and John B. Trafton of Fort Fairfield, Oliver Smith of Mars Hill, Milton Welch, Theophilus Cary and Z. P. Wentworth of Houlton, Hon. P. P. Burleigh of North Linneus, George H. Downing of No. 5, R. 3 of Aroostook county, ——— Sylvester of Parlin Pond, ——— ——— of Forks Hotel, (Forks of the Kennebec,) J. H. Eveleth of Greenville, Capt. Thomas Robinson of the "Fairy of the Lake," Seward Dill and Mr. Russell of Phillips, E. Darwin Prescott of Sandy River Plantation, D. M. Benjamin of East Livermore, Prof. D. T. Smith of Bangor, Messrs. Best and McAdam of the Woodstock Charcoal Iron Co., ——— Batchelder of Union, and Dr. John DeLaski of Vinalhaven.

Very substantial assistance was furnished the Geologist by the superintendents of most of the railroads in the State; insomuch that enough funds were saved by this means to authorize the excursion to Aroostook county when the valuable properties of the iron ore in No. 13, R. 6 were discovered. We do not see, then, but that the credit of this discovery—the most important yet made by the survey—is to be ascribed to those gentlemen who so kindly furnished these passes. From Edwin Noyes, the Superintendent of the Penobscot and Kennebec, and the Androscoggin and Kennebec Railroads,—from B. H. Cushman, the Manager and Superintendent of the Kennebec and Portland Railroad, and from W. W. Sawyer, the Superintendent of the Calais and Baring, and Lewy's Island Railroad, complimentary tickets for the season were received; while for occasional passes for particular trains over the Grand Trunk Railway and the Somerset and Kennebec Railroad we are indebted to the kindness of their superintendents.

To Mr. Sawyer, Superintendent of the Lewy's Island Railroad, we are not only indebted for season tickets for our whole party, but also for the use of special engines and a steamboat—the "Gipsey,"—which plies over Lewy's Long and Big Lakes. These favors assisted us very materially, both in time and money.

We cannot close without alluding to the wisdom your Honorable Assembly used in the selection of the commissioners to whom we are responsible, viz: His Excellency the Governor, and the Secretary of the Board of Agriculture. They have fully appreciated the value of scientific explorations in their instructions to us; they have been careful to see that the appropriations were expended to the greatest possible advantage; they have managed for us difficult questions of a pecuniary nature; have borne with our failings, and assisted us all that was in their power. If the explorations are not conducted so as to secure the greatest possible benefit to the State, it is not their fault.

With these preliminaries, we now present our reports.

Respectfully submitted,

EZEKIEL HOLMES,
C. H. HITCHCOCK

PART I.

REPORTS UPON THE ZOOLOGY AND BOTANY OF THE STATE OF MAINE.

DR. HOLMES' REPORT

ON THE FISHES OF MAINE, INCLUDING SOME OF THE ELEMENTARY PRINCIPLES OF ICHTHYOLOGY.

PART I.

To the Hon. Senate and House of Representatives
in Legislature assembled, January, 1863:

Gentlemen:—In accordance with the resolves passed by your honorable body at the last session, providing for a continuance of the "Scientific Survey of the State," I herewith submit the following report on the Ichthyology of Maine.

This report, I have divided into two parts, and you will find it somewhat anomalous in its plan as usually followed in such cases, inasmuch as it is not confined to a mere detail, or catalogue of the fishes which are found in the waters of Maine, but embraces also some of the more important elementary principles of the science of Ichthyology in general. My reasons for this, I trust, will, on a candid consideration of the subject in all its bearings, be fully appreciated and approved.

A dry, formal catalogue of the fishes found in our State and vicinity, with their technical names and synonymes, would interest the experienced scientific Ichthyologist for a few moments, but the people none at all.

As Naturalist to the survey, it becomes my duty to give in detail, as far as observation and facts will warrant, the natural history of fishes which frequent our coast, and streams, and lakes.

Now, the natural history of fishes treats "of their structure and form—their habits and uses—their classification," and territorial distribution. To do this, appropriate use must be made of the peculiar language—nomenclature, or terms and phrases used, both among practical fishermen, and scientific men in this department; for Ichthyology, like every other science, profession and occupation, has its own particular language. As a general thing, however, the terms and phrases which scientific men have, from time to time adopted as appropriately descriptive of Ichthyological subjects and facts, are, to the majority of us, "*heathen Greek.*" It is not strange that it should be so, for there has been little, or no effort to familiarize the people, whether old or young, with them. Books on this subject are comparatively rare and costly, and the few in existence are better calculated for those already adepts, than as guides to the first rudimentary principles, leading the young inquirer to a full knowledge of the science in all its branches.

Hence arises the fact that, although the fishermen of Maine are among the most enterprising, intelligent and shrewd of the people—are well versed in all the technicalities of practical seamanship, whether in calm or in storm, understand thoroughly the habits of most of the fishes they pursue, and are perfectly acquainted with the best modes of taking, curing and preparing them for dietetical, commercial and economical purposes, but very few of them could point out the true scientific distinctive characteristics between a *cod* and a *sculpin*.

As before remarked, they cannot be blamed for this, so long as the avenues to knowledge in this department of natural history are virtually closed against them. Indeed, I found it sufficiently difficult, during the writing of this report, to obtain some of the standard authors on Ichthyology, for the purposes of reference, and the clearing up of points on which the mind was in doubt.*

To obviate, in some degree, this difficulty, I have thought it advisable to incorporate into this report, some of the general elementary principles of Ichthyology, by which the subject matter might be more clearly elucidated and better understood. What I have given is designed to aid the student in his investigations of this branch of natural science, so that his "pursuit of knowledge"

---

* But one work of the kind could be found in the State Library, where, it is but fair to expect the student would be enabled to find all the more expensive works on every subject, and this belonging to the select class which bears the inscription, "*not to be taken from the Library.*" As further proof of the difficulty of obtaining elementary instruction in this department of natural history, Worcester's Dictionary may be cited. The publishers of that work boast of it as being the best of the kind in use for giving and defining scientific terms and phrases. On examination, you will find that, in Botany, Ornithology, and some other branches, nearly all the terms used are put down well explained and also illustrated by small neat wood cuts ; but for the strictly Ichthyological terms you will seek in vain. It is true a few of the more common ones may be found, but no wood cut illustrates them to the eye.

shall lie "under" less "difficulties" than they otherwise would be. In doing this, I have found it a difficult thing, as have undoubtedly many others who have essayed a similar task, to steer a middle course, so that what was written should not be so *dryly* scientific as to repulse and discourage the beginner—nor so purely elementary as to afford no interest to a deeply scientific man.

If some of the rising generation, who are fond of deep sea fishing, with lead and line—or of Isaak Walton's "gentle art" of angling by brook and lake-let, should be led by it to turn their amusement to incentives for more thorough research into the works of nature, as manifest in the "finny tribes," the writer will be amply compensated for his labor, and the State expenditure will not have been made in vain.

The Early Settlers of Maine drawn thither by its Excellent Fisheries. Progress and Importance of Them.

The fisheries of Maine constitute one of the oldest and most valuable interests of the community. No person can sail along our coast, or explore our bays and creeks, without being struck with the uncommon facilities offered for marine fisheries. No person can travel over our territory, and examine the innumerable lakes, rivers and smaller streams he meets with, without also being struck with the uncommon chances and advantageous localities offered by nature for interior fisheries.

These advantageous sites, and the facilities for such pursuits were quickly observed by the very first discoverers of Maine, and were among the principal inducements which drew so many pioneers and adventurers to our shores, and made them so persevering and determined to establish settlements on the coast and islands, notwithstanding the social privations that at first attended, and the Indian hostilities that soon surrounded them.

Pring, in 1603, and Weymouth, in 1605, both foresaw the uncommon advantages here offered for fishing and trading. Weymouth, during his voyage came to anchor near Monhegan island, and lay for a time in what he called "Pentecost Harbor," from which he sailed, as his historian says, "to enter a newly discovered river, swept by strong tides, and *enlivened with fish, some of which were seen great leaping above water judged to be salmon.*" Around the island where they first landed (Monhegan) they relate that *great lobsters—rock fish and plaice were fished—all the fish being well fed, fat and sweet.*"

The natives, whom they found there inhabiting the seaboard, obtained an easy and luxurious living from the ocean, and were expert fishermen, as well as bold and skilful hunters. Rosier, the historian of Weymouth's voyage describes those whom he (Weymouth) basely kidnapped and carried with him on his return to England, as peaceable, kind-hearted, generous, truthful and honest"—as "expert whalemen by profession, often capturing this mammoth fish(?) in our waters." The account which these captives gave of the resources of their native country in these things, heightened the already roused excitement in England respecting the newly discovered regions, and two years after (1607) we find Popham, with true British enterprise, planting a village of fifty houses to accommodate his colony on the shores of the Kennebec, and erecting a fort for their defence. The objects of this settlement were principally fishing in the adjacent waters, and trade with the natives. In all the early descriptions of our State, fishing and fisheries always stood out in strong relief. Speaking of "Norembega," it was described by Purchas as being an island at the mouth of a goodly river,* "*very fit for fishing * * * and that the region that goeth along the sea doth abound in fish.*"

The death of Popham, and the hostilities of the Indians, exasperated probably by the imprudence of his men, brought his settlement to an early destruction, but the inducements were too great to allow the country and fishing grounds to remain unoccupied, and accordingly we find the French settling Mount Desert in 1609, and a few years after (1614) we find the celebrated Capt. John Smith exploring along our coast with two ships. He, as did Weymouth, anchored at Monhegan. This island had been more or less a resort for fishermen since its discovery by Weymouth.

Smith made it the centre of his operations during the summer. "Whilst the sailors fished," says he, "myself with eight others ranged the coast in a small boat. We got, for trifles, 11,000 beaver skins, 100 martins, and as many otters, and the most of them within the distance of 20 leagues. We ranged the coast east and west much further."†

---

* Supposed to be Damariscotta, see Sewall's Ancient Dominions of Maine, p. 32.

† Sewall's Ancient Dominions, p. 175.

He returned to England in September following. Besides the furs, they carried home 47,000 dry and core fish made at Monhegan.

Plymouth Colony saved from Starvation by Maine Fisheries.

From this time the fisheries and settlements began to increase, proving a source, not only of profit to those who engaged in the business, but of food and life to many, especially to the Pilgrims of Plymouth, who obtained provisions from them to sustain their starving families, and without which relief, it is doubtful if they too would not have suffered the fate of Popham's colony and become annihilated. Some historians have stated, and many people believe, that the landing of the Puritan Pilgrims on Plymouth Rock, in 1620—an event whose anniversary is annually celebrated—constituted the first settlement in New England, and the germ of all its population and prosperity. Not so. Several settlements had been made and were in existence, years before, and to these were the Plymouth Pilgrims indebted for relief and timely succor. Monhegan and Damariscove, and Saco, and other places, were settled before that. In 1622, "thirty sail of vessels entered at Damariscove—which was now the granary of the embryo settlements of New England—whose name (Damariscove) an English corruption of Indian words signifying a "place for fish," indicates its early importance as a fishing depot.

The ship Swallow, from here, sent her shallop to Plymouth, and to Damariscove came Winslow of the Plymouth plantation (the Governor of the colony) to draw supplies for his settlement famishing on the shores of Cape Cod—who says—" I found kind entertainment and good respect, with a willingness to supply our wants —which was done as far as able—and would not take any bills for the same, but did what they could freely"—which certainly indicates that the inhabitants of Damariscove were a thrifty and generous people.*

A trading house was early established on Monhegan, where fish and furs were purchased and stored until shipped to the mother country. This house was broken up in 1626, when the goods, being offered for sale, Governor Bradford and Mr. Winslow of the New Plymouth colony, and Mr. Thompson of Piscataqua, went

---

* Sewall's Ancient Dominions, p. 105.

thither and purchased them. The moiety of the Plymouth planters being £400. The island was sold that year by Mr. Jennings of Plymouth, England, to the future Pemaquid patentees, and continued to be a favorite resort of fishermen.*

Sir Fernando Gorges, who had taken a deep interest in the discoveries in this country, and who had given a shelter and a home to the natives whom Weymouth had kidnapped, while they were in England, and listened attentively to their descriptions of their country, fitted out, in 1616, an expedition under the command of Richard Vines, Esq., to explore the country still more, with a view to settlement.

He came to anchor at a place which, in consequence of his wintering there, he called "Winter Harbor," a spot near the mouth of Saco river, which river had been previously described by Champlain, a French early voyageur and explorer, as being "three or four fathoms in depth, *and is well stored with* fish."

The Plymouth Colony Purchase a Fishery in Maine.

The Plymouth colony, finding the fisheries in this section of the coast much more productive than further south, purchased, in 1628, of Monquine, Sagamore of Kennebec, a large tract of country on both sides of that river *from Cusenock (Cushnoc) up to Wesserunskick.* It is conveyed by deed, which is still to be seen in the Register's office in Lincoln county, to William Bradford, Edward Winslow and others, in behalf of Plymouth Company. This grant was enlarged and confirmed to them in 1629-30. This patent gave to them the control of the fisheries and trading sections on that tract. The monopoly, or exclusive right to fishing in these waters, for one or more years, used to be sold to the highest bidder at the expiration of each lease, and thus the colony derived a profitable income from their domain. This was continued until the lands began to rise in value, when the colony sold out to a company of individuals known as the Plymouth Company, or proprietors.

The Fishing Business becomes more Systematic and Increases Commerce.

As population increased, the fishing business began to assume more systematic arrangements and regular business forms. In

* Folsom's History of Saco.

1636, Folsom observes* that fishing was the most common occupation, as it was both easy and profitable to barter the proceeds for corn from Virginia and other stores from England.

The trade with the planters of Massachusetts soon became considerable. At this time, Mr. Vines had a consignment of bread and beef from that quarter. Jocelyn remarks that "Winter Harbor is a noted place for fishes; here they have many stages." He describes the mode of pursuing this business in the following manner. "The fishermen take yearly on the coast many hundred quintals of cod, hake, haddock and pollock, and dry them at their stages, making three voyages in a year. They make merchantable and refuse fish, which they sell to Massachusetts merchants; the first, for 32 ryals ($4.00) per quintal—the refuse for 9 and 10 shillings ($2.00 to $2.25.)

The merchant sends the fish, the first to Lisbon, Bilboa, Marseilles, Toulon, Bourdeaux, and other cities of France—to Canaries, pipe staves and clapboards; the refuse fish to the West Indies for the negroes. To every shallop belong four fishermen—a master, a steersman, a midshipman and a shoreman, who washes the fish out of the salt and dries it upon hurdles pitched upon stakes, breast high, and tends their cookery. They often get in one voyage eight or nine barrels a share a man. The merchant buys of the planters, beef, pork, peas, wheat, Indian corn, and sells it to the fishermen."

Thus, step by step, grew the fisheries of Maine into a business of magnitude and importance, and in proportion as the fisheries prospered, grew the maritime portions of our State in population and corresponding strength. During the vexatious and bloody Indian wars which soon after this began, and continued with but occasional cessation for more than a hundred years, and kept the settlers in constant watch for the safety of their property and lives, their principal reliance for sustenance and supply of other comforts was on their fisheries. But for these, many a family, and many a hamlet on the seaboard, would have been reduced to actual starvation. It was this never failing resource which gave them life and energy and the means to resist the assaults of their wily foes, until they finally conquered and exterminated them.

---

* History of Saco, p. 37.

Laws began to be called for to Regulate the Fisheries.

As population increased, and settlements multiplied, and competition on the sea, and improvements on the land become more prevalent, separate interests of individuals and communities began to clash, and we find the government, whatever it was at the time, often called upon to make laws and regulations for restraining encroachments on the one hand, and resistance on the other. As governments therefore, whether proprietory or colonial, began to be established for the preservation or the protection of property, the expenses accruing thereby, were met by some sort of taxation. This was often paid in fish.

Thus, we find that as early as 1684, when Pemaquid, and the "region round about" were formed into a "Ducal State," under the Royal Grant to the Duke of York, a duty, or tax was put upon the fishermen for the purpose of revenue. "All vessels, not of the Ducal State, were ordered to pay into public revenue—if a decked vessel, *four quintals*—if an open boat, *two quintals of merchantable fish.*" In 1732, we find that the people of Saco met with trouble in regard to their river, or interior fishery, by reason of the practices of the officers and soldiers of the "Truck-house" (Block-house or fort,) and the town voted "that Mr. John Gordon lay a memorial before his Excellency the Governor, and the Honorable Council of the difficulties that the inhabitants and residents on Saco river sustain by those in the public pay of this Province, by setting of nets and drifting with nets to the disturbing of the common course of the fish, and any other difficulties that are not for the honor of this Province."*

From this date to the present time, legislative enactments have been frequently called for, and an examination of our statute books will prove the fact, that if the natural history of fishes had been more thoroughly understood, some of the laws would have been very differently framed and much better executed.

Massachusetts commences Legislative Encouragement to Fisheries.

Previous to this (in 1639) Massachusetts, whose government had become more stable than that of some of its sister provinces, seeing the great importance of this branch of industry, began a system of encouragement to it by legislative protection. It was provided by

* Folsom's History of Saco.

law, that all vessels and other property employed in "taking, making and transporting of fish, should be exempt from duties and public taxes for seven years; and that all fishermen, during the season of their business, should be dispensed from military duty. This so stimulated the business that in 1641, the mariners of that colony followed the fishing so well, that there was above *three hundred thousand* dry fish sent to market."*

They become a Practical School for Seamen.

This system of encouragement also resulted, not only to the increase of the fisheries, but also, by consequence, led to the business of ship building, and to a more extended commerce. It also proved itself to be one of the very best practical schools for seamanship and a source from which, in process of time, the merchant service derived their most expert and skilful sailors and shipmasters. At the breaking out of the revolution, these men formed the nucleus of our navy, which, though small, did essential and effective service in the cause of their country, and in the war of 1812, constituted a formidable rival to the greatest naval power then on earth.

The United States adopt and continue a System of Encouragement.

When peace was established and the Federal Congress was organized under the new constitution, they remembered this service, and to aid in continuing a school productive of such good results, they adopted a system of national bounty to those fishermen who embarked in the business to a certain extent. This bounty continues in operation to this day, and under its provisions more than three millions of dollars have been received since its commencement by the hardy fishermen of Maine alone.

Aided by the stimuli and encouragement, which we have mentioned, and the profitable character of the business itself, the Maine fishermen have continued to increase from the humble beginnings we have related, until in 1850, the product of the shipping so employed amounted to $569,876; the capital employed to $496,910; and the number of hands to 2,783, being third State in rank.

One would be lost in trying to estimate the myriads of fishes, of various kinds, that have been drawn from our waters, both sea and

---

* Palfrey's History of New England, vol. 11, p. 55.

inland, since the day that Weymouth found them so plentiful "fat and sweet" at Monhegan, until the present time. Every year, and every season since, has the sea yielded to the industry of the adventurous fisherman, a life-giving, exhaustless harvest, and that harvest has been as continually replenished and nurtured in the coral fields of the ocean by an unseen but Almighty hand.

CLASSIFICATION.

By classification in natural history, is meant the arranging, or grouping into classes, orders and genera, the several objects to be described, which have properties and characteristics similar and common to each other. Some system of this kind was found necessary at a very early day, and some not very successful attempts of the kind were made by the older naturalists. The most successful systematizer in natural science was Linneus, the Swedish Philosopher, whose researches and writings opened a new era in studies of this kind. His keen observation and talent of discrimination enabled him to develope a more simple, and at the same time more practical arrangement, than any writer before him had done. By his writings and lectures he rendered all the departments of the science popular, and awakened an enthusiasm among the scientific of every nation, that has continued to this day and been of incalculable benefit to mankind.

A theory had long obtained belief that God had created every thing in nature according to a natural gradation, or natural orders; or, in other words, that there is a continuous series, or chain of creation from the least to the greatest, and from the most simple to the highest and most complicated organizations—that a perfect knowledge of the whole range would enable us to place any particular object under consideration, unerringly into the exact place or link in the great chain of created beings or things—that by searching out the resemblances and affinities of the objects in question, they could all be grouped into true natural orders, each order sufficiently definite and distinct to warrant a specific name, or designation, and yet its extremities or borders (so to speak) so nearly resembling those on either side as to enable the student to see and point out where they meet and blend into each other.

It is evident, that, in order to designate and accurately describe these natural orders (admitting their existence,) a perfect knowledge of the whole would be needed. This is impossible for one man to attain.

Linneus, therefore, while he conformed as far as he was able to what he considered natural orders, thought it advisable to adopt what has been called an artificial system, establishing classes, orders, genera and species upon certain organs which are always present and uniform in form, position and structure. By these means the study and description of natural objects have been admirably systematized and facilitated. These important aids have wonderfully promoted investigation and research, and increased the knowledge of natural history in all its branches.

Different individuals, following their taste and "bent of their genius" have devoted themselves to different departments of the science. Some making Botany, some Ornithology, some Entomology, some Ichthyology, others Mineralogy and Geology, and so on, their speciality, thus becoming adepts in their favorite science. New discoveries have brought new changes in grouping or classification, and it will be found in tracing back the progress that has thus far been made, that the arrangement of the present day is very different, in many important respects, from the systems adopted and promulgated by our predecessors—each of which had its day. In Ichthyology this has been especially the case, and as many of the terms, and some of the orders and genera are still used by modern writers on this science, it may be useful to look briefly over some of the several classifications which the older Ichthyologists adopted in their works. We shall thus more understandingly pursue the science as arranged by the more modern writers. As long ago as 1555, Belon, a French physician, wrote a work entitled "*The nature and diversity of fishes with their portraits.*" He was the first who divided or grouped fishes into two grand divisions of cartilaginous and osseous fishes.

Willoughby and Ray were among the earliest authors who reduced the study of Ichthyology to something like a systematic arrangement. Their work made its appearance in 1686, in four Books, folio.

WILLOUGHBY AND RAY'S SYSTEM.

II. BOOK—cetaceous fishes.

III. BOOK—concerning cartilaginous fishes. This is divided into three sections.

IV. BOOK—concerning oviparous fishes which have spines. This is divided into five sections.

Ray, in 1713, published an improvement on this system in a work entitled "A Synopsis of Fishes."

A classification by Samuel Dale, was published in 1739, who made several improvements upon former arrangements, showing that there had been some advance in the science. He based his method upon the respiratory organs, dividing them into two grand classes.

DALE'S SYSTEM.

CLASS I.—Fishes breathing by gills (*Branchiis respirantes,*) and having but one ventricle to the heart. This class was divided into two orders, viz:

I.—OVIPAROUS.

II.—VIVIPAROUS.

These orders were sub-divided into families, genera and species.

CLASS II.—Fishes breathing by lungs (*Pulmone respirantes,*) and having two ventricles to the heart as the whales.

About the same time, Linneus, who had collected the manuscripts and writings of his deceased friend, Artedi, published two volumes entitled *Philosophia Ichthyologiæ,* (Philosophy of Ichthyology.)

Artedi was a true and thorough naturalist for that day. He had confined his researches to the natural history of fishes with indefatigable zeal.

He had systematized and arranged the classification of the science in accordance with the advice of Linneus—established new genera, gave rules for their formation and descriptions, assigned to them their proper limits, and gave the methods of separating different species, so as to render their descriptions clear and simple. So correct was he in the description and distribution of genera, that many of those he established are still retained in the nomenclature of the science on his authority to the present day.

Artedi's System.

Class I.—Fishes with tails placed perpendicularly (*Pisces cauda Perpendiculari.*)

This class is divided into sections, orders and genera, as follows:

A.—Fish with bony rays to the fins and bony gills.

1.—Fins unarmed, *Malacopterygii*, (soft fins.)

	Genera.
Fish with one fin almost in middle of back.	1. Syngnathus, *Pipe fish.*
	2. Cobitis, *Loche.*
	3. Cyprinus, *Carp.*
	4. Clupea, *Herrings, &c.*
	5. Argentina, *Argentine, Silver fish.*
	6. Exocætus, *Flying fish.*
One fin nearly in middle of back, and one adipose fin near the end of back.	7. Coregonus, *White fish.*
	8. Osmerus, *Smelt.*
	9. Salmo, *Salmon and Trout.*
One fin in further end of back.	10. Esox, *Pickerel.*
	11. Echeneis, *Sucking fish.*
One or more fins extending whole length back.	12. Coryphæna.
	13. Ammodytes.
	14. Pleuronectes, *Halibut, Flounders.*
	15. Stromateus.
	16. Gadus, *Cod, Haddock.*
	17. Anarrhichas, *Wolf fish.*
One long fin scarcely distinct from the tail.	18. Muraena, *Eel.*
	19. Ophidion.
One very small fin, or none at all, on extreme part of back.	20. Anableps.
	21. Gymnotus, *Electrical Eel.*

2.—Fish with bony fins, some of which are spines, *Acanthopterygii*, (thorny fins.)

Fish with smooth heads.	22. Blennius, *Blenny or Slime fish.*
	23. Gobius, *Goby.*
	24. Xiphias, *Sword fish.*
	25. Scomber, *Mackerel.*
	26. Mugil, *Mullet.*
	27. Labrus, *Bass.*
	28. Sparus, *Gilt Head.*
	29. Sciæna.
Fish with rough heads.	30. Perca, *Perch.*
	31. Trichiurus, *Weever.*
	32. Trigla, *Gurnard.*
	33. Scorpæna.
	34. Cottus, *Bull-head.*
	35. Zeus, *Dory.*
	36. Chætodon, *Beard-teeth.* [*belly.*
	37. Gasterosteus, *Stickle-back, Bony-*
Fish destitute of bony gills; with gill-coverings.	38. Balistes, *File fish.*
	39. Ostracion, *Coat of mail, Trunk fish.*
	40. Cyclopterus, *Sucker.*
	41. Lophius, *Angler.*

B.—Fish with fins having cartilaginous rays hardly distinct from a membrane.

Chondropterygii, cartilaginous or leathery fins.	42. Petromyzon, *Lamper Eel.* 43. Acipenser, *Sturgeon.* 44. Squalus, *Sharks.* 45. Raja, *Skates.*

Class II.—Fish having tails placed horizontally, including whales.

We have thus given the system of Artedi in detail, because, although Willoughby and Ray, by their improvements in classification of this branch of natural history gave it a more respectable position among the sciences than it before had, Artedi placed it on a firm basis and gave permanency to its nomenclature.

Other writers on this subject followed from time to time and added, by their researches and discoveries, to the fund of knowledge on these matters. Among them, Klein published an interesting work in 1740, on the "Natural History of Fishes," in which he endeavored to improve upon Artedi.

He increased the number of genera partly from the number of newly discovered fishes till then undescribed. He divided them into three grand orders, viz:

I.—Cetaceous or whales.

II.—Fishes with concealed gills as in Lamprey Eels.

III.—Fishes with open gills. These were again separated into many sub-divisions or groups, and then again into genera.

Artedi's arrangement, however, continued to take the lead until 1766, when Linneus himself, who at first had adopted his friend Artedi's system, came out with a new one of his own, which then, from the popularity of its author as well as from its simplicity, kept the ascendancy for several years.

Linneus' System.

Order I.—Apodes—fishes having no ventral fins.

Order II.—Jugulares—fishes having their ventral fins placed before or forward of the thoracic or pectoral fins.

Order III.—Thoracici—fishes having the ventral fins placed directly below the pectorals.

Order IV.—Abdominales—fishes having their ventrals placed behind the pectorals.

This was a very simple and concise arrangement, but nevertheless a purely artificial one, and by following it out fishes of very diverse form and characteristics were brought into the same order, as the eel and the sword fish for instance, individuals having but few properties similar or in common. He omitted the cetaceous tribe (whales, &c.,) altogether from the arrangement; placing them, inasmuch as they breathed air with lungs, among the mammalia.

The assertion from such high authority that a whale was not a fish, made some stir at the time, but its correctness has long since been established.

Ichthyology now began to be pursued as a well founded science, and, as a pursuit, contributing largely by the researches and discoveries of its followers, indirectly at least, to the comfort and wealth of the people by its suggestive aids to the practical economy and wants of mankind.

Fisheries began to be established more understandingly and to better advantage as a branch of national industry, than they had hitherto been, and that too in proportion as the knowledge of the habits and instincts of fishes became better known and more widely disseminated. Other writers upon this subject appeared from time to time, some of whom adopted the classification of Linneus wholly, and some only in part,* but his system kept the ascendancy until Cuvier, the celebrated French comparative Anatomist and Naturalist introduced his method, based in part upon differences of anatomical structure.

Cuvier's System.

He divided, as did Belon, all fishes into two great divisions, viz: those whose skeletons were made up entirely of bone which he called Osseous fishes: The other included those whose skeleton or frame work instead of being composed of bone was principally made

* Among them, Gronovius, a contemporary and friend of Linneus, published his work (*Museum Ichthyologicum*,) a few years before the latter had made public his system. He adopted Artedi's two natural orders of fishes with horizontal tails and those with tails in a perpendicular position, while his other characters were derived from those of Ray and Linneus. Brunich, in 1771, published a work on the "Principles of Zoology" (*Zoologia Fundamenta*,) in which he united as far as he could, the natural system of Ray with the artificial system of Linneus. Prof. Gowan followed, adopting Brunich's method. Scopoli, in a work published in 1777, adopted a new method which was never followed. Bloch, in 1785, published a work with excellent plates. It was written in French and German. He followed the Linnean system. Bonnaterre, who wrote the article on Ichthyology, in that great work, the *Encyclopedie Methodique*, in 1788, also adopted Linneus' method.

up of cartilaginous matter. These he called cartilaginous fishes, such as the Lampreys or lamper eel—some of the sharks, &c. His next sub-divisions were based upon the structure of the gill apparatus and the Linnean plan based on the position of the fins. These again divided into families or groups dependent upon the formation of mouth or other appendages. A tabular view of his first formed system would read thus:

Cartilaginous fishes.	Fixed branchiæ or gills chondropterygii.	Round mouth at end of nose.	
		Transverse mouth under snout.	
	Free branchiæ with branchiostegi or gill covers.	do. do. do. teeth.	
		do. do. do. no teeth.	
		Mouth at end of nose; no teeth.	
		do. do. teeth.	
		Bones of jaw answering for teeth.	
		Mouth very wide; number small teeth.	
Osseous fishes.	Apodes	Mouth at end of nose.	
		Mouth under the nose.	
	Jugulares	Head unarmed.	
		Head armed.	
	Thoracici	Dorsal fin partly spinous; head armed.	
		do. do. do. head unarmed.	Two dorsals.
			One do.
		Bones of the jaws naked; used as teeth.	
		Two eyes on the same side.	
		Body very long.	
		A furrowed disc on the head.	
	Abdominales	No operculum to the branchiæ (gills.)	
		No teeth.	
		Sharp teeth; no cirri or beard.	
		Head depressed; cirri.	
		Spines free on the back.	
		Mouth at the end of the nose.	

It will be seen on careful comparison that the two first divisions are those of Belon; the secondary characters are derived from Artedi and Linneus; the third are groupings of his own, which he has since improved upon, establishing several natural families which have been used by subsequent writers on Ichthyology. He subsequently, with Valenciennes, published a work on the "Natural History of Fishes," in 13 volumes. I have given his improved system on 33d page.

Lacepede's System.

A few years afterwards (1803) Lacepede published his celebrated work, "The Natural History of Fishes," in five volumes. From the accumulations of facts in the several authors that had preceded him, and the fact of his having access to the best museums or cabinets of natural history in Europe, he was enabled to make a very valuable work, and one which for many years was considered as standard authority. The following is a tabular statement of the classification which he adopted by which it will be seen that the first or primary divisions are those of Belon—the second his own, and the third those of Linneus :

Sub-Classes.		*Divisions.*	*Orders.*
CARTILAGINOUS FISHES.—The spine composed of cartilaginous vertebræ.	1.	1. No operculum nor branchial membrane.	1. Apodal. 2. Jugular. 3. Thoracic. 4. Abdominal.
	2.	2. No operculum, but a branchial membrane.	5. Apodal. 6. Jugular. 7. Thoracic. 8. Abdominal.
	3.	3. An operculum; no branchial membrane.	9. Apodal. 10. Jugular. 11. Thoracic. 12. Abdominal.
	4.	4. An operculum, and a branchial membrane.	13. Apodal. 14. Jugular. 15. Thoracic. 16. Abdominal.
OSSEOUS FISHES.—The spine composed of bony vertebræ.	5.	1. An operculum and a branchial membrane.	17. Apodal. 18. Jugular. 19. Thoracic. 20. Abdominal.
	6.	2. An operculum; no branchial membrane.	21. Apodal. 22. Jugular. 23. Thoracic. 24. Abdominal.
	7.	3. No operculum, but a branchial membrane.	25. Apodal. 26. Jugular. 27. Thoracic. 28. Abdominal.
	8.	4. Neither operculum nor branchial membrane.	29. Apodal. 30. Jugular. 31. Thoracic. 32. Abdominal.

Practical application of these Systems as aids to Study.

We will not spend any more time in the review of the several systems of classification which have heretofore been adopted by the several writers in question. They are all of them based partly upon the existence or non-existence of certain organs without reference to their structures or uses. This, as has been remarked, while it did much better than none, and aided, not only the writer in establishing something like method in his work, also aided the student in his researches or studies, as far as merely ascertaining the place in the system of the individual specimens found, and the names given them. This may be thus illustrated in reference to either of the foregoing systems given. Take for instance the last named one (Lacepede's.)

Suppose the student to have before him the five volumes of that author, and has also a fish, the name and distinctive characteristics of which he is desirous of knowing. If there had been no more system in classification than is found in Pliny's writings, he would have to read the volumes all over in course, until he came to a description that tallied with the specimen in hand; but by the aid of the classification adopted, it will be a much less difficult task to find its name and description here.

In the first place, he will look to its skeleton or frame, and ascertain whether it be cartilaginous or bony. He finds it bony—it therefore belongs to the osseous class. The ascertaining this fact abridges his labor materially. He may pass over the pages describing cartilaginous fishes and confine himself to the description of osseous fishes only. Again he will look over the gill coverings, and observe the appearance of those organs. Suppose he finds that it has an operculum and branchial membrane. On reference to the arrangement it will be seen that it belongs to the sixth division. His examination will therefore be still more curtailed, it being unnecessary to search for the description in any part of the volume than that which treats of fishes belonging to the sixth division. Next, he will turn his attention to the fins, particularly to the ventrals. It either has or has not any ventrals. If it has none, it belongs to the apodal order of the sixth division, and the further search will be directed to that part of the work treating of such. If it has ventrals, the position of them will determine the order—if they are placed before the pectoral fins it comes under the order of jugulares—and will be found in the description of that order, and so of the others.

Remarks on the difference between Artificial and Natural Orders.

It may be asked, if such classifications answer the purpose of convenient arrangement?—if they aid the author in giving method and system to his work?—if they aid the student in his research into the distinctive points of the subjects under examination? Why is it not sufficient?—and why should it be stigmatized as artificial? For the mere and single purpose of descriptive aid, it is enough. It is called artificial, because nature has never been guilty of grouping animals so diverse into the same order as these arrangements do—a proof that they are not arranged according to the natural order of things. We alluded to this fact while speaking of the system adopted by Linneus, which brought into one of his groups or orders, the eel and the sword fish, animals exceedingly different in form, anatomical organization and habits. Such classification does not, therefore, accord with nature, and will not satisfy the student who wishes to investigate the harmony displayed in God's work, and realize if possible the range and extent of the natural orders as they came from the hand of the Creator. The nearer we can come to this—the more clear and satisfactory and practically useful will be our views and knowledge of such things. The deeper man penetrates into the arcana of nature the more evidence does he find of method and system—of a beautiful and harmonious classification and skilful connection of manifold groupings, throughout the range and chain of animal existence, and indeed of all organic life. To follow nature, and to unfold as far as possible this classification, and the wise adaptations of means to ends and to render the discovery of these facts applicable to the practical business of our lives, has been, and still is, the legitimate work and desire of the true naturalist.

Modern Classification more conformable to Natural Orders.

Hence, modern Ichthyologists, using the experience of their predecessors and profiting by facts almost daily brought to light are establishing an entirely different classification, one much more conformable to natural orders, based on the unvarying anatomical structure, and physiological functions of the general organs of the fishes hitherto discovered.

The classification I *at present* follow in the description of the fishes found in Maine, is the arrangement as given by Dr. Girard in his general report of the fishes found during the exploration of the Pacific Railroad route. One still more modern may be followed in the final report.

This classification comprises ten orders. Each order is grouped into families, and the families into genera and species.

Enumeration of the Orders.

I. ACANTHOPTERI—*Thorny fins.* This order is made up of fishes that have one or more dorsal fins. If more than one, the rays of the forward one are stiff, sharp, inarticulated bony spines. If there be only one fin the anterior portion has spine rays. The remainder have soft articulated rays. The common perch affords a good illustration of this order.

II. Anacanthini—*No spines, thornless fins.* The fins of the fish have no bony spiny rays. The general anatomical structure, however, is similar to the preceding. The cod-fish is one of the representatives of this order.

III. Pharyngognathi—*Throat bones united.* Fishes of this order have the inferior pharyngeal bones united into one piece. The tautog belongs to this order.

IV. Malacopteri—*Soft fins.* The fishes of this order have their fins made up of soft articulated rays. The trouts are good representatives of this order.

V. Plectognathi—*Soldered jaws.* Fishes in this order have the outer or premaxillary bone and the jaw united into one continuous immovable bony piece. The balloon fish belongs to this order.

VI. Lophobranchii—*Tufted or crested gills.* This order comprises fishes that have their jaws united into a tube or pipe and have tufted or crested gills, a small fish found in the Hudson river, called the river sea-horse (Hippocampus,) illustrates this order.

VII. Ganoidei—*Plated Scales.* Fishes in this order have their bodies covered with enamelled plate like, or shield like scales. The sturgeon is a good example of this order.

VIII. Holocephali—*Solid heads.* This order is made up of the few fishes that have the jaw bones and the bones of the head all united into one. The very rare fish called the Northern Sea Monster (Chimera monstrosus,) represents this order.

IX. Plagiostomi—*Skew mouths.* Fishes of this order have their mouths transverse to their heads. The sharks and scates, &c., belong to this order.

X. Dermopteri—*Skin fins.* The fins of fish in this order are of a skinny or cartilaginous texture. The lamper eel affords a good illustration of them.

A more full explication of these orders and of the families into which they are divided, together with a description of the genera and some of the species of the fishes which have been found thus far in our waters, will be given in part II of this report.

SYNOPSIS OF THE FISHES OF MAINE, IN PART.

ORDER I. ACANTHOPTERI—*Thorny fins.*

Family.	
Blennidæ.	Gunellus mucronatus, *Butter fish.* Zoarces anguillaris, Blenny, *Eel shaped.* Sticheus subbifurcatus. Annarrhicas vomerinus, *Wolf fish.*
Lophidæ.	Lophius Americana, *Angler.*
Batrachidæ.	Batrachus tau, *Toad fish.*
Percidæ.	Perca flavescens, *Yellow perch.* Labrax lineatus, *Striped bass.* Labrax rufus, *White perch.* Pomotis vulgaris, *Bream.* (I think 3 species of them.)
Scorpenidæ.	Sebastes Norwegicus, *Norway haddock.*
Gastoridæ.	Gasterosteus DeKay,(?) *Stickle back.*
Sciænidæ.	Otolithus regalis, *Squeteague.*
Sparidæ.	Pagrus agyrops, *Porgee.*(?)
Scombridæ.	Scomber vernalis, *Mackerel.* Argyreiosus unimaculatus, *One-spot Dory.* Thynnus secundo-dorsalis, *Horse Mackerel.* Cybium maculatum, *Spotted Mackerel.* Temnodon saltator, *Blue fish.* Rhombus anacanthus, *Skip Jack.*
Atherinidæ.	Atherina notata, *Dotted Silver-side.*
Triglidæ.	Dactylopterus volitans, *Sea Swallow.* Acanthocottus variabilis, *Greenland Sculpin.* Acanthocottus Virginianus, *Common Sculpin.* Aspidophorus monopterygius, *Single Fin Bull-head.* Cryptocanthodes maculatus, *Spotted wry mouth.* Cryptocanthodes inornatus. Hemitripterus Acadianus, *Deep-water Sculpin.*

ORDER II. ANACANTHINI—*No thorny rays.*

Gadidæ.	Morrhua Americana, *Cod fish.* Morrhua aeglefinus, *Haddock.* Morrhua pruinosa, *Tom Cod.* Merlangus purpureus, *Pollock.* Phycis Americanus, *White Hake.* Phycis filamentosus, *Squirrel Hake.* Brosmius flavescens, *Cusk.*
Ophididæ.	Ophidium marginatum,(?) *Freshwater Cusk.*
Pleuronectidæ.	Hippoglossus vulgaris, *Halibut.* Pomabopsetta dentata, *Gill.*

ORDER III. PHARYNGOGNATHI—*Pharyngeal bones united.*

Family.	
Labridæ.	Ctenolabrus ceruleus, *Cunner or Connor.* Tautoga Americana, *Tautog.*

ORDER IV. MALACOPTERI—*Soft fins.*

Siluridæ.	Pimelodus atrarius, *Horned Pout,* (perhaps two species.)
Cyprinidæ.	Cyprinus auratus, *Golden Carp,* (introduced.) Leucosomus Americanus, *Shiner.* Catostomus communis, *Sucker.* Catostomus gibbosus,(?) *Chub.*
Cyprinodontidæ.	Fundulus pisculentus, *Minnow,* (2 species.)
Esocidæ.	Esox reticulatus, *Pickerel.*
Salmonidæ.	Salmo salar, *Salmon.* Fario fontinalis, *Brook Trout.* Fario erythrogaster,(?) *Red-bellied Trout.* Fario Sebago, *Girard.* Fario confinis,(?) *Togue or Lake Trout.* Fario tsuppitch,(?) *Salmon Trout.* Fario ——,(?) *Blue-back Trout.* Osmerus viridescens, *Smelt.* Coregonus albus, *White fish.* Coregonus clupeiformis,(?) *Shad-Salmon.*
Clupeidæ.	Clupea elongata, *English Herring.* Alosa præstabilis, *Shad.* Alosa tyrannus, *Alewife.* Alosa menhaden, *Menhaden.* Alosa cyanonoton, *Blue back.* Engraulis vittata,(?) *Anchovy.*
Scomberesocireæ.	Scomberesox storeri,(?) *Bill fish.*
Anguillidæ.	Anguilla Bostoniensis, *Common Eel.*

ORDER V. PLECTOGNATHI—*Soldered jaws.*

Balistidæ. Gymnodontidæ.	Not certain about genera of either family being found in Maine.

ORDER VI. LOPHOBRANCHII—*Bearded or tufted gills.*

Hippocampidæ. Syngnathidæ.	Not certain about genera of either family being found in Maine.

ORDER VII. GANOIDEI—*Enamel or plated scales.*

Sturionidæ.	Acipenser sturio, *Sturgeon.*

ORDER VIII. HOLOCEPHALI—*Solid heads.*

Chimaeridæ. { Probably none of these in Maine.

ORDER IX. PLAGIOSTOMI—*Oblique mouths.*

Sub-order I. Squali. { Genera in one or two families (*sharks*) are found in Maine, but am not well posted yet about them.

Sub-order II. Rajai. { Ditto of the rays, and skates, and flounders.

ORDER X. DERMOPTERI—*Skin fins.*

Petromyzontidæ. { Petromyzon marinus, *Lamprey, or Lamper Eel.*

—

CUVIER'S LAST SYSTEM.

On page 26th was given Cuvier's system as first promulgated. He afterward enlarged and arranged it in the following manner:

FIRST DIVISION—BONY FISHES.

ORDER I. ACANTHOPTERYGII.

Family Percidæ, (Perch family.)
Hard cheeks.
Sciænidæ, (Maigre family.)
Sparidæ, (Sea Breams,)
Menidæ,
Squammipennes, (scaly fins.)
Scomberidæ, (Mackerels.)
Tænidæ, (Ribbon shaped.)
Theutyes, (Lancet fish.)
Pharynginæ labyrinthiformæ, (Pharyngeal labyrinths.)
Mugilidæ, (Mullets.)
Gobiodæ, (Gobys.)
Pectorales pedunculati, (wrists to pectoral fins.)
Labridæ, (Rock fish, thick lips.)
Fistularidæ, (Pipe mouths.)

ORDER II. MALACOPTERYGII ABDOMINALES. *Soft or jointed fins. Ventral fins behind the pectorals.*

Family Cyprinidæ, (Carp family.)
Esocidæ, (Pickerel or pikes.)
Siluridæ, (Sheat fish or smooth skins.)
Salmonidæ, (Salmons and trouts.)
Clupeidæ, (Herrings.)

ORDER III. MALACOPTERYGII SUBBRACHIATI. *Ventrals under the pectorals.*

Family Gadidæ, (Cod fishes.)
Pleuronectidæ, (Halibuts and flounders.)
Discoboli, (ventrals formed into suckers.)

ORDER IV. MALACOPTERYGII APODA. *Ventral fins wanting.*
Muræenidæ.

ORDER V. LOPHOBRANCHII. *Tufted gills.*

ORDER VI. PLECTOGNATHI. *Soldered jaws.*

Family Gymnodontes, (naked teeth.)
Sclerodermi, (rough skins.)

SECOND DIVISION—CARTILAGINOUS FISHES.

CHONDROPTERYGII.

This Division is divided into two orders.

ORDER I. CHONDROPTERYGII BRANCHIIS LIBERIS. *Free or open gills.*
Sturgeons.

ORDER II. CHONDROPTERYGII BRANCHIIS FIXIS. *Gills fixed with separate openings for water to pass through.*

Family Selachii, (Sharks and Rays.)
Cyclostomata, (Sucker-formed mouths.)

—

AGASSIZ SYSTEM.

Agassiz, the eminent Professor of Natural History in Harvard College, has given the following arrangement of fishes, which seems to accord better than any other with that which geological researches among fossil remains point out as the order of succession in the formation or existence of fishes in early epochs of the world. The distinctive characters are founded upon the structure of the scales.

1. PLACOIDS.—Embracing those with cartilaginous skeletons or a skin covered with enameled plates or scales, as the *shagreen* of sharks.

2. GANOIDS.—Those, whether bony or cartilaginous, covered with

a continuous armor of angular scales, or bony plates fitting into each as in the sturgeons.

3. Ctenoids.—Those having scales with their posterior edges comb-like (pectinated,) as in the perches.

4. Cycloids.—Those with scales entire and of a circular form, as in the Salmons. It seems, from this, that the first fishes formed, judging from the oldest fossil remains were Placoids, and the last Cycloids.

—

GLOSSARY:

OR EXPLANATION OF SOME OF THE TERMS USED IN ICHTHYOLOGY.

A.

Abdominal—Belonging to the abdomen, or belly. Abdominal fins are those attached to the belly part of the fish, (*see fins.*) An order established by Linneus, having the abdominal fins placed behind the pectorals.

Acanthopterygii—Thorny or spiny fins; from *akanthos*, a thorn, and *pterugion*, fin or little feather. Fishes having such fins on their backs form one of the natural orders of fishes.

Acanthopteri—Same as the above.

Acerated—Pointed; sharpened to a fine point; needle-shaped.

Adipose—Fatty; fleshy. An adipose fin is one without any rays, but made up of a fleshy or fatty substance, and is generally placed on the back near the tail, as in the salmon and trout.

Anal—Relating to the vent, placed near the vent, (*see fins.*)

Anacanthini—Literally, this means no thorns, or thornless, (from the Greek *ana* none, and *acanthus*, thorn.) It is the name of the second order of fishes in some systems.

Anterior—Going before; fore-part; forward of something else; the opposite to posterior.

Apodal—Without ventral fins; from *a* (*privative*) no, and *pous*, feet; no feet, the ventral fins being considered, in comparative anatomy, as in the place of feet, (*see fins.*)

Arcade—An arch or series of arches.

B.

Barbel—A beard; a slender soft filament attached to the lips, chin or snout of some fishes.

Bifurcate—Divided into four points or two forks; twice forked.

Blenny—Mucus fish; from Blennius, mucus.

Branchii—The gills or breathing organs of fishes, (*see gills.*)

Branchiæ—Same as Branchii.

Branchiostegal—Pertaining to the gill covering, or bony part of the gills, (*see gills.*)

Branchiostegous—Having gill covers; from *branchia*, gills, and *stegos*, covering. The Branchiostegi were an order of fishes in Artedi's system, the rays of whose fins were bony but whose gill covers are destitute of bony rays.

C.

Carinated—Keeled; having appendages in form like a keel.

Carpus—The small bones of the wrist. In Ichthyology, the joint bones of the pectoral fins.

Carpal—Pertaining to the carpus, or bones of the pectoral fins in fishes.

Cartilaginous fishes—Fishes whose spinal column or vertebræ are made up of cartilage.

Caudal—Pertaining to the tail. The caudal fin is the tail fin of a fish, (*see fins,*) from the Latin *cauda*, a tail.

Centronotus—Thorny back; from *kentros*, a thorn, and *notos*, back.

Centropomus—Thorny gill, having thorns or spines on the gill covering or opercle; from *kentron*, thorn, and *poma*, an opercle.

Chætodon—A beard or bristle; from *kaitee*, a bristle.

Chondropterygii—Fish having cartilaginous fins; from *chondros*, cartilage, and *pterygion*, a fin.

Cirri—Filaments, or beard-like appendages.

Coecum—The commencement of the large intestine; sometimes the large intestine. It is also applied to the vermiform or worm-like appendages to the intestines, as in birds and fishes.

Corpus papillæ—The villous surface of the skin.

Crenated—Notched or cut into circular or curved shaped notches.

Crescentic—In form like a new moon; also growing or increasing.

Ctenoid—Comb formed; from *kteis*, a comb, and *eidos*, form. Ctenoidians formed the third order of Agassiz classification of fishes, containing those having jagged or comb-edged unenamelled scales.

Cuirassed—Covered with shield-like covering or scaly plates.

Culdesac—The bottom of a bag; an appendage to the stomach or intestines, like a small bag or sac, with no opening except the passage into it from the stomach.

Cycloid—A peculiar curve or circle formed by any point in a circle while it is rolling on a plane surface as does a carriage wheel.

Cycloidal—Pertaining to or resembling a cycloid.

Cycloidians—The fourth order of Agassiz system of Ichthyology containing those having circular smooth edged scales, as the herring.

D.

Deciduous—Falling; when applied to the scales of fishes it means those that adhere but slightly and fall easily.

Dentary—The place or places where teeth are inserted.

Denticulated—Having small teeth.

Dermic—Skinny; pertaining to the skin.

Dermopteri—Skin fins; (from *derma,* the skin, and *pteron,* wing or fin;) the tenth order of fishes in some systems.

Dichotomized—Separated into two forks or branches.

Diphyllous—Divided into two-leaf like divisions.

Discoid—Having the form of a disc.

Dorsal—Belonging to the back.

Ductus pneumaticus—A tube or passage by means of which the air bag communicates with the throat.

E.

Echeneis—(Sucking fish,) or a fish that often attaches itself to ships; from Greek *eko,* I have, and *naos,* ship.

Emarginated—Notched on the margin with circular notches.

Enoplus—Armed; from Greek *enoplos,* armed.

F.

Fins—The organs of motion and position of fishes. They are formed for the most part of a membrane spread over a number of rays which are either bony or cartilaginous and which are jointed, and thus expand or fold up the membrane at the will of the fish. They are considered by comparative anatomists to be analagous to the wings of birds, or the arms and feet of mammalia.

They vary in *number*, situation and structure in the different genera and species. The *size* of the fins is also equally various in the different species, as it bears no constant proportion to the figure or magnitude of the fish, nor to its habits or instincts.

The *situation* of the fins furnishes to the Ichthyologist some of the most obvious and useful distinctive characters, and have therefore received names expressive of their respective locations.

Ventrals placed forward of the pectorals he termed *Jugulars*. Those that had the ventral fins beneath the pectorals he called *Thoracic*. Those that had the ventral fins placed behind the pectorals were called *Abdominal*. This system has now given way to one more in accordance with natural orders; but these distinctions are of great use in marking characteristic differences between genera and species.

In regard to the structure and operation of the fins of fishes, we make an abstract of remarks by a writer in the Ed. Encyclopedia on this subject. In general these organs consist of numerous pointed rays, which are sub-divided at their extremities. These are covered on each side by the common integuments, which form in some instances soft fibres projecting beyond the rays. These fins, with articulated rays, were considered by the older Ichthyologists as furnishing characters for systematic arrangement of great importance. Fishes possessing these were termed *Malacopterygii*.

Besides these articulated rays, there exists in the fins of some fishes one or more rays made up of a single bony piece, enveloped like the former by a common membrane. Some fishes have one or more fins consisting entirely of these bony rays. Fishes with such rays are called *Acanthopterygii*. In a few genera the posterior dorsal fin is entirely destitute of rays and has obtained the name of *finna adiposa*, or adipose fin, as in the salmon and trout.

As these rays serve to support the fins, and are capable of approaching or separating like the sticks of a fan, we may conclude that they move upon some more solid body as a fulcrum. Accordingly we find in the sharks, for example, that the rays of the pectoral fins are connected by a cartilage to the spine. In the osseous fishes the pectoral fins are attached to an osseous girdle, which surrounds the body behind the *branchiæ* (gills), and which supports the posterior edge of their apertures. This osseous girdle is formed of one bone from each side, articulated at the posterior superior angle of the cranium and descending under the neck, where it unites with the corresponding bone. Between the rays of the fin and this bone, which resembles the *scapula* (shoulder blade,) there is a range of small flat bones separated by cartilaginous intervals, which may be compared to the bones of the *carpus*. The rays of the *ventral* fins are articulated to bones corresponding to the *pelvis* in the higher classes of animals. The pelvis is never articulated with the spine, nor does it form an osseous girdle round the abdomen.

In the *Jugular and Thoracic* fishes it is articulated to the base of the osseous girdle which supports the pectoral fins.

In the *Abdominal* fishes, the bones of the pelvis are never articulated to the osseous girdle and are seldom connected with each other. They are preserved in their situation by means of certain ligaments. The rays of the caudal fin are articulated with the last of the caudal vertebræ which is in general of a triangular form and flat.

The rays of the dorsal fin are supported by little bones, which have the same directions as the spinous processes, and to which they are attached by ligaments.

As connected with the fins we may here take notice of those organs which are termed *cirri* or *tentacula* (barbel which see) according as they are placed about the mouth, or on the upper part of the head. They are in general soft but often contain one jointed ray. They do not differ in structure from the fins, and are so closely connected with them, that it is difficult to point out their use.

The motions of a fish are performed by means of its fins. The caudal is the principal organ of progression. Those fins which are situate on the back are termed *Dorsal*. These vary greatly in shape and number.

Fig. 1.

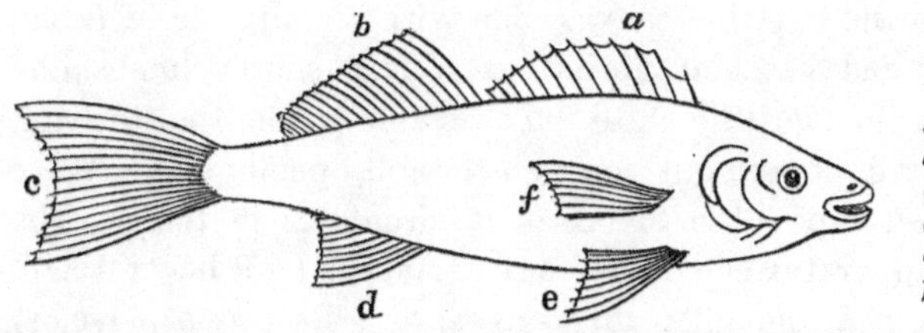

a. First dorsal.
b. Second dorsal.
c. Caudal.
d. Anal.
e. Ventral.
f. Pectoral.

The fin which surrounds the extremity of the tail is termed *Caudal* fin, and is always placed perpendicularly. It is forked in some and in some even or rounded. Between the caudal fin and the vent is placed the *Anal* fins. These vary in number and shape according to the species. Between the vent and the throat are placed the *Ventral* fins. When these are found, they are always parallel to each other and never exceed two in number. A short distance behind the gill openings are the *Pectoral* fins, so called, one on each side.

Linneus and some other writers on Ichthyology, considered the ventral fins analagous to the feet of quadrupeds. He used their varying positions as the basis of his classification of this branch of natural history, a brief abstract of which may be acceptable to some of our readers. Those fishes which had no ventrals he called *Apodal* or no feet. Those which had ventral fins placed nearer to the anterior extremity than the pectoral fins, or in other words those that had the ventral fins forward of the pectorals he called *Jugulars;* if beneath the pectorals they were called *Thoracic*; and if behind the pectoral *Abdominals*.

FULIGINOUS—Dark, dusky or sooty.

FUSIFORM—Spindle shaped.

G.

GANOID—Pertaining to the order of fishes called ganoids or ganoidei; from ganois, bright, and eidos, form.

GANOIDEI—The seventh order of fishes in some arrangements, the fishes having enamelled and plate like or shield like scales.

GANOIDIAN—The second order of fishes in Agassiz classification, having angular scales covered with bright enamel, as in the sturgeons.

GILLS—(From the Swedish word *gel.*) In Ichthyology, by gills are meant the organs of respiration in fishes, consisting of a cartilaginous or bony arch, attached to the bones of the head, and furnished on the exterior convex side with a multitude of fleshy leaves or fringed vascular fibrils resembling short plumes, and of a red color in a healthy state. The water is admitted through the mouth and poured out through the gill openings over these fibrils, and acts upon the blood as it circulates in them. The whole gill apparatus consist of four parts, viz: gill lid, gill flap, gill opening and the gills themselves, or gills proper, (*see gill* proper) as may be seen in the head of the common trout.

FIG. 2.

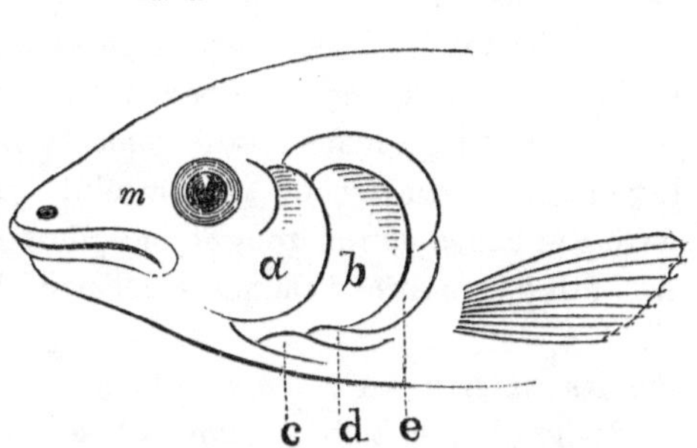

GILL LID, (*a*)—This is called the preopercle or preoperculum, and is situate behind the eye on each side. In its structure it is scaly, membranaceous or bony, and is articulated to the bones of the head. It consists of one or more pieces, and is therefore termed monophyllous when but one, diphyllous when two, triphyllous three, and so on. Its use is to support the gill flaps and act as cover to the opening of the lids. It is absent in fishes which have fixed branchiæ, and in a few with free branchiæ.

GILL FLAP, (*b*)—This was called by Linneus *membrana branchiostega* (*branchiostegal membrane, or gill covering membrane.*) He (Linneus) also considered it a true fin. It is also called opercle or operculum. It consists of a number of curved bones, or cartilages with a membrane. Its posterior or hindmost edge is generally free, and its anterior edge or base is united with the gill lid.(*a*) It is capable of extension and contraction, and when at rest is folded up partly under or beneath the gill lid (opercle). In some fishes it is wanting. When present it appears to assist the mouth in promoting the current of water over the gills. The rays of this organ (*c. d. e.*) sometimes called *Branciostegal rays*, furnish the Ichthyologist with valuable distinctive characters.

GILL OPENING—In the osseous (bony) fishes, and among the Branchiostegi (of some authors) this opening is a simple aperture behind the gills on each side formed by the lifting up the *gill flap*. It is sometimes round or semilunar, and in relative position differs according to the genera and species. In cartilaginous fishes as the "Lamper Eel," for instance, the gill opening is a round hole or aperture over each gill.

GILLS (PROPER)—Lifting up the gill flap you will see directly beneath them the true gills, several organs consisting, as has been before said, of a multitude of red colored, fringed vascular fibrils, or delicate fleshy plumes attached to curved bones jointed on to the head. In fishes with gills, or *branchiæ* as they are often

6

called, these organs are generally eight in number, four on each side. Each of these consists of three parts, viz: a cartilaginous or bony support and its convex side, and concave side. The bony or cartilaginous support consists of a crooked bone or cartilage generally furnished with a joint. At its base it is united with the bones of the tongue and above with those of the head. At both extremities it is movable, and throughout flexible like a rib.

From its exterior or convex side issues the multitude of fibrils or fleshy plumes before named closely connected with its base. The internal or concave side next the mouth varies in appearance. It is always more or less furnished with tubercles. It is of a white color.

In the cartilaginous fishes (Chondropterygii of some authors,) the gills are not so perfect. They are fixed to partitions which serve the purpose of the bony arches in bony fishes just described. These partitions extend from the mouth to the gill openings and vary in number in different genera. They are destitute of the inner or concave white side, but the fibrils or fleshy plumes are of the same structure with those on the convex side of the bony fishes.—*Ed. Ency.*

Gular—Pertaining to the throat or gills.

H.

Helicoid—A spiral curve; (from *helix* winding, and *eidos* form.)

Helical—Spiral.

Hermophrodite—Both sexes in one animal. This is alleged by Dr. Home to be the fact with the Lamprey Eel.

Heterocereal—Having upper lobe of the tail larger than the lower; from *heteros*, different, and *kerkos*, tail.

Holocephali—Solid heads; from *holos*, whole or solid, and *kephale* head, the eighth order of fishes in some arrangements.

The jaw bones and bones of the head united.

Homocereal—Both lobes of the tail equal.

I.

Ichthyology—The science which treats of the natural history of fishes. From Greek *ikthus* fish, and *logos* discourse.

Interradial—Between the rays. The membrane between the spines or rays of the fin is said to be interradial.

ISTHMUS—A narrow connecting space.

INTERMAXILLARY—Applied to bones situate between the upper jaws or cheek bones.

J.

JUGULAR—Pertaining or relating to the throat or neck.

L.

LATERAL LINE—A line extending from the head of a fish to the tail on each side of the body. It contains numerous pores or ducts from which proceeds the mucus or slimy matter covering the body of fish.

LOPHOBRANCHII—Tufted gills or crested gills; from *lophoura* a crest, and *branchii* gills. The sixth order in some systems.

M.

MALACOPTERYGII—Soft fins. Fish, the rays of whose fins are soft and not spinous, belong to the Malacopterygii or Malacopteri; the fourth order of fishes; from *malakos* soft, and *peterygii* little wing.

MALACOPTERI—Same as above.

MILT—The roe of fishes. The spermatic part of the male fish corresponding to the roe in the female.

MONOPHYLLOUS—Literally having one leaf. In Ichthyology, when the gill lid is composed of but one piece, it is said to be monophyllous.

O.

OPERCLE—The gill flap, (*see gills.*)

OPERCULAR—Belonging to the opercle.

OPERCULUM—The opercle; which see.

OSSEOUS FISHES—Fishes whose spinal column and skeleton are made of bones.

OVOVIPAROUS—Fishes whose eggs are hatched in the uterus and are excluded with the fry or young fish.

P.

PALATINES—The bones belonging to and forming the back part of the roof of the mouth.

PECTORAL—Near to or pertaining to the breast. The fins on the side of the fish near the gills are called pectoral fins.

PECTINATED—Resembling the teeth of a comb.

PEDUNCLE—A stem or stalk.

PETROMYZON—A stone sucker; from *petros* stone, and *muzo* to suck. The Lamprey Eel has received this generic name from the fact of its having an apparatus on its lip whereby it attaches itself to rocks and stones.

PHARYNGEALS—Belonging to the pharynx. Bones belonging to the pharynx or throat of fish are called pharyngeals.

PHARYNGONATHI—The name of the third order of fishes in some systems and refers to the union of the bones of the inferior pharyngeal bones into one.

PHARYNX—The upper part of the oesophagus or gullet. The cavity back of the tongue and above the oesophagus and wind pipe.

PLACOID—Plate like. The first order of fishes in Agassiz system is called Placoids, from their having plate like scales often elevated in the middle, and sometimes a point or spine as in the shark, rays, &c.

PLAGIOSTOMI—Oblique mouth or transverse mouth; from *plagios* oblique or transverse, and *stoma* a mouth, the ninth order of fishes.

PLECTOGNATHI—Solid or soldered jaws; from *plectos* woven, and *gnathos* cheek or jaw. The fifth order of fishes. The outside bone is united to the main jaw immovably.

PLEURONECTES—Broad swimmers, as the flounders; from *pleuros* broad, and *nektœs* swimmer.

PLURISERIAL—Made up of many serials.

POLYMORPHIC—Many forms.

PREMAXILLARIES—Bones on the outside of the jaws of fishes resembling bony lips. (*See fig.* 2, *m.*)

PREHENSILE—Adapted to seizing or laying hold of.

PREOPERCLE—Before or forward of the opercle; the gill lid. (*See gill.*)

PSEUDOBRANCHII—False branchiæ or false gills.

PTERYGOID—Wing shaped.

PYLORIC—Pertaining or belonging to the pylorus.

PYLORUS—The lower orifice of the stomach.

R.

RAPTATORIAL—Fitted to seize or snatch any thing.

RAY—A slender, bony or cartilaginous spine or filament supporting the membrane of the fins of fishes.

RAY FORMULA—Each species of fish has the same number of rays in their fins and this affords one of the characters to distinguish them. The number of rays is expressed in the following abbreviated method. Take for instance the common yellow or brindle perch. Its Ray formula would be expressed thus:—B. 7: D. 13; 2—13; P. 15; V. 1—15; A. 2—8; C. 18—which must be read as follows: Branchial rays, *seven*. Dorsal rays in first fin, *thirteen*, in second *thirteen*. Pectoral rays, *fifteen*. Ventral fins, *one* spiny ray, *fifteen* soft ones. Anal fin, *two* spiny and *eight* soft. Caudal ray, *eighteen*. The rays of the common brook trout, which has no spiny rays, would be expressed thus, omitting the branchial ray:—D. 11: P. 13; V. 8; A. 11; C. 19.

ROE—The eggs of fishes. In bony fishes these consist of two long bodies resembling the soft roe or milt of male fish, except that they are of a firmer consistency, and are filled with a prodigious number of spheric ovula. They are situated side of the intestinal canal, and near the liver and swimming bladder and extend as far as the vent. The ovula (little eggs,) composing these hard roes are so numerous that nearly 350,000 have been counted in a carp eighteen inches long, and in a sturgeon weighing one hundred and sixty pounds there was discovered nearly 1,500,000. They are enveloped in a delicate membrane which forms a peripheral part of the ovary, and joining that of the opposite ovary near its sacral extremity forms with it a common tube that opens behind tho vent for the passage of the egg.—*Ency.*

S.

SCALES—The small thin plates which cover the bodies of fishes.

SERRATED—Notched like saw-teeth.

SOUND—The air-bag of fishes.

SPIRACULA—Small holes through which air is passed.

SPINIGEROUS—Spine bearing.

SUB-FUSIFORM—Somewhat spindle-shaped.

SUB-CRESCENTIC—Somewhat crescent or moon-shaped.

SUB-ORBITAL—Under the eye or orbit.

SUPRA OCULAR—Over the eyes.

T.

TAENOID—Ribbon-shaped.

TENTACULA—A filiform or thread-like process or organ fitted for holding or fastening the object to which it belongs to some particular place. It may also be an organ of feeling or of motion.

THORACIC—Pertaining to the thorax or chest. One of Linneus' order of fishes was called Thoracic, in which the ventral fins were placed below the pectorals.

TRIPHYLLOUS—Divided into three leaf-like divisions.

U.

URANOSCOPUS—Star-gazer; from *uranous*, sky or heavens, and *skopeo*, I see.

V.

VENTRAL—Belonging to the belly. In Ichthyology the ventral fins are those situate between the anus and throat, (*see fins.*)

VILLOUS—Covered with soft hairs or fine soft fibrils.

VOMER—Literally a ploughshare; the point or end of the nose or snout of a fish.

DR. HOLMES' REPORT

ON THE FISHES OF MAINE.

PART II.

Descriptive Ichthyology.

We come now to an enumeration of some of the fishes of Maine, including their arrangement into families,—the distinctive characteristics of their respective genera, together with their specific descriptions, and such general remarks as facts and observations have suggested.

It has been seen by a perusal of Part I, that the classification of fishes has always been a rather difficult task, and that from the time attention was first turned to the subject, to the present, continued changes have been made. Increase of knowledge on this subject authorized these changes—successive increase of knowledge authorizes a continuance of changes. Indeed, we may consider this part of Ichthyological science as yet in a transition state, and still progressing towards completion.

The researches of experienced and enthusiastic devotees to this branch of natural history are continually developing new and interesting facts, each of which either corroborate the correctness of former arrangement or point out such new variations as shall help to a consummation of the work in progress. Even since this survey was in contemplation, discoveries have been made leading those high in authority as to this science to follow out changes already begun with a view to the ultimate perfection of the classification proposed.*

As before stated, on page 30, I proposed, without considering

* Those who will examine the recent publications of the Smithsonian Institute, of the Philadelphia Academy of Arts and Sciences, Memoirs of American Academy of Arts, and other scientific periodicals, will be convinced of this.

myself pledged to follow it in the final report, to adopt, for the present, the arrangement followed by Dr. Girard, in his general report of fishes found during the exploration of the Pacific Railroad route, published by order of Congress in 1855, an abstract of which has already been given. On further consideration I have concluded to change this plan, and follow the more modern classification of Prof. Gill of the Smithsonian Institute, it being more consonant with natural structure of fishes.

Only a part of the species of fishes found in our waters can be at this time described. Many specimens are on hand, waiting further time for examination and their assignment to their true and proper place in the catalogue. Many more will undoubtedly be obtained by further search.

There may be some question in regard to the limits of the geographical distribution of the fishes of Maine. Whether this term is to be applied to those only which are constantly found here, and those also which come regularly by periodical emigration to breed and feed their young to an age and size sufficient to enable them to follow their parents; or whether those also which are only occasionally found among us—chance visitors—should be claimed and included in the list.

The fishes found on the north shore of Massachusetts bay, and those that frequent the coast of the British Provinces on the Atlantic as far as Newfoundland, may be fairly, I think, enumerated among Maine fishes. Our coast occupies a middle ground between them, and fishes are not restrained in their movements from one haunt to another, or in their instinctive explorations for food, &c., by any national or conventional boundaries.

For the same reason some of the species which sometimes wander from their more southern localities to the extremity of Cape Cod, may be considered as coming into the region or confines of the Maine fishing grounds, as Kittery point is but about one-third of a degree more north than Race point, and but between twenty and thirty leagues distance from it on a straight line.

If, therefore, some of the fishes which make the Cape Cod waters their particular "*habitat*," should occasionally be taken by the hook, or stray into the net of some of the Maine fishermen, and thereby be placed on our catalogue, I trust no accusation of breach of the "*fishing treaty*" or of poaching on our neighbors "*aquarium*" will be brought against me.

I will here very cheerfully acknowledge the aid I have received from the published works of Dr. Girard in his general report before referred to—the late Dr. DeKay on the fishes of New York, and Dr. Storer on the fishes of Massachusetts. Few Ichthyologists describe a fish with more accuracy and precision than Dr. Storer. I have found his description of the color exhibited by different species remarkably correct, and have taken the liberty to often avail myself of this part of his description in this report. I am also under special obligations to Prof. Theodore Gill, of the Smithsonian Institution, who kindly revised a list of our fishes sent him and furnished information of more species not yet obtained by me. These, with the proffer of other favors, have been gratefully received.

The system of classification devised and promulgated by Prof. Gill and which we propose to follow, is a modification of, and as we think, an essential improvement on that laid down by the celebrated European Ichthyologist, Johannes Muller. It is based upon the idea, as well as the fact of a gradual rise from a very simple form and frame work of fishes, up to that which is more complicated and more perfect in anatomical structure and physiological organization, classifying and arranging them into classes, orders, groups and families step by step, along the whole line as you ascend from base to pinnacle. These are established according to similarity of permanent organs, and the affinities of those subordinate parts and appendages which are more variable in form and location.

To use his own words, "fishes appear to be constructed according to four different sub-plans which are characterized by their correspondence to different stages or grades of developement of a typical or model osseous fish" To elucidate this more clearly we will begin at the lowest and simplest form and organization of fishes, as the "Lamper eels" for instance, which exhibit scarcely any organs of a complicated kind or remarkable developement. This sub-class * is called Dermopteri, or skinfins. Their fins, instead of being fitted out with rays and membranes, are merely a duplicature or folding over of skin.

The next step upwards brings us among the sharks and rays, or skates, a class whose frame-work or skeleton, though made up of

* As Fishes, as a whole, constitute one of the grand classes of the animal kingdom, the next division of them must be into sub-classes.

cartilage like those below, is a little more complicated and their organs more perfect in form and adaptation. This sub-class receives the name of "ELASMOBRANCHII," or plated gills, from the fact of their having separate and detached breathing apertures on each side of their necks.

Passing upwards, we come to the sturgeons, where we find a still greater "advance of developement," and more perfect and complicated organs than in those below them. This class is called GANOIDEI, from the enamelled plate-like scales with which its representatives are covered.

The next, and fourth step upward brings us among fishes with a perfected bony skeleton—showing a complete vertebrated organization, and their several parts constructed with reference to carrying out higher physiological functions,—in short a perfect fish. This sub-class is called TELEOSTEI, perfected bones; in refer-ence to the completeness and finish of its bony skeleton.

He.e we have four sub-classes, founded on what appears to be a natural series of orders, each representing different but connected stages in the creation of the animals under examination.

These several sub-classes are further divided into orders, the orders into sub-orders, these again into groups and families, the families into sub-families, and these into genera, and the genera into species. To elucidate this system so as to make it clear and plain, let us begin at the "*top of the heap*," and go downwards. Commencing, therefore, with the sub-class Teleostei, or perfected bones, let us examine their several distinctive characteristics more minutely and technically, and we shall find them as follows: using in part, Prof. Gill's description of them. *

SUB-CLASS TELEOSTEI, *Muller.*

The endo-skeleton (inside frame) is almost always osseous. The scapular arch is suspended from the skull; the supra scapula generally connected with the mastoid, and paroccipital bones. The exo-skeleton (outside frame) is generally in the form of cycloid, or ctenoid scales, but sometimes the body is naked, and sometimes covered with bony scales, plates or spines. The optic nerves cross each other in their passage from their respective lobes to the eyes.

* Catalogue of fishes of the eastern coast of North America from Greenland to Georgia, by Theodore Gill.

The *bulbus arteriosus*† has almost always only two opposite semilunar valves. The branchial apertures are represented by simple fissures on each side. There are four pairs of true and well developed branchial arches, each of which generally supports free branchiae; an air bladder is generally present. The neutral fins vary in position, and are sometimes absent. This class embraces by far the largest proportion of existing fishes, and have been divided by Prof. G. into five "natural and easily distinguished orders." The first order he calls TELEOCEPHALI, or perfect heads. Let us examine this order more in detail.

ORDER TELEOCEPHALI, *Gill.*

The endo-skeleton is almost always perfectly developed. Body generally covered by ctenoid or cycloid scales, branchiæ pectinated. The supramaxillaries and intermaxillaries, are always present and separate from each other. The sub-opercular bone is almost invariably present. Many of the rays are articulated and branched. Nearly all the fishes most esteemed as food, belong to this order; it is divisible into several sub-orders. Let us next examine these.

SUBORDER PHYSOCLISTI, *Bona.*

It has been found that a large number of fishes have an air bladder that is closed, and has no visible duct or communication out of it. Such of the fishes of the order Teleocephali are put in this sub-order which, from this fact, is called *Physoclisti*, from the Greek *phusa*, bladder, and *kleisos*, closed. The scales on fishes of this sub-order, when present, are either ctenoid or cycloid; there are rarely osseous plates. The anterior rays of the dorsal and anal fins, and the first ray of the ventrals are simple, or spinous. The ventrals are generally more or less anterior. The lower pharyngeal bones are small and triangular, sometimes united, but generally distinct, the teeth are implanted on the plain surface.

SUB-ORDER HETEROSOMATA, *Bona.*

Next, on further examination, we find some fishes that have

† Fishes are destitute of a double heart, such as the mammalia have. In some it appears to be merely an expansion of the aorta provided with one or two valves within.

their bodies apparently all askew from their eyes, the eyes being both on one side of their heads, and their mouths distorted, and no air bladder, as the halibuts and flounders or flat-fish. These have been gathered into a sub-order and called *Heterosomata* (from *heteros*, another, and *soma*, body.) The side on which the eyes are situated is dark, or variously colored, while the eyeless side is almost always white; the scales are either ctenoid or cycloid. The dorsal and anal fins are very long and composed mostly of articulated rays.

Sub-Order Physostomi, *Muller*.

Proceeding in our examination of the *Teleocephalic* order, we find many that have an air bladder that communicates by means of a duct or passage with the mouth or intestinal canal. These are put into another order called *Physostomi* (from *phusa*, bladder, and *stoma*, mouth.) The scales are generally cycloid, there being but one or two exceptions. The fins are mostly sustained by branched rays, only the first rays being sometimes simple. The ventrals are always abdominal. The lower pharyngeal bones are separate and almost always small and triangular, with the teeth on a plain surface. The salmons and herrings are embraced in this order.

Sub-Order Eventognathi, *Gill*.

Continuing our investigation, we find fishes whose air bladders are divided by constriction into two or three portions, and communicate by a duct with the throat. With few exceptions they are covered with cycloid scales. All the rays of the fins except the first of each, are branched; the ventrals always abdominal. The lower pharyngeal bones are more or less falciform (sickle shaped,) greatly developed, nearly parallel with the branchial arches as provided on the internal surface of the curved portion, with large teeth of various forms.

In allusion to the developement of the pharyngeal jaws, the Prof. has named this order *Eventognathi*. This concludes the division of the first order Teleocephali into sub-orders. We will now consider the details of another order.

Order Apodes, *Kaup*.

We find in our researches, many fishes that have a snakelike form of body,—skin generally naked or rarely covered with scales imbedded in the epidermis. The branchiæ, or gills are pectinated (comb-like.) The supramaxillaries and intermaxillaries are small and rudimentary. The teeth are planted on the palatine and vomerine bones. With the vomer, the nasal and ethnoid bones are coalescent. The pectoral fins are often absent, and the ventral fins always wanting. Hence the name *Apodes* is given to this order. The dorsal, anal and caudal fins, when present, always run together. The common eel and the conger eel belong to this order. This order has no sub-orders.

Order Lemniscati, *Kaup*.

Continuing the examination we find a few fishes of rather doubtful affinity to the above order or sub-orders, but which for the sake of convenience are placed, for the present, in an order by themselves until further light is thrown upon the doubts in regard to them. They are small, destitute of ventral fins and are generally diaphanous, or they are greatly elongated and compressed, or ribbon formed. The skull and vertebral column are incomplete and cartilaginous. The blood is colorless, and there is no spleen; the body is entirely naked and the arrangement of the muscles is very apparent. It has received the above name (from *Lemniskos*, a ribbon, or crowned with a ribbon.) No sub-orders.

Only one species has been found on the Atlantic coast,—the *Leptocephalus gracilis* (thin head) of Storer.

Order Nematognathi, *Gill*.

(*Threaded Jaws.*)

Further search brings us to a grade of fishes with either naked bodies, or else protected with ganoid plates. The branchiæ are pectinated and supported on four arches as they are in the order Teleocephali. The supramaxillary bones are little developed, and are enveloped in the integuments which terminate in longer or shorter barbels,—hence the name Nematognathi, threaded jaws, (from *nematos*, thread or threaded, and *gnathi*, jaws.) The sub-opercular bone is always absent. The rays are mostly articulate and branched.

The "catfishes," "hornpouts" and "bullheads" are embraced in this order.

Order Plectognathi, *Cuvier.*

(*Soldered Jaws.*)

We next come to fishes that have some peculiarities in the structure of their jaws. The supramaxillary and intermaxillary bones are united together in a continuous piece, and the palatine arch and cranium are connected by immovable sutures. Hence they are called Plectognathi, (soldered or woven jaws.) The internal skeleton is less perfectly developed than those of the perfect head, (Teleocephali.) The exterior is covered with ganoid plates, granulations, or spines. The branchiæ are pectinated and the branchial apertures small. The air bladder has no duct.

The "balloon fish," "puffers" and "blowers" belong to this order.

Order Lophobranchii, *Cuvier.*

(*Tufted Gills.*)

All the fishes hitherto examined, have their gills or branchiæ fringed like the teeth of a comb, but we now come to a few whose gills are in small round tufts, disposed along the arches in pairs. The branchial apertures small, and on each side of the nape. The jaws are produced into an elongated tubular mouth. The internal skeleton is less perfectly developed than in those with fringed gills. The external skeleton is composed of polygonal plates of a bony, or horny nature, which are joined to each other but permit considerable motion in the animal.

In warm latitudes, there are quite a number of genera of this order, but there are but two in the Eastern Atlantic coast, viz: the "Syngnathus," or pipe fish, a "Hippocampus," or "River Horse," a small fish of the Hudson river.

This closes up the orders and sub-orders of the first, or upper class of this sytem. Another step down the descending series brings us into the

Sub-Class Ganoidei, *Agassiz.*

Here we find fishes of a mixed organization, if we may so speak, some of them exhibiting many of the characteristics of the class above them (Teleostei,) the more like the class below, and others intermediate. As revised by Muller, it embraces forms in which the vertebral column and skull are either osseous or cartilaginous.

The scapular arch is suspended directly from the skull. The exo-skeleton is generally deposited in the form of ganoid plates, but there are in representatives of some families, oval or cycloid scales, and the body is still more rarely naked, and the bony plates absent. The optic nerves, like those of the plaigiostoms are only connected by a commissure, and do not cross or decussate. The *bulbus arteriosus* is muscular and provided with two or more rows of valves, which, in one order are replaced by two spiral and longitudinal valvular folds. The intestine has frequently, — but not always, — a spiral valve. The branchial apertures are simple fissures, or spiracles on each side as in ordinary fishes; the branchiæ are free. An air bladder is present and communicates by a duct with the intestinal anal. The ventrals are abdominal.

This sub-class has been, for the present, divided into four orders, as follows:

ORDER HOLOSTEI, *Muller*.

(*Whole Bones.*)

This order embraces fishes which are provided with plates which are either rhomboid and tiled, or oval and imbricated. The hyoid (tongue) apparatus has one or many branchiostegal rays. The centre of the vertebræ are either ossified or represented by a persistent notochord, (spinal chord.) The neuro-apophyses (Ganglions) and haemapophoses) arterial bulb, are always ossified. The dorsal and anal fins are sustained by true dermo neural spines articulated with the inter neural spines. The scapular arch is suspended by two processes to the paroccipital and the mastoid bones, and sustains well developed pectoral fins which are provided with many rays. The abdominal ventral fins, are also supported by several rays. The bulb of the aorta has several longitudinal rows of valves.

ORDER PLACOGANOIDEI, *Owen*.

This order embraces those fishes that were, rather than those of the present day, and their characteristics are studied from the organic, or fossil remains, discovered by the geologist in his researches into the formations of early epochs of the earth on which we live.

Order Chondrostei, *Muller*.

Ossified Cartilage.

The fishes embraced in this order, among which are the sturgeons, have a skeleton, or frame work made up of both cartilage and bones. The body is sometimes naked, but in most of the species is covered with more or less interrupted rows of long or ganoid plates of irregular form.

There also many smaller plates and tubercles scattered on different parts of the body. There are no true branchiostegal rays. The vertebræ and their elements are cartilaginous. The skull is also cartilaginous, but it is sometimes imperfectly ossified in front. The scapular arch is suspended by two processes of the paroccipital and mastoid bones. It supports two well developed ventral fins. The ventral fins are also furnished with several rays. The bulb of the aorta is furnished with several longitudinal rows of valves.

Order Dipnoi, *Muller*.

(*Double Spiracles, or breathing holes.*)

The bodies of the fishes embraced in this order are elongated and covered with regularly imbricated cycloid scales. The centre of the true vertebræ are cartilaginous, the notochord being persistent. The continuous vertical fin, or fold encircling the posterior part of the body, is sustained by *articulated* rays, immediately connected with the spinous processes of the neurapophyses and haemapophyses, which spines are osseous. The scapular arch is suspended only to the exoccipital bone and supports on each side a simple unjointed, or articulated ray on each side. The bulb of the aorta is furnished internally with two spiral ridges, or valves.

This brings us through the second class from the highest and to the commencement of the third stage in the descending series of the structural arrangement under consideration. The next is

Sub-class Elasmobranchii, *Bona*.

(*Plated gills or branchiæ.*)

This class embraces the sharks and the bony skates. They have the endo-skeleton or vertebral column, and skull cartilaginous, or, very imperfectly ossified. The exo-skeleton is developed in the form of placoid granules. The brain is much more complex and highly developed, than in the true fishes; the optic nerves are connected by a commissure, but do not cross each other. The *bulbus arteriosus* or *aorta* has a thick muscular coat, and is provided with at least two rows of semi-lunar valves. The intestine has a spiral valve. The males are provided with the so called "claspers," which are present as appendages to the posterior edge of the ventral fins. Fecundation is effected by copulation. The branchial apertures are usually five in number, and are generally *all* external. There is no air bladder.

This sub-class was divided into two orders, viz: Plagiostomi, (oblique mouths,) and Holocephali, (solid heads.) The Plagiostomi are distinguished by several separate branchial apertures on the side of the neck, and the Holocephali with only one branchial aperture on each side, as in the true fishes. The Plagiostomi order was divided into four sub-orders, viz: Squali, Rhinæ, Pristes and Raiæ; and this arrangement is so laid down in Prof. Gill's catalogue of East Coast fishes before referred to. Subsequently, further investigation of the Sharks and Rays, has induced the Professor to modify this arrangement by raising the SQUALI to a full order, dividing it into several families and sub-families, and reserving the Rhinæ and Raiæ (Rays) as sub-orders. The divisions, therefore, of this sub-class, (Elasmobranchii,) are

ORDER SQUALI; *Gill.*

Several branchial apertures on the side of the neck,—pre-supplementary eye-lids. An incomplete scapular arch; naso-pectoral cartilage absent.

SUB-ORDER RHINÆ; *Gill.*

This sub-order is distinguished from the Squali by the depressed head and body, and the dorsal position of the eyes. The branchial apertures are situated on the sides, but are placed in a furrow which separates the large expanded pectoral fins from the body. The mouth is at the extremity of the snout.

SUB-ORDER RAIÆ; *Muller & Henle.*

The *Raiæ* have the branchial apertures beneath the body under the pectoral fins. The body is flattened until from its union with the large and fleshy pectorals it forms a disc. These pectorals are united before the snout. Eyes and spout holes are always above on the dorsal aspect of the head ; the scapular arch complete ; naso-pectoral cartillages present ; no eye-lid, or only an adnate upper one.

Order Holocephali ; *Bona.*

There are but one or two genera of fishes belonging to this order. The *Chimæra,* or sea monster of the high northern seas, is embraced by it. None of them, to our knowledge, have been found on the coast of Maine. They have but one branchial, or gill aperture on each side of the head, as in the perfected or true fishes.

The next step in the descending series brings us to the

Sub-class Dermopteri ; *Owen.*

This is the lowest and most simply constituted of all the classes of the fishes. The body is very much elongated and worm-like, either sub-cylindrical or compressed. The endo-skeleton is very rudimentary and cartilaginous, and in one of the orders, (Pharyngobranchii,) there is no distinct head. The pectoral and ventral fins are both absent. The skin is entirely naked and mucous, and the fins are only folds of the skins. There is no pancreas and no air bladder. The olfactory organ and nostril are single.

There are three orders to this class, viz :

Order Hyperoartii ; (*Bona.*) *Muller.*

The body of the members of this order is invariably greatly elongated, and sub-cylindrical or anguiliform. The head is distinct. The "myelon," or medulla spinalis, is described by Owen as being flattened or depressed, "of opaline sub-transparency, ductile and elastic." The *bulbus arteriosus* is absent, but there are two opposite valves at the organ of the branchiæ vessel, as in the Teleostei. The branchiæ are purse shaped and without opercula. There are seven in number on each side. Each receives the streams of water for the correction of the blood, through short tubes, entering from a medium canal, which is below and distinct from the oesophagus and which terminates behind in a closed wall, and according to Professor Owen, communicates with the fauces anteriorly "by an opening guarded by a double membranous valve." This order includes the well known "Lamper Eel," and is equivalent to the Marsipobranchii, (purse gills,) of other authors.

ORDER HYPEROTRETI; (*Bona.*) *Muller.*

The representatives of this order resemble those of the other, (Hyperoartii,) except in the respiratory organs. The branchiæ are bursiform and fixed, receiving the streams of water directly from the throat or oesophagus, through short tubes communicating with each sac. The water is discharged through tubes which either severally open externally, or into two lateral and longitudinal canals, directed backwards and discharging by as many orifices on each side of the medium line of the ventral surface.

ORDER PHARYNGOBRANCHII; *Muller.*

There is but a single genus embraced in this order, and this is the very lowest in point of organization of all fishes. It has no distinct head, and no heart; body elongated and compressed. This brings us to the foot of the structural series following it, according to the anatomical characters as laid down by Muller and the great comparative anatomist, Professor Owen of England, and modified in its arrangement by Professor Gill, upon whose description we have drawn pretty freely.

That it is perfect the authors themselves do not pretend, but it comes nearer to the scheme of natural orders than anything that has yet been promulgated, and will help greatly to the desired consummation of a perfect classification, according to the true plan of nature, so much desired and sought for by every naturalist.

CATALOGUE

OR SYNOPSIS OF A PART OF THE FISHES OF MAINE, ARRANGED ACCORDING TO PROF. GILL'S CLASSIFICATION.

SUB-CLASS TELEOSTEI, *Muller.* ORDER TELEOCEPHALI, *Gill.*

Sub-order Physoclisti, Bona.

PERCOID FAMILY.

Perca flavescens, *Mitch.*, Yellow or brindle perch.
Roccus lineatus, *Gill*, Striped bass.
Morone Americana, *Gill*, White perch.
Pomotis vulgaris, *Cuv.*, Bream, flat fish.
Pomotis appendix, *Mitch.*, Pumpkin seed.

SPAROIDS.

Pagrus argyrops, *Cuv.*, Big porgee.

SCIÆNOIDS.

Cynoscion regalis, *Gill*, Weak fish, squeteague.

SCOMBROIDS.

Scomber vernalis, *Mitch.*, Spring mackerel.
Scomber grex, *Mitch.*, Fall mackerel.
Orycnus secundi-dorsalis, *Gill*, Tunney or Horse mackerel.
Apodontis maculatus, *Gill*, Spanish mackerel.

CARANGINOIDS.

Vomer setipinnis, *Ayres*, Blunt-nose shiner.

SERIOLINOIDS.

Pomatomus saltatrix, *Gill*, Blue fish.

SCOMBERESOCOIDS.

Scomberesox scutellatus, *Lesueur*, Bill fish.

GASTEROSTOIDS.

Pygosteus DeKayii, *Brevort*, Many-spined stickle back.

ATHERINOIDS.

Chirostoma notatum, *Gill*, Silverside.

SCORPENOIDS.

Sebastes Norvegicus, *Cuv.*, Norway haddock.
Hemitripterus Acadiensis, *Storer*, Sea raven.

COTTOIDS.

Acanthocottus Groenlandicus, *Girard*, Sculpin; Greenland Bull-head.
Acanthocottus octodecim-spinatus, *Gill*, Common sculpin or bull-head.
Acanthocottus Labradoricus, *Girard*, Labrador or northern sculpin.

AGONOIDS.

Aspidophorus monopterygius, *Storer*, American aspidophore.

TRIGLOIDS.

Dactylopterus volitans, *Lac.*, Sea swallow.

BATRACHOIDS.

Batrachus tau, *Linn.*, Toad fish.

BLENNIODS.

Stichaeus subbipinnatus, *Gill*, Radiated shanny.
Muraenoides mucronatus, *Gill*, Butter fish.

ZOARCEOIDS.

Zoarces anguillaris, *Storer*, Thick-lipped eelpout.

ANARRHICHOIDS.

Anarrhicas vomerinus, *Storer*, Sea wolf.

CRYPTOCANTHOIDS.

Cryptocanthodes maculatus, *Storer*, Spotted wry mouth.
Cryptocanthodes inornatus, *Gill.*

LOPHIOIDS.

Lophius Americanus, *Val.*, Angler; goose fish.

—

Sub-order Anacanthini, Muller.

GADOIDS.

Gadus Americanus, *Gill*, Cod fish.
Gadus pruinosus, *Mitch.*, Frost fish; tom cod.
Melanogrammus aeglefinus, *Gill*, Haddock.
Merlangus purpureus, *Storer*, Pollock.
Merlucius vulgaris, *Reinh*, Hake.
Brosmius flavescens, *Lesueur*, Cusk.

PHYCINOIDS.

Phycis Americana, *Cuv.*, American codling.
Phycis DeKayii, *Kaup.*

OPHIDIOIDS.

Ophidium marginatum, *Mitch.*, New York ophidium.

—

Sub-order Pharyngonathi, Muller.

LABROIDS.

Tautoga Americana,(?) *DeKay*, Tautog; blackfish.
Ctenolabrus Burgall, *Gill*, Cunner.

—

Sub-order Heterosoma, Bona.

PLEURONECTOIDS.

Hippoglossus Americana, *Gill*, Halibut.
Pomatopsetta dentata, *Gill*, Toothed flat fish; summer flounder.
Pleuronectes Americanus,(?) *Walbaum*, Flounder.

—

Sub-order Physostomi, Muller.

CYPRINOIDS.

Carascira auratus,(?) *Fitz.*, Gold fish; golden carp.
Leucosomus Americanus, *Storer*, Shiner.
Plargyrus cornutus,(?) *Storer*, Red fin.

Catastomoids.

Catastomus Bostoniensis, (?) *Lesueur*, Common sucker.
Moxostomus oblongus, *Ayres*.

Cyprinodonts.

Fundulus pisculentus, *Val.*, Minnow ; killifish.

Esocoids.

Esox reticulatus, *Leseur*, Common pickerel.

Salmonoids.

Salmo salar, *Linn.*, Salmon.
Salmo fontinalis, *Mitch.*, Brook trout.
Salmo Gloveri, *Girard*.
Salmo Sebago, *Girard*, Salmon trout.(?)
Salmo hamatus, *Cuvier*.
Salmo Tomah, *Hamlin*, Togue.
Salmo erythrogaster, *DeKay*, Red-bellied trout.
Salmo oquassa, *Girard*, Blue back.
Coregonus albus, *Linn.*, White fish.
Argyrosoma clupeiformis, *Ayres*.
Osmerus mordax, *Gill*, Smelt.

Clupeoids.

Clupea elongata, *Linn.*, Herring ; English herring.
Alausa sapidissima, *Storer*, Common shad.
Alausa tyrannus, *DeKay*, Alewife.
Alausa cyanonoton, *Storer*.
Brevoortia menhaden, *Gill*, Menhaden ; moss banker.

Engraulinoids.

Engraulis vittata, *Baird* and *Girard*, Anchovy.

—

Order Apodes.

Anguilloids.

Anguilla Bostoniensis, *Les.*, Eel.

—

Order Lemniscati, *Kaup*.

Leptocephalus.

Leptocephalus gracilis, *Storer*, Thin head.

—

Order Nematognathi, *Gill*.

Pimeloids.

Amiurus pullus, *Gill*, Hornpout.

SUB-CLASS GANOIDEI, (*Ag.*) *Muller*. ORDER CHONDROSTEI, *Muller*.

STURIONOIDS.

Acipenser oxyrhynchus, *Mitchell*, Sturgeon.

SUB-CLASS ELASMOBRANCHII, *Bona*. ORDER PLAGIOSTOMI, *Cuv*.

Sub-order Squali, (*Muller*) *Gill*.

CETORHINOIDS.

Cetorhinus maximus (?) *Blainville*, Basking shark.

SCYMNOIDS.

Somniosus brevipennis, *Lesueur*, Sleeper.

SUB-CLASS DERMOPTERI, *Owen*. ORDER HYPEROARTII, *Bona*.

PETROMYZONTOIDS, *Bona*.

Petromyzon Americanus, *Les.*, Lamper eel.

Familiar and Scientific Description of some of the Maine Fishes named in the foregoing Synopsis.

SUB-CLASS TELEOSTEI. ORDER TELEOCEPHALI.

Sub-order Physoclisti.

FAMILY PERCOIDÆ.* SUB-FAMILY PERCINÆ.

FAMILY CHARACTERS. *Body* more or less elongated ; in most cases protected by pectinated scales, generally rough to the touch, occasionally smooth, owing to the deciduous nature of their prickles.

Head.—The preopercle and opercle exhibit various spinous or serrated edges,—in a few they are smooth. Jaws, front of vomer, and often the palatine bones furnished with teeth of various kinds, velvet-like, card-like, or of the canine type ; the canines occurring, occasionally intermixed with the former two kinds.

Fins.—Dorsals always well developed, sometimes single, at oth-

* In this description it is thought best to commence with the more perfected orders and proceed to the lower grades. Some of them I have described in full, in the form and manner that it is intended all of them shall be in the final report. Others, merely the family and generic characteristics are given, and in a few even these are omitted because either some doubtful points in regard to them need further investigation, or I have not the proper authorities to which I could refer in the matter.

ers subdivided into two distinct fins; the anterior portion, or anterior fin, as the case may be, being spinous,—that is, composed of bony or rigid rays. The anal fin has a variable number of spiny rays in its anterior margin, but in a few cases they are wanting. The caudal is either truncated posteriorly or more or less emarginated. Ventrals are inserted posteriorly to the base of the pectorals or composed of an external stoutish spine or fine soft dichotomized rays.

GENERA. The following genera belong to this family:

GENUS PERCA, *Cuv.*

GEN. CHAR. Two dorsal fins distinctly separated; the rays of the first spinous, those of the second flexible; tongue smooth, teeth in both jaws, in front of the vomer, and on the palatine bones; preoperculum notched below, serrated on the posterior edge; operculum bony, ending in a flattened point directed backwards. There are branchiostegous rays. Scales roughened, and not easily detached.

Perca Flavescens, Mitchell.

Brindle perch. Yellow perch.

One of the most common and abundant of the fishes in our ponds and lakes, and one which every schoolboy recognizes as among the trophies of his earliest fishing expeditions, is the yellow or brindle perch. The perches breed in great numbers in the fresh waters throughout the State, preferring sandy bottoms and clear waters to any others. They take bait readily, and offer fine sport to anglers, especially in the months of July and August, when they congregate together in great numbers in the comparatively shoal and warm waters of the ponds where they breed. In winter they are sometimes caught, but they do not bite so readily as in summer. They are in the colder parts of the season most generally found near where warm streams enter the ponds or lakes and thereby furnish water of warmer temperature. They associate freely with other fish, which very seldom molest them. Indeed, they oftentimes take entire possession of their favorite grounds, to the exclusion of larger and stronger fish. Even the voracious pickerel, at times, has to give way to them, when they dash in among them, the sharp spines of their front dorsals erect and scraping their bellies as they *scoot* under and around with great dexterity. The brindle perch are a very good "panfish," their flesh being firm and sweet, though not very high flavored. They vary much in size, seldom weighing a pound, though instances are known of those being caught which weighed three pounds.

Distinctive Characteristics. The ground color of the sides is yellow. There are from six to eight vertical dark colored bands on the sides and over the back. The pectoral, ventral and anal fins are yellow.—*DeKay.*

Specific Description. The head is of a darker color than the other parts. Jaws equal. Pupils of the eye black and the irides of a bright gold color. Above and between the eyes you will find it smooth. The top of it is broad and somewhat depressed, or flattened. DeKay says that in the old fish "the rostrum becomes more elongated, producing a concavity of the facial angle," or, as some term it, "*dish faced.*" Back of the eye, the raised striæ radiating from centres make it rough to the touch. The preoperculum is scaly and notched with fine serratures along its edge, except a portion of the posterior superior angle, which is bare.

The opercle is subtriangular, of a green color in its centre, and has a few scales on its upper margin, is notched or serrated beneath, and has a sharp spine on its posterior angle. The length of the head is not quite one-fourth the length of the body.

The body is somewhat compressed and elongated, with a subcircular or gibbous outline. Sides of a golden color, crossed by "seven transverse bands," of a dark color. These bands are located in the upper part—those of the middle broadest of all. The abdomen is white. Chin or lower jaw pink or flesh-colored. The lateral line is a series of tubular orifices, and runs parallel with the curve of the body from the humerus to the tail. The scales are small and pectinated or ciliated on their posterior edges, giving a rough feeling to the hand.

There are two dorsal fins—the first with spinous rays, thirteen in number, and the second with mostly articulated or soft rays, also thirteen in number. These fins are of a yellowish-brown color, the first or anterior one tinged more with a light yellow. The spines strong—the first shorter than the second, and the fourth and fifth longest, the last much shortest. The whole height of this fin is about one-third of its length. The distance between the first and second is small, seldom more than three-tenths of an inch.

The second dorsal is of a subquadrangular shape, a little more than half the length of the first, and has oftentimes two spinous rays, the first one quite small.

The pectorals rise very near the shoulder, or humeral bone, are rather long, somewhat fan-shaped and rounded slightly on the margin, and have fifteen rays. They are of a yellow color.

Ventrals rise a short distance behind the pectorals, are a little triangular in shape, and have the outside ray spinous. Color orange, tinged with scarlet.

Anal rises about opposite the middle of the second dorsal, has ten rays; its first two outside rays are spinous, the first one shortest, color like that of the ventral.

Caudal is emarginate. Ray formula is as follows:

B. 7, D. 13, 2-13, V. 1-5, A. 2, 8, C. 18.

Whole length 10 to 15 inches.

SYNONYMES.—*Bodianus flavescens*, Mitchill, Kirtland.
Morone flavescens, Mitchill.
Perca acuta, Cuv. and Val.
Perca gracilis, Rich.
Perca granulata, Linsley Cat. Fishes of Conn.

GENUS MORONE, (*Mitch.*) *Gill.*

GEN. CHAR. Body oblong ovate. Gibbous as far as the commencement of first dorsal fin. Maxillary teeth and those of the palatines and vomer villiform. Lingual teeth situate on the margin of the tongue—none on the base. Scales on the head pectinate. Preopercule denticulate, or serrate on its lower posterior margin. Operculum two spined. The two dorsal fins are connected by a slightly elevated membrane at their base. First dorsal is composed entirely of spinous rays. The anal has three spines, of which the second is often the largest. The lateral line slightly convex anteriorly—nearly concurrent with the body. The chief distinctive characteristics of this genus are the presence of strongly pectinated scales on the cheeks and opercular bones, and the band of villiform teeth on the sides, and more scattered over on the tip of the tongue.—*Gill.*

Morone Americana, Gill.

White perch.

Wherever you find the yellow perch you also find the white perch, but in many of our lakes and ponds in Maine the white perch in some seasons are the most abundant, and furnish the best sport to the angler. Although not altogether similar in form and structure to the yellow perch, they are nevertheless very much alike in their habits and modes of life. Like them they prefer clear waters and sandy bottoms.

In the latter part of summer they congregate together in such waters, and afford fine sport to fishing parties in many sections of the State. They are then active and bite greedily, and we have known of as many as 150 being caught in a single day by one individual. They breed in great numbers in the Cobbosse Contee, and adjacent ponds in Winthrop and Hallowell, as well as in many others of our numerous sheets of water in different sections of the State. They are easily transferred from their native haunts, and are found to become easily habituated to their new locations, provided they are of suitable character, and soon multiply rapidly. They are much esteemed by many sportsmen, as making an agreeable ingredient in chowders and other savory dishes of the fisherman's and hunter's camp fare. They are of a silvery gray color, and like their cousins, the yellow perch, vary in size from a half a pound to a pound and a half, although the latter size is seldom caught.

Distinctive Characteristics. Body compressed, front part gibbous. First ray of the posterior dorsal nearly as long as the second. Opercle with a single spine. General length three to six inches.—*DeKay.*

Specific Characters. Head rather small and slopes gradually to the snout, with a suture behind the eyes, and is about one-fourth the length of body. Jaws about equal, with rose tints on the lips and underneath lower jaw. Upper jaw protractile ; tongue and lips spotted with very small black dots. A number of small teeth on each jaw, and a row of velvet like teeth on sides of the tongue, but none on the tip or centre. Eyes circular, pupils black, iris silvery ; space between the eyes covered with scales. The nostrils double, hindmost one oval and transversely situated to the other. Preopercle gives metallic reflections ; serrated on its posterior and lower margin. Operculum scaled ; has two points, the one in the posterior edge sharp, and the one above more blunt.

Body rather deep beneath the first dorsal fin, compressed, curved, or gibbous, in front, and tapering with a slight curve back of the first dorsal ; scales silvery, and covered with small black dots, like the tongue and lips. Their free edges are serrated ; lateral line follows the curve of the back ; first dorsal rises on a line a little back of the insertion of the pectorals; has strong spinous rays, the fourth of which is longest. The second dorsal is connected with the first by a low membrane; it is quadrangular in shape.

The pectorals rise just below and on a line with the spine of the operculum; are broad, the upper ray longest, and the edges rounding downwards. They are brown at the base, but yellowish above; ventrals rise a little back of the pectorals, and are reddish at the base.

Anal rises on a line with the middle of the second dorsal; second spine very strong. Caudal forked or "deeply emarginate."

Ray formula—

D. 9, 1, 12; P. 15; V. 1, 5; A. 3, 9; C. 17.

Length six to ten inches.

SYNONYMES.—*Perca americana*, Grd., Block., Lacepede.
Perca mucronata, Raf., Sw.
Morone rufa and *Morone pallida*, Mitch.
Bodianus rufus, Mitch; *Centropomus albus*, Raf.
Labrax mucronatus, Cuv. and Val., Storer, Ayres, Linsley, Baird, Hill.
Labrax rufus, DeKay, Gill.
Labrax pallidus, Mitch., Storer, Perley.
Labrax nigricans, DeKay, Storer.

GENUS ROCCUS, (*Mitch.*,) *Gill.*

GEN. CHAR. Body slender oblong-ovate. Back anteriorly curved or gibbous, Maxillary teeth and those of the vomer and palatines villiform. Lingual teeth villiform, deposited in two lateral bands, and separated at their base in two longitudinal series, either separated or coalessing. Scales from the nape to the nostrils and for the most part cycloidal. Preoperculum at the lower posterior part serrated. Operculum two spined. Dorsal fins not joined at the base by a raised membrane. Anal has three spines, increasing in size. Lateral line rectilinear.—*Gill.*

This genus has been separated recently by Prof. Gill from the genus *Labrax*, from which it differs chiefly in the character of the armature of the preoperculum, and by the absence of teeth at the anterior extremity of the tongue; the whole margin of the tongue in the latter genus (*Labrax*) being provided with a band of villiform teeth, and the spur-formed teeth of the lower margin of the preoperculum.

Roccus lineatus, Gill.

Striped bass. Rock fish.

This fish seems to posses rather a migratory character. Late in winter and early in spring it comes up from the sea into the rivers along the coast, where it is taken by seines or nets. It formerly ascended quite high up our rivers, but the dams have now shut them out from many of their old haunts, and they are now found only in those fresh waters near the ocean that are unobstructed. On the Kennebec they are now seldom seen higher than Eastern river, in Dresden, where they used to be taken in considerable numbers. The experiment has been tried of transferring them into lakes above the dams to breed and increase, but with rather imperfect success thus far. On the seaboard on some parts of the coast they come up into the creeks and arms of the sea, generally during the flood tides in the night, to feed and return to deeper water at ebb. They prefer rather shoal waters of the sea, with rocky bottoms. They bite the hook at times very readily, and many are thus taken. Fishermen use for bait, crabs or clams, but they prefer, of all things, *squid* to anything else. DeKay says that the largest individuals, called green heads, never ascend the fresh water streams. He also observes that in New York harbor, as the weather grows cold, they penetrate into the little bays and ponds connected with the sea, and imbed themselves into the mud.

The bass, or rock fish is highly prized by some, though the larger ones are of rather coarse flesh. Those found in the market vary greatly in size, from one pound upwards. Storer mentions one that weighed 84 pounds.

Specific Description. The head as well as the body of the striped bass or rock fish is covered with large and pretty strongly adherent scales. It is rather blunt, lower jaw a little the longest. The eyes are rather large, and their distance apart equal to a little less than two of their diameters. The pupils are black and the irides of a golden hue. The operculum is of a golden shade and has two spines on its posterior edge, the lowest of which is largest. The preoperculum is of the same color, and its posterior margin is finely serrated, the denticulations being largest on its lower edge. Teeth in the jaws and palatines, no teeth at the extremity of the tongue. Holbrook says "there are two bands of minute teeth at the root of the tongue, separated slightly from each other in the mesial line; the sides of the tongue are also armed with small teeth." The body is cylindrical, tapering to the tail, and about four times as long as the head. It is of blueish brown above, and near the abdomen commences a bright silvery hue. There are eight or nine blackish bands or streaks passing from near the operculum, the middle ones terminating in the caudal fin. Sometimes these bands are curved and sometimes interrupted. Scales large and somewhat quadrangular, showing concentric lines on their sides, as also very minute striæ diverging from the centre. They continue more or less on all the fins. Storer enumerates sixty-two scales along the lateral line. The lateral line is nearly straight, arises near the superior spine of the operculum, and extends through the fourth stripe of the body to the tail.

The first dorsal fin arises behind the line from the posterior half of the pectoral. The spinous rays are nine in number. The first smallest, the fourth and fifth longest, and the others decreasing in length to the last. The second dorsal has the first ray spinous, and twelve branched ones—second ray longest.

Pectorals arise a short distance behind the gill opening and a little below the lower spine of the operculum, rather short, and have sixteen rays. The ventrals are a little behind the pectorals ; the first ray is spinous and short, the second ray is longest. The anal arises on a line midway of the second dorsal. Its three first rays are spinous, rather short and stout.

Caudal deeply forked, and twice as wide at its posterior edge, when spread, as it is at its base.

Ray formula—

D. 9, 1-12 ; P. 16 to 18 ; V. 1-5 ; A. 3, 11 ; C. 17 to 18.

Length from six inches to four feet.

SYNONYMES.—*Sciæna lineata*, Block.
Perca saxatilis, Schnæd.
Perca Septiontralis, Schnæd.
Centropome Raye, La. Cep.
Perca Mitchilli, Mitch.
Labrax lineatus, Storer, DeKay, Rich.

GENUS POMOTIS, *Raf.*

GEN. CHAR. Body subcircular or elliptical, very much compressed. Head small or moderate ; mouth proportionate to size of head. Jaws generally equal, lower one longest. Velvet or card teeth upon the jaws and front of the vomer only. Tongue smooth, cheeks and opercular apparatus scaly. Branchial apertures continuous under the throat. Spinous portion of dorsal fin longer and less elevated than the soft portion. Three anal spines. Insertion of ventrals situate posteriorly to the base of pectorals.

Posterior margin of caudal fin emarginated or subcrescentic. Scales small, developed and pectinated.

Pomotis vulgaris, Cuv.

Bream, Roach, Sunfish, Pumpkin-seed, Pondfish, Kiver.

If you should take a pumpkin seed and carve out a fish's head on the blunt end, attach dorsal fins to one side or edge of it and a fish's tail to the small end, you would have no mean miniature representation of this species of fish. This resemblance has given rise to one of the many names given it, "*pumpkin seed*," which is quite expressive of its form.

The *Pomotis* or Bream is found in nearly all the fresh waters of New England. It is found in nearly all the brooks and margins of lakes and ponds in Maine. The avidity with which it seizes the hook, and the ease with which it is taken, render the fishing for it a capital subject to initiate boys into the craft and mystery of the angler's art, being often caught by them with the simple and rude apparatus of a worm on a pin-hook, tied to an alder twig for a rod. Many an urchin in the country, thus equipped, becomes quite an expert at brook fishing, and exhibits as much zeal and prowess, and as many trophies of his piscatory victims as the most skillful angler, with reel and gaff, does over the subtle trout and salmon. In the spring of the year the Bream repairs to the margin of ponds, or to the eddies and still waters of brooks, where there is a gravelly or sandy bottom, and cleaning off any vegetables that may be in the way, they scoop away the sand by swimming around and stirring up the surface, and form a basin shaped cavity, sometimes two feet across. Here they deposit their spawn and carefully watch the premises until their eggs are hatched.

Characteristics. Green mixed with olive, and dull reddish spots over the body. Appendix of the opercle black, bordered with scarlet. Length four to eight inches.—*DeKay.*

Specific Description. The head of the *Pomotis* is rather small and less than a quarter part as long as the body. Eyes are large and circular, and near the "facial outline." Nostrils double, the forward one tubular. Mouth is small, and the teeth sharp and rather thickly set upon the jaws, vomer and pharyngeals.

The edge of the preopercle very finely serrated. The operculum terminates in a blunt point, which is prolonged by a membranous appendage, which is black with a bright scarlet spot at its extremity. The gill covers present beautiful blue, or azure lines running lengthwise across them.

Body is oval and very much compressed and flattened, of a greenish color above, with irregular spots or blotches of red or rusty colors rather irregularly disposed along the sides. The abdomen is of a whitish silvery color. The lateral line curves with the outlines of the back. The scales are quite large and are toothed at their base.

The dorsal fins are made up of spinous and soft rays, and the portion of soft rays is highest and rounded; the anterior portion of it is spinous, and this portion has ten rays, and the soft portion twelve.

The pectorals are pretty long, of a triangular shape, and generally reach to the soft portion of the dorsal. The ventrals have one spinous ray. The anal fins extend as far as the dorsal. The caudal fin is emarginate.

Ray formula—

D. 10, 12; P. 1, 5; A. 3, 16; C. 17.

Very seldom over eight inches long.

SYNONYMES.—*Labrus auritus*, Lin.
Morone maculata, Mitch.

Pomotis appendix, DeKay.

Red-tailed Bream.

This species of *Pomotis* is not so abundant as the *Pomotis vulgaris*. DeKay describes it as more robust in body, thick and chubby than the above named. The prolongation or appendix to the operculum broader and longer, and of a black color. Mouth and gape larger. Storer describes the dorsal fin as being anteriorly dark brown and the posterior membranous portion red; ventrals red at their base. Pectorals yellowish brown. Anal yellowish at the base and fuliginous at its margin. Caudals of a blood red color. After death he observes that the body becomes of a bluish-gray color, abdomen orange, extremities of the ventrals purple, and the tail a rust color.

Ray formula—

D. 10 to 11; P. 11 to 12; V. 1, 5; A. 3 to 10; C. 18.

Length six inches.

SYNONYMES.—*Labrus appendix*, Mitch.

FAMILY SPAROIDÆ, *Cuv.* SUBFAMILY SPARINÆ, *Bona.*

CHAR. No spines or denticulations on the opercular bones. No teeth in the palate. Mouth not protractile, sides large. No scales on the fins; muzzle not thickened, nor the bones of the head cavernous. Pylorus has coecal appendages.

GENUS PAGRUS, *Cuv.*

GEN. CHAR. Four to six stout teeth in front of each jaw, and two series of round teeth on the sides. Many species have, behind the front teeth, numerous small granular or cardlike teeth. Body generally deep.

Pagrus argyrops, Cuv.

Big Porgee. Scapaug.

Characteristics. Brilliant metallic reflections on the sides. A short recumbent spine in front of the dorsal fin. The second and third dorsal rays often filamentous.—*DeKay.*

DeKay gives the fin rays—

D. 9, 3, 22; P. 18; V. 1, 5; A. 3, 18; C. 16.

Storer's formula is different, as follows:

D. 12, 12; P. 15; V. 6; A. 3, 11; C. $16\frac{2}{3}$.

Perhaps it is an error to call this a Maine fish. It is sometimes brought in during summer by fishermen who have been out Cape Cod way.

SYNONYMES.—*Sparus argyrops*, Lin.
Spare xanture, La. Cep.

FAMILY SCIÆNOIDÆ, *Cuv.*

CHAR. Body similar to that of the Percoids, protected with pectinated or else ctenoid scales, extending over the head and a portion of the fins.

Head peculiar, owing to a convexity of its upper surface and the bluntness of the snout; the bones of the skull being cavernous and otherwise provided with crests or ridges. They may be distinguished from the Cataphracti tribe by the suborbital bones not extending across the cheek.

Mouth little protractile. Barbels sometimes under the lower jaw, near the mouth. Maxillary teeth various.

Vomer and palatines toothless, which distinguishes them from Percoids. The operculars have spines or serratures. The preopercule is occasionally smooth. Fins—Either one or two dorsals, having the general modifications as the Percoids.

Air-bladder peculiar by the horn-like process it exhibits.

SUBFAMILY SCIÆNINÆ, *Bona*. GENUS OTOLITHUS, *Cuv.*

GEN. CHAR. The bones of the anal fin are weak and there are no barbels ; some of the teeth terminate in elongated hooks, or are of the canine form. Their natatory bladder has a horn on each side, projecting forwards. (Storer.) Two small pores on the lower jaw are entirely wanting. Two dorsal fins. Body elongated.—*DeKay*.

Otolithus regalis, Cuv.

Weak fish, Squeteague, Squetee, Checouts.

Prof. Gill has transferred this species to a different genus, "Cynoscion," and terms it *Cynoscion regalis*, but I have not seen his generic characters of it.

It is a fish of very delicate structure, and makes such feeble resistance to the hook that it breaks from it very easily.

Characteristics. It is of a bluish color above, veined with dusky. Ventrals and anal, orange. Ventrals with fine branched rays. Length one to two feet.—*D. K.*

Ray formula—

D. 8, 1, 28 ; P. 1g ; V. 1, 5 ; A. 13 ; C. 17.

SYNONYMES.—*Labrus squeteague*, Mitch.
Sciæna regalis, Rich.

FAMILY CORYPHÆNOIDÆ, *Lowe*. SUB-FAMILY PEPRILINÆ, *Gill*.

GENUS PRORONOTUS, *Gill*.

Poronotus triacanthus, Gill.

Harvest fish.

This is the Rhombus triacanthus described by Storer and DeKay as belonging to the Scombridæ or mackerel family, but removed by Prof. Gill into the above-named family and genus. As we have not their descriptive characters at hand, we will give those under which they were formerly described. For description of Scombroid family see further along.

GENUS RHOMBUS, *La. Cepede.*

GEN. CHAR. Head and body compressed. Body covered with minute scales. A small trenchant and pointed blade before the vent. A horizontal partially connected spine before the dorsal and anal fins.

Rhombus triacanthus, DeKay.

Short-finned Harvest fish. Skip Jack.

DeKay says of this fish that it is equally remarkable for the splendor of its coloring and its excellence as an article of food, although many fishermen consider them unfit for eating on account of the unpleasant odor which they emit when opened. They are believed to feed chiefly on marine plants. He found the oesophagus of many which he opened, filled with pebbles about the size of a pin's head. When taken from the water at night, it is said to emit vivid phosphoric flashes.

Prof. Peck describes them on the New Hampshire coast as long ago as 1794. They are sometimes used on Cape Cod as a manure. They make a good bait for mackerel.

Ray formula—

D. 3, 45; P. 19; A. 3, 42; C. 19.—*DeKay.*

Storer has it—

D. 45; P. 21; A. 43; C. 20.

SYNONYMES.—*Poronotus triacanthus*, Gill.
Stromateus triacanthus, Peck.
Stromateus cryptosus, Mitchill.
Peprilus cryptosus, Cuvier.

FAMILY SCOMBROIDÆ.

Mackerel.

This family has been recently sudivided into several subfamilies, among which are enumerated *Scombrinæ, Bona.; Orycninæ, Gill.; Caranginæ, Bona.; Seriolinæ, Gill.* We cannot now give the characters of each, and therefore recite the character given to the family before its subdivision.

CHAR.—Body exceedingly diversified in form and aspect, covered with minute scales, giving a smooth appearance to the skin.

These scales are both cycloid and ctenoid. Many genera are provided with a crest or ridge on the sides of the tail, often protected with a series of keeled, bony, scale-like shields.

Head.—Sides of head smooth. Operculars have neither spines nor serratures.

Fins.—Dorsal, caudal and anal fins scaleless and varied in structure, according to different genera.

Stomach.—Numerous pyloric appendages to the intestines.

Air-bladder wanting.

SUB-FAMILY SCOMBRINÆ, *Bona.* — GENUS SCOMBER, *Cuv.*

GEN. CHAR. — Body fusiform, covered by scales which are uniformly small; sides of the tail not connected but merely raised into two small cutaneous crests; dorsal fins widely separated; some of the posterior rays of the second dorsal and anal free, forming finlets. One row of small conical teeth in each jaw.

Scomber vernalis, Mitch.

Spring Mackerel.

The mackerel, or spring mackerel, is one of those migratory fishes that everybody knows something of, because they form so important a branch of our fisheries and enter so largely into the commercial and dietetical uses of the community.

They appear on our coast about the middle of May, and their numbers gradually increase until into June. The first comers are males, and are rather lean and do not rank so high in the inspection as those later comers, or those caught later, that have become fatter.

They are caught with the hook, in large quantities, but they are subject to what the fishermen call "freaks" in this respect, sometimes taking the bait eagerly, and at other times pass along without taking the least notice of the bait and rejecting all the allurements which the fishermen can devise. To obviate this, many have adopted the custom of catching them with seines and drift nets by which great numbers are caught during the season of their stay in our waters. I have gathered some statistics in regard to the amounts taken in different years, from 1850 to 1860, but as they are quite defective and unsatisfactory I omit them at present. They give, however, some interesting facts showing the great importance of this branch of our industry.

Characteristics. A dark spot at the base of the pectoral and anal fins.—*DeKay.*

Specific Description. The mackerel when first drawn from the water is a beautiful fish. De Kay describes its colors as exceedingly vivid; dark steel blue above, becoming lighter on the scales and mixed with metallic green near the lateral line. From 24 to 30 vertical deep blue half bands which are sometimes angular, curved, interrupted and occasionally forming irregular circles. Below the lateral line, and parallel with it, is a longitudinal, dull, brownish line, often interrupted and sometimes forming a series of inequidistant, irregular spots, occasionally both line and spots wanting. Beneath, silvery, with greenish and yellowish metallic reflections, a black blotch of the pectorals and ventrals. Pectorals, second dorsal and caudal, dark colored; the remaining fins lighter. Iris of the eye white with a slight tinge of yellowish; pupils black.

A careful examination of these colors would be sufficient to identify this species. The head is somewhat pointed and about a sixth of the length of the body; mouth of a moderate size; eyes large. A single row of small teeth on the jaws, a single row on the palatines; tongue black. The lower margin of the operculum has a row of mucus pores. The body is cylindrical, tapering to the tail. The lateral line is waved or undulating, passing from the humeral bone to the tail.

Dorsal fin commences over the ventrals; contains in the first 13 simple rays. The second dorsal is one third as high as the first; it has ten rays. Behind these are six finlets composed of one ray and equidistant. Generally there are five corresponding finlets on the opposite side beneath; pectorals acute, fourth ray longest; ventrals subdivided at their middle and again at their extremities, and opposite the commencement of the second dorsal, at its commencement is a short spine, and next to it the finlets before named; caudal deeply forked—on each side are ridges or carinæ; no air bladder.

Ray formula—

D. 13, 10—V. 1; P. 17; V. 6, A. 12; V; C. 15.—*DeKay.*

Storer has it—

D. 10, 12; P. 17, V. 5, A. 12, C. 20.

Length 15 to 17 inches.

SYNONYME.—*Scomber scomber*, Schoepf.
La maquereale printanier, Cuv. et Val.

Scomber grex. DeKay.

Fall Mackerel.

Characteristics. Small, a black spot at the base of the pectorals and tip of lower jaw. Dorsal bands very tortuous. Length 8 to 10 inches.

There have been doubts in the minds of some Ichthyologists in regard to this species being a spring mackerel (*Scomber vernalis*) of a different age.

GENUS THYNNUS. *Cuv.*

GEN. CHAR. Form of the body like that of the Scomber, but less compressed. A kind of corselet around the thorax, formed by scales larger and coarser than those of the rest of the body; a long and elevated crest on each side of the tail. The anterior dorsal reaching almost to the posterior one. Numerous finlets behind the dorsal and anal fins; a single row of small, pointed, crowded teeth in each jaw.—*Storer.*

Of this genus we have one species on our coast the

Thynnus secundo dorsalis. Storer.

American Tunny, Horse Mackerel, Albicore.

It comes on to our coast in the early part of summer, when it is very lean; by autumn, before it diappears, it is very fat and affords a large quantity of oil from its head and belly. It frequently grows to a large size, weighing from 500 to 1000 pounds.

Its *characteristics* according to DeKay are—very large and long pectorals, corselets pointed behind, no colored lines or spots. Length 9 to 12 feet.

Ray formula —

D. 14, 1, 23 — X; P. 34, V. 1, 5; A. 2, 12 — IX; C. 19.

This fish is placed by Prof. Gill into the sub-family *Orycninæ*, and into Cuvier's genus (Orycnus.) This species he terms

Orycnus secundo dorsalis. Gill.

GENUS CYBIUM, *Cuv.*

GEN. CHAR. An elongated body without a corselet, and large compressed sharp teeth. The palatines have only short teeth.

We have one very rare species of this genus, occasionally found on our coast, viz :

Cybium maculatum, Storer.

Spanish Mackerel.

This is characterized as large,—numerous greyish brown spots distributed along the sides. Length one to two feet.

This species has been placed by Prof. Gill into the Orcyninæ sub-family and into the genus *Apodontis* of Bennet. Its specific name under this arrangement is *Apodontis maculatus, Gill,* and is synonymous with the *Scomber colias,* (Storer) and *la maquereau colias* of Cuvier and Valenciennes.

GENUS VOMER, *Cuv.*

GEN. CHAR. Body compressed. No filaments or prolongations of the fins Profile nearly vertical.

SUB-FAMILY CARANGINÆ. VOMER SETIPINNIS, *Ayres.*

Blunt-nosed Shiner.

This species is inserted here on the authority of another. I have never met with it in our waters and have considered, on the statement of DeKay, that New York, or the southern coast of Massachusetts was its northern limit. It has been described as the body of a lustrous silvery tint, passing into a leaden tint on the back. Iris yellow ; membrane of the second dorsal minutely dotted with black, tinged at its base with light yellow. Pectorals olive green, verging to dusky.

First dorsal composed of short isolated rays deeply hidden in a groove.

Ray formula —

D. 7, 1, 22 ; P. 1, 18 ; V. 1, 3, A. 1, 18.

air bladder very large with two horns behind.

This has very recently been placed into the sub-family *Caranginæ* (Bona.) Its synonymes are

Vomer Brownii. Cuv.
Zeus setapinnis. Mitch.

SUB-FAMILY SERIOLINÆ. GENUS TEMNODON, *Cuv.*

GEN. CHAR. The tail unarmed, the little fins or detached spines are before the anal as in Seriola. The first dorsal, fragile and long, the second and the anal covered with small scales; but the principal character consists in a row of separated, pointed, cutting teeth in each jaw; behind the upper ones is a row of smaller teeth, and there are some fine as velvet on the vomer, palate and tongue. The operculum terminates in two points, and there are seven branchiostegous rays.

Temnodon saltator. Storer.

Bluefish. Skip-jack.

This species of fish seems to have an historical interest on account of its great abundance, at times, and then disappearing as to great numbers for many succeeding years. It is a great scourge to the herring and mackerel fisheries when they appear among them, as it eats them voraciously and becomes fat upon them. It It is rather a handsome fish and sometimes grows to the weight 14 pounds. Storer describes its colors as bluish on upper part of the body; greenish tinge upon the side and abdomen. Iris is yellow. Pectorals of a greenish yellow with a deep black blotch at their base. Second dorsal and caudal fins are likewise of a greenish brown color. Ventrals and anal fins are of a bluish white color.

Fin rays —

D. 7, 26, P. 17, V. 6, A. 28, C. 20.

This species has been placed recently in the sub-family *Seriolinœ, Gill.* Genus Pomatomus of Lacepede, and is termed *Pomatomus saltatrix, Gill.* Its synomyms are

Gasterosteus saltatrix Lin.
Scomber plumbeus Mitch.

FAMILY SCOMBERESOCOIDÆ.

CHAR. *Body* elongated.

Head.—Gills fully developed—last branchial aperture extant. Pseudo branchiæ glandulous, and covered by the mucus membrane of the branchial aperture, and therefore concealed.

Scales cycloidal—a row of keeled ones on either side of the body, distinct from the lateral line.

Dorsal opposite the anal; rays soft and articulated.

Ventrals abdominal in position; rays soft, articulated.

Air-bladder has no duct leading to the throat.

Stomach has no *culdesac* and no pyloric appendages—straight and hardly distinguished from the intestines passing gradually into them.

GENUS SCOMBERESOX, *Cuvier*.

GEN. CHAR. Snout greatly attenuated and elongated, as in the preceding. Teeth in both jaws; more on the palatines and tongue. Dorsal and anal fins divided behind into numerous finlets.

Scomberesox scutellatus, Les.

Bill fish.

The Bill fish is a very handsome species, with a body shaped like that of a mackerel, and the head, or rather snout, elongated much longer than that of a pickerel. Hence the common name of Bill fish. It is found all the way from the shores of Newfoundland to those of Cape Cod. At the latter place it is most abundant in October.

Its *characteristics* are, "dark green above. Lower jaw longest. Body with a broad silvery band."—*DeKay*.

The head is narrow and long; eyes small; gill covers large, smooth; lower jaw one quarter of an inch longer than the upper. Minute teeth on the base of it. Nostrils are large. Body is somewhat eel-formed; scales small. The lateral line straight and near the back. A furrow extends from the lower edge of the operculum to the base of the caudal fin. Storer describes this as consisting of two yellowish lines, which are a continued series of scales. When raised, they resemble serrations; when not erect, they look like sinuses (furrows.) Between these rows are situate the ventrals, the anal, and the anal finlets.

Ray formula—

D. 10, v. or vi; P. 14; V. 6; A. 12, v. or vi; C. 20.

The caudal is deeply forked, the lower lobe slightly the longest.

SYNONYMES.—*Scomberesox Storeri*, DeKay.
Scomberesox equirostrum, Lesueur.
Esox longirostra, Mitch.

FAMILY GASTEROSTEOIDÆ, *Bona*. SUB-FAMILY GASTEROSTEINÆ, *Bona*.

CHAR. *Body* diminutive in size—no scales; sometimes naked, sometimes plated entirely or in part.

Head.—Gills four in number on each side, composed of two perfect branchial combs. Four branchiostegal rays; gill openings being separated beneath by an isthmus—the last gill opening situ-

11

ate between the fourth gill and inferior pharyngeal bones, is fully developed.

Fins.—The spinous rays of the dorsal region, instead of being united by a membrane into a dorsal fin, are isolated from each other. Each spine has a very small membrane at its posterior base. Spines variable in number, transversely flattened upon their base, and acerated upon their extremities, either smooth upon their edges or denticulated. Can be laid back in a horizontal position. Ventral fins situate about the middle of the abdomen, composed almost exclusively, with few exceptions, of one stout spine.

Pelvic bones are external and united to the thoracic belt.

GENUS GASTEROSTEUS, *Artedi*.

GEN. CHAR. Upper surface of head plane either smooth or corrugated. Opercular apparatus without spines. Mouth rather small, oblique; posterior extremity of maxillary not extending as far as a vertical line drawn in advance of the anterior rim of the orbit.

Minute velvet-like teeth upon the dentaries and premaxillaries; none on either the vomer or the palatines. Gill openings separated by a narrow isthmus; branchiostegal rays three on either side. First dorsal represented by a series of isolated spines, varying in number. Caudal fin subtruncated, or subcrescentic posteriorly. Insertion of ventrals situate opposite the second dorsal spine, therefore abdominal; has one strong spine. Body either covered with a smooth skin, or partly, or totally covered with tranversely elongated plates. Lateral line very obsolete. Bones of the pelvis forming a shield to the belly, pointed behind.—*Grd.*

Gasterosteus biaculeatus, Mitch.

Two-spined Stickleback.

The stickleback is among fishes what the Bantam is among poultry—small, active, smart and pugnacious—often attacking and driving much larger fishes than themselves.

This little species is found in the small pools and creeks in many parts of the seashore and in salt marshes. It is characterized by two distinct spines on the back and a third near the dorsal; a strong serrated spine on each side, representing the ventrals.

The head is somewhat depressed or flattened above, with numerous punctures in rows, or as it is said by some, to be granulated; jaws equal; teeth minute and numerous. Eyes large for the size of the fish—pupils black, iris silvery. Opercles of a silvery color, spotted with dusky, covered with striæ. Mouth protractile. Nostrils large and placed half way between eyes and point of the jaw. A broad silvery plate back of the gill opening. Body somewhat oblong, compressed, and very slender at the base of the tail.

It is covered with twenty-eight or thirty bony plates, narrow and plated vertically, with slight serratures on their posterior margins and with perpendicular striæ or markings. It is of a green color above and silvery below.

Lateral line pretty high up. It has the power of raising or depressing the spines on the back; the dorsal is longer than it is high. The forward rays longest, one spine and twelve soft rays; pectorals fan-shaped. Two sharp serrated spines stand in the place of the ventrals, and between them is a kind of bony lance-shaped plate, rough upon its surface, serrated on its edges and on central keel. It seems to be made for a support to the abdominal point which is anterior to the vent.

Anal fin commences on a line a little posterior to the dorsal, and ends on a line with its own preceded by very minute spines.

Caudal fin is very slightly emarginate and has twelve rays.

Ray formula—

D. 2, 1, 11; P. 10; V. 1; A. 1, 6; C. 12.

Length two and a half inches.

Gasterosteus DeKayi, Agassiz.

Many Spined Stickleback.

This little fish is an occupant of both fresh and salt water, often in the brackish water about salt meadows. It has more than seven spines in front of its dorsal fin.

Specific description. The head of this species is small, about one-fifth the length of the body. Gape of the mouth obliquely downward; nostrils round, and near the orbit; jaws full of small teeth. The opercula are of a silvery color; eyes circular black, iris metallic.

Body is elongated, a little compressed, and tapers gradually from the dorsal. Upon the back there are generally ten sharp, slightly curved spines, inclined from right to left. The first are placed midway between the operculum edge and dorsal fin.

The dorsal is of a triangular shape; has one stout spine and seven soft ones; anal beneath dorsal and like it in shape, with four rays; ventral spines sharp, curved, with a slight membrane at base, and a bony plate between them pointed posteriorly.

Caudal rounded, and the tail is keeled or carinated with from twelve to fourteen distinct plates.

Ray formula—

D. 80, 9 or 10; P. 11; V. 1; A. 1, 9; C. 13.

Length from one to two inches.

SYNONYMES.—*Gasterosteus pungitius*, Storer.
Gasterosteus occidentalis, DeKay.

This species has been placed by Brevoort into a new genus, PYGOSTEUS, and terms this species *Pygosteus DeKayi*.

FAMILY ATHERINOIDÆ, *Bona*. SUB-FAMILY ATHERININÆ, *Bona*.

CHAR. *Body*, covered with cycloid scales, provided laterally with a silvery band. Mucous pores wanting.

Head. Upper arcade of mouth formed by the pre-maxillaries. The maxillaries which are situate behind, are tapering towards its free or posterior extremity instead of being dilated.

Upper jaws very protractile. Six branchiostegal rays on either side. Four gills on either side but no pseudo-branchiæ. Branchial apertures continuous under the throat. Pre-maxillar, maxillar and other teeth, so small as to require being magnified to be seen.

Fins. Two dorsals, widely separated. Ventrals are abdominal.

Stomach, a simple membranous pouch, no culdesac, or pyloric appendages.

Air bladder is extant.

GENUS ATHERINA, *Linn*.

GEN. CHAR. Body elongated; two dorsals, widely separated; ventrals further back than the pectorals; mouth highly protractile, and furnished with very minute teeth. A broad silvery band along each flank on all the known species.—*Storer*.

Atherina notata, Mitch.

Dotted Silver Side. Sand Smelt.

In some parts of our seaboard, especially at the mouth of the Piscataquis river, and the creeks around Kittery during the smelting season, there is a beautiful little fish caught, known in some places as the "*Sand Smelt*," "*Silver Side*," "*Capelin*," "*An-*

chovy" and "*Young Smelt.*" It is taken in company with the smelt and sold with them as such. It is from three to five inches in length, and when first taken from the water quite semi-transparent, especially along the back.

Characteristics. Body, slender; dorsal fins some distance apart, the second one over the middle of the anal.

Specific Characters. Head small, pupils somewhat straight above and curved beneath, smooth on the top, greenish and covered with small black spots; upper jaw slightly the longest, and both jaws well furnished with very small teeth; mouth, when shut, a little oblique, the angle being lower than the point of the lips, "mouth protractile." Eyes black and slightly oval, iris silvery.

Body slender, a little compressed, of a greenish color on the upper part, and "marked into diamonds by dotted lines." These dotted lines also disappear on the edge of the scales of the upper portion of the body.

The lateral line is dark, commencing near the upper angle of the operculum, extending parallel with the back. Beneath it, and parallel to it, is a broad band of a bright silvery color starting at the root of the pectorals to the tail. This is a very distinguishing mark. Below this belt the body is of a higher color than above it. It has two dorsals, the first arising about one inch back of the extremity of the pectorals, and is smaller than the second and of a triangular shape. The second dorsal commences about one inch back of the first. It is square or quadangular in shape, its last rays being longest.

Pectorals commence near the upper angle of the operculum and cover a portion of the silvery band. The ventrals arise on a line with the extremities of the pectorals. Anal commencing a little back of a line of the beginning of the first dorsal and is the largest fin of all. Caudal forked. All the fins have very delicate, colorless semi-transparent membranes.

Ray formula—

Branchial 6; D. 5, 9; P. 12; V. 5; A. 25; C. 18.

Prof. Gill has separated this species from the genus Atherina and places it into "Chirostoma," (*Chriostoma notatum, Gill.*)

SYNONYMES.—*Atherina menidia*, Lacepede,
L'atherine de Bosc. *A. boscii*, Cuv. et Val.
Chirostoma notatum, Gill.

FAMILY SCORPENOIDÆ, *Swainson*. SUB-FAMILY SCORPENINÆ, *Bona*.

CHAR. *Body*, large, and more or less spinous. Some genera exhibit flaps or membranous appendages ; in others there are scales all over the head, as also to the tips of the snout and along the jaws.

Head. There are seven branchial rays in all, and the gill openings are continuous under the throat. Gills, three and a half on either side, the fourth having but one branchial comb developed, consequently the last branchial split does not exist.

Fins. The dorsal is unique, the spinous portion combining closely with the soft.

GENUS SEBASTES, *Cuv*.

GEN. CHAR. Body rather short and contracted. Head largely developed ; upper surface covered with scales, and with or without spines. Mouth large ; eyes large ; inferior jaw the longest ; velvet or cardlike teeth upon the premaxillaries, dentaries, the front of the vomer and the palatines. Surface of the tongue smooth ; spines on the preopercle and opercle. Gill openings continuous under the throat. Branchiostegal rays seven on either side. Dorsal fins united at base and resemble one fin. Caudal posteriorly subcrescentic, or concave. Insertion of ventrals posterior to the pectorals. Body covered with well developed pectinated scales, opercular apparatus, cheek and jaws, and also over portion of the fins.

SEBASTES NORVEGICA, *Cuv*.

Norway Haddock, Red Sea Perch, Rose Fish, Snapper, Hemdurgan.

This fish is more abundant in the northen portion of our waters than in the southern, and is still more plenty on the shores of Newfoundland. Not having an opportunity to examine a good specimen, I here copy a part of Storer's description of it.

Color. In the recent fish the entire body, together with the fins is of a beantiful bright red, with the exception of a blotch upon the posterior portion of the operculum. After death the color partially disappears upon the throat and abdomen, and the space between the ventrals becomes nearly white, and at the posterior base of the soft portion of the dorsal a dull blotch is observed. Pupils black, iris yellow.

Description. Body oblong, compressed, covered with small rough scales. Head flattened above between the eyes and upon the occiput. The operculum is armed with three spines, one pointing upward and backward at its posterior upper angle; a second, beneath this, directed obliquely backward and downward; and a third, much smaller, at its inferior angle, and the pre-operculum is rounded at its edge and furnished with five spinous processes; the three posterior of which are the larger. Two spines upon the scapular bones and two upon the suborbitars. Four spinous projection upon the supraorbitars, all of which are pointed backwards; one at the anterior angle of the eye; a second with its base continued under the greater portion of the ridge; and the two smaller ones behind. Two elevated sharp ridges upon the occiput which bifurcate posteriorly into spinous points. Eyes circular, very large; the diameter of the orbit equal to one-third the length of the head, when the jaws are closed; uostrils just in front of the eyes, the posterior one largest. The jaws, pharynx, vomer, and palatine bones are armed with numerous minute teeth; upper jaw very protractile, and has an emarginature at its centre, into which the extremity of the lower jaw shuts when the mouth is closed.

The chin prominent. The lateral line above the operculum, and taking the curve of the body, terminates at its caudal ray. About thirty-six tubes are seen in the curve of the line.—*Storer's Hist. of Fishes of Mass., Mem. Am. Acad., Vol. 5, p.* 86.

Ray formula—

D. 15, 15; P. 18; V. 1, 5; A. 3, 7; C. 19

Length one foot.

SYNONYMES—*Perca marina*, Pennant.
Serranus Norvegicus, Fleming.
Scorpæna Norvegica, Jen.

FAMILY COTTOIDÆ, *Rich.* SUB-FAMILY COTTINÆ, *Bona.*

CHAR. *Body.* Very thick anteriorly, tapering rapidly posteriorly. An absence of true scales; spine not always smooth. In some perfectly smooth; in others studded with minute prickles, simple or compound. Others with series of longitudinal long shields; others with parallel rows of small scale-like plates, various in form and structure.

Head. Very large. Opercular apparatus provided with large spines. (Hence the name scull pin.) Three complete gills of two branchial combs, and a half one with but one comb. Teeth of velvet type on the jaws. Palate smooth, or velvet teeth on front of vomer and palatines.

Fins. Two dorsals sometimes continuous, sometimes separated. Anterior one always composed of spiny rays. Anal fin is opposite second dorsal. Ventrals under the pectorals, posteriorly to base of the latter with small number of rays. Pectorals very large, broad and expanded; their inferior rays undivided, (though articulated,) and projecting beyond the interradial membrane which is emarginated.

GENUS ACANTHOCOTTUS, *Girard.*

GEN. CHAR. Spines upon each of the opercular bones. Surface of head, and often the circumference of the orbits either serrated, or notched, or armed with spines. Mouth more deeply cleft than those of the Cottus genus. Lateral line uninterrupted.

Acanthocottus octodecim spinosus, Gill.

Sculpin, Pig Fish, Bullhead, Sea Toad, Sea Robin.

This is a handsome oddity, and a plague to fishermen who are fishing for better fish, bnt often have to haul it up, to their no small vexation when they see its head and horns emerging from the water. It is rightly named "*sculpin*" for its scull is full of spines and thorns as sharp as pins, and as strong as so many ten-penny nails, projecting point foremost in every direction. If irritated when first taken from the water it shows fight—swelling out its gill membranes, pushing out the horns, and erecting the spines of its dorsal like the bristles on the back of a wild boar. It is said to be eaten by some, but little or no use is made of it among our fishermen.

Characteristics. Spine of the pre-opercle reaching the point of the opercle. Pectorals very broad and rounded.—*DeKay.*

Specific Description. The common sculpin has a large, broad head with channels or furrows on the top, apparently made by a continuation of the spinous ridges. Storer says there are twenty spines upon and about the head. These are strong and generally curved backwards. Each nostril is armed with one of these. Each eye has one, and on the nape of the neck is one on each side of it. The pre-operculum is also armed with them. The one on the posterior angle is a formidable one, very stout and sharp, partly curved at its base, but may be laid bare its whole length. Just below this is another much smaller, and pointing backward and downward. Below this a smaller one pointing a little forwards. The operculum has but two. The larger on the upper anterior angle pointing backwards, and a small one on the inferior portion pointing downward. On the shoulder blade, just above the pectoral fin is another, pointing upward and backward, and above the commencement of the fleshy part, a membrane of the operculum, is another, but rather short. The mouth is large, and capable of being very much distended bythe fish itself. Card-like teeth are upon the jaws in pretty compact order; also upon the pharyngeal and palatine bones. Eyes large and the orbits projecting from the skull.

The body variegated in color, with a mixture of black and greenish yellow. Light colored in the belly, with a yellowish or browned tint occasionally. There are four dusky bars commencing on the back extending downward a short distance irregularly. It tapers gradually and regularly to the tail.

First dorsal smaller than the second with rounded margin on the top. Nine spinous rays, third ray largest, and all the rays are above the edge of the membrane. It is of a sooty color with brown bands.

The second dorsal has articulated rays, and is nearly twice as long as the first. It is brownish, with three brown bands nearly horizontal across it. The pectorals are very large with rounded or circular edge when expanded. Yellowish above, flesh colored below, and crossed with six brown circular bands. Ventrals rather slender but long, yellowish white, the rays projecting above the membrane. Anal fin nearly even and opposite second dorsal; yellowish color, with two dark bands. Caudal long and even at its end.

Ray formula—

D. 9, 16; P. 18; V. 3; A. 14; C. 12.

Length ten to twenty inches.

SYNONYMES.—*Scorpius Virginianus*, Willoughby.
Cottus octodecim spinosus, Mitch.
Cottus Virginianus, Storer.
Cottus scorpius, Schroep.
Acanthocottus Virginianus, Girard.

Acanthocottus Groenlandicus, Girard.

Greenland Sculpin.

This is the handsomest of the sculpin *beauties*, but is not so abundant as last described species. It can be distinguished from it by the different markings of its colors, and by a " quadrangular area in the head, bounded by four tubercles ; circular white spots upon the abdomen ; dark brown color of the body, with large clay-colored blotches on the top of the head and upon the gill covers, with a few smaller ones on the back and sides ; and small circular spots on the sides toward the abdomen."

It may also be distinguished by the feel of the sides of the body, both above and below the lateral line, which are "roughened by granulated tubercles, which seem like spines when the finger is drawn over them."

Ray formula—

D. 9 or 10, 16 or 18 ; P. 17 ; V. 3 ; A. 13 ; C. 16.

Length ten to fourteen inches.

SYNONYMES.—*Cottus scorpius*, Fabricius.
Cottus quadricornis, Parry's Voyage.
Cottus variabilis, Ayres.
Acanthocottus variabilis, Girard.

Richardson and Bonaparte have separated the Triglidæ family to which this genus formerly belonged, into several new ones—placing the above into the newly family Cottoidæ and sub-family Cottinæ.

GENUS HEMITRIPTERUS, *Cuv.*

GEN. CHAR. The head depressed ; two dorsals ; no regular scales on the skin, but teeth in the palatines ; head is bristly and spinous, and has several cutaneous appendages. The first dorsal is deeply emarginate, a circumstance which has led some authors to believe there were three dorsal fins.—*Storer.*

Hemitripterus Acadianus, Storer.

Deep water Sculpin.

One of the most elegantly colored though at the same time uncouth of the sculpin tribe, is the "Deep-water Sculpin," as it is called ; called also by fishermen Sea Rover, which though not exactly coming under the same genus is nevertheless closely allied to it in form and habits. It is taken often times when fishing for codfish in deep water, and is an unwelcome intruder upon the lines and bait and time of the fishermen. It is easily distinguished from the other Sculpin by the odd fringes or cirrhi about the eyes and ranging from the lower jaw, which, with the indentations, ridges and spines give it a singular expression of countenance, if a Sculpin can be allowed to have a countenance.

Specific Description. Head large and broad and measuring from the point of the lips to the point of the operculum nearly or quite a quarter of the whole fish. Sometimes it is of a blood red, sometimes of a yellowish or pink purple with variegated markings of a brown or some other darker color with white. The mouth is very large when opened. Jaws of the same length, from the lower one there hang about a dozen fleshy fringes or cirrhi, giving it a singular appearance, cardlike teeth are found on the jaws, vomer palatine bones and the pharynx, but the tongue is quite large and smooth. The snout presents, a little above it, a ridge on each side with several spinous projections. Eyes moderate in size but the orbitar projections are large, and a fringe is attached to them composed of the fleshy cirrhi before mentioned They are also marked with white vertical lines. Pupils black, Iris yellow, tinged with brown.

The preoperculum has two spines on its posterior angle, the superior one curving upward and backwards. The operculum is sub triangular in shape, rather small, ending in a blunt point and has ridges on its surface.

The body is oblong, in outline cylindrical, and as Storer observes "granulated, and studded with innumerable tubercles which are quite large upon the back and very small or almost imperceptible below the lateral line." The dorsal has somewhat the appearance of three fins. It rises just behind the spinous processes of the head. The first ray is long while the next, fourth, fifth and sixth are much shorter, and the next following rise again. Small appendages like tentacula are suspended from the tops of those rays. The second dorsal rise immediately behind, the first ray shortest and the others gradually increasing in length giving a sub quadrangular form to the fin. The rays project above the membrane. The Pectorals are large and the rays very marked and distinct. There are 18 of them in all, the first lowest one shortest and increasing in length as you count upward, giving the fin an oval outline. It rises close to the gill opening. The ventrals rise just back of the pectorals, are small composed of one short strong spinous ray, and two sometimes three soft ones. Anal is opposite the second dorsal and similar in form to it. Caudal fin nearly even in width and slightly rounded at its extremity.

Ray formula—

D. 16, 13; P. 18; V. 1, 3; A. 15; C. 12.

SYNONYMES.—*Cottus hispidus*, Bl..
Scorpena flava, Mitch.
Hemitripterus Americanus, Cur.
Cottus Acadianus, Penn.

This has also been removed from the Triglidae family into the Cottoidae of Richardson.

GENUS ASPIDOPHOROIDES, *Lacep.*

GEN. CHAR. Body octagonal covered with scaly plates; head thicker than the body, with points and depressions above, flattened below; teeth in both jaws only, none on the vomer; snout with recurved spines, branchiostegous rays six; body tapering to the tail; one or two dorsal fins distinct.—*Storer*.

Aspidophoroides monopterygius, Cuvier.

Asphidophore, One fin Aspidophore.

I place this species on the list of Maine fishes from hearsay evidence, never having seen a specimen. It is a very rare fish and most of the individuals obtained have been taken from the stomach of fishes caught on our coast, or further south. Storer gives a very fine engraving of it and a full description in detail.

It is small, with a slender body divided longitudinally into eight rows of scaly plates which give it an octagonical shape. One dorsal fin on the last half of the body at the extreme portion of the dorsal furrow.

Ray formula—

D. 5; P. 10; V. 1, 2; A. 4; C. 16.

Length 5 inches.

SYNONYMES.—*Agonus monopterygius*, Bl.
Cottus monopterygius, Richardson.

This genus has been removed from the Triglidae to the family Agonoidæ of Swainson, and sub-family Agoninæ of Gill.

SUB-FAMILY DACTYLOPTERINÆ, *Lac.* GENUS DACTYLOPTERUS, *Lacepede.*

GEN. CHAR. The rays under the pectorals are numerous and large and are united by a membrane into supernumerary fins, larger than the fish itself, and which will support it in the air for some length of time. The muzzle which is very short appears to be cleft like the lips of a hare ; the mouth is situated beneath ; there are in the jaws only, certain rounded teeth, arranged like pavement ; the head is flat, rectangular, and granulated. The preoperculum is terminated by a long and strong spine. All the scales are carinated.—*Storer.*

Dactylopterus volitans. Cuv.

Sea Swallow. Flying Finger Fin.

This Sea swallow is one of those singular fishes that have the power of springing into the air, and by means of its long and wide spread pectoral fins supporting or buoying itself up some little time, and thus scaling along quite a distance forward. It is one of those varieties called flying fish, though there is no flying done, the large fins acting only as a sort of parachute to let them down gently as the momentum of the spring they take just before they leave the water, ceases. It is thus enabled to elude its numerous enemies, though it undoubtedly often performs the act for the sport of it. They have a wide territorial range, according to some, from Newfoundland to Brazil. They swim together in large schools (scholes?) and their frolics, in sea and air, often enliven the dullness and monotony of a sea voyage. Sometimes, as they are not very well able to steer, or vary their course, they fall on board of vessels in their way.

Specific Description. Head somewhat foursided and wider than its height, and flatish above and of a darker color than its body, and there is a furrow between the eyes descending down in front, granulated. Mouth rather small, lower jaw shortest, lips fleshy. Teeth small, conical, three or four rows on the jaws. Small teeth exist on the pharyngeal bones but none on the palate—nostrils double, lower one smallest—snout very blunt, upper jaw of a yellowish color. Eyes large and circular. " Suborbitar bones are pushed forward nearly joining in front, their posterior upper angle passes upward and the opposite inferior angle continued back to the preoperculum where it terminates in a sharp point. The preoperculum has a long stout spine extending to the base of the pectoral fins. Operculum is small, covered with scales and is triangular in form. The body, forward of the vent, cylindrical and flattened, or compressed back of it. It is of a slaty color mottled with darker spots—abdomen yellowish, sides silvery. Scales rough, hard, toothed finely on their outer margins, with a ridge on each of those on back and sides.

"First dorsal is composed of two nearly free and flexible filaments nearly abreast of each other and united near the base by a low membrane ; closely contiguous to these, but not united to them by a membrane, follow four feebly spinous rays, united together by a membrane and the rays diminishing in length backwards. Both this, and the following fin, are lodged in a groove. Between this and the second dorsal is a short immovable triangular crest, the "stiff spiny stump "of Mitchill." [DeKay."] Second dorsal of a quadrangular shape with a very delicate membrane.

Pectorals large, and when spread very wide and extend to the base of the tail. They are made up of two parts ; the first having six rays in part free at their tips, and the posterior or main fin. Ventrals beneath the pectorals, and beneath the second dorsal. The caudal is fanshaped and concave at its extremity. It has two elevated scales resembling finlets near the base.

Ray formula—

D. 2, 4, 1, 8 ; P. 30, 6 ; V. 1, 4 ; A. 6 ; C. 10.

Length 4 to 6 inches.

SYNONYMES—*Trigla volitans*, Shaw.
Polynemus sex radiatus, Mitch.

FAMILY BATRACHOIDÆ, *Rich.*

CHAR. *Body* more or less tapering, subdepressed anteriorly, and compressed posteriorly ; in some protected by ctenoid scales, others scaleless.

Head branchial, apertures continuous under the throat in some genera, and widely separated by an isthmus in others. Four branchial combs in some, and three only in others. Some have the suborbital bone, other genera want it. Carpus much developed, but all the carpal bones contribute to its developement.

Fins. Pectorals not pediculated but exhibit a broad and fanlike base. Ventrals inserted in advance of the thoracic belt.

Stomach has pyloric appendages.

Air bladder wanting.

SUB-FAMILY BATRACHINÆ, *Bona*. GENUS BATRACHUS, *Linn*.

GEN. CHAR. Head depressed, broader than body, ventrals jugular with three rays; the first elongated. First dorsal small, second low and long. Base of the pectorals elongated. Branchial aperture small, with six rays. Sub-opercle as large as the opercle, and both spinous. No suborbital. Teeth on the jaws, in front of the vomer and palatines.

Batrachus tau, *Cuv*.

Toad fish.

Whoever, in the summer season, looks carefully among the rocks where the eel grass is abundant in shoal water, will frequently see a "*queer*" looking fish peeping out from under the stones or among the grass. I have noticed it while standing on the bridge which connects the Navy yard at Kittery with one of the islands. This is the Toad fish, named by Linneus, Batrachus tau. The *tau* (being the Greek word for the letter T) refers to a fancied resemblance of the ridges of bone on the top of the skull of this fish when dried. It is not used for food nor put to any economical use, but it is, nevertheless, interesting to the naturalist on account of its habits and parental affection it manifests for its young.

Dr. Storer, to whom we are all greatly indebted for much valuable information in regard to the fishes on the coast of Massachusetts and Maine, has attentively studied the habits and characteristics of this fish, and we copy the following remarks in full from the memoirs of the American Academy.

"The particular situations which it chooses vary with the nature of the coast. Thus along our southern shore it is found in the shallow bays. The sandy or muddy bottoms of these are overgrown with eel grass (*Zostera marina*) under cover of which it lives in security, and finds abundant sources of food. When the coast, on the contrary, is more or less rocky, we meet with it chiefly under the stones. Examining the places where the water is but a few inches deep at low tide, we see that under many of the stones or smaller rocks the sand has been removed, leaving a shallow cavity, perhaps a foot in width, and extending back beneath the stone. If we approach it cautiously, we shall probably distinguish the head of a Toad fish, very much in the position of that of a dog as he lies looking out of his kennel. The fish is at rest, and might be overlooked by a careless observer. A close attention, however, readily distinguishes the curve of its broad mouth, the delicate laciniated furrows with which its jaws and other parts of its head are ornamented, its truly beautiful eyes and sometimes the anterior portion of its body. At the slightest alarm, it retreats beneath the stone, but presently reappears. It is lying here perhaps merely as a safe resting place, perhaps on the watch for its prey.

But during the month of June, July and August, we shall, in many instances be able to discover another purpose, it is apparently guarding its eggs or young. We shall then find on the interior surface of the stone the young Toad fish adhering, to the number of several hundred. They will be in different stages of developement according to the season of our examination.

We may see the eggs not larger than very small shot. A little later they are increased in size, and the young fish plainly visible through their walls; a little later still, the young have made their escape but are still attached to the stone. The attachment now is accomplished in a different manner. The yolks not being yet absorbed, occupy a rounded sac protruding by a narrow orifice from the abdomen, and the part of the sac near its outer border, being constricted, leaves external to it a disc, by means of which, acting as a sucker, the young fish adheres so firmly as to occasion difficulty in detaching it. They remain thus until they have attained the length of half or three quarters of an inch, or until the yolk sac is entirely absorbed. During this period an adult fish occupies the cavity beneath the stone, and if driven from it speedily returns.

* * * * During the winter season, in our colder latitudes, the Toad fish in some instances, perhaps, retire into deep water; it is true, moreover, that many of them become nearly torpid. They are found buried beneath the mud, in the same manner as the eels, and are sometimes taken with the spear thrust down in search of their more valuable neighbors."

Specific Description. The Toad fish has a broad flattened head, as broad as it is long. Its mouth very large; lower jaw longest; several rows of conical, blunt pointed teeth on the jaws—thicker in front; smaller teeth on the inter-maxillaries and vomer; palatines have none; scarcely any tongue; lips fleshy; cirrhi about the head, and a row of from five to seven suspended from lower jaws—one or more over each eye; eyes moderate in size, and guarded by a gelatinous or membranous covering. Numerous mucous pores are seen about the head, about the body, and under the eyes. Preopercle has three spines partially concealed. Branchial apertures the size of the base of the pectorals.

Body thick at fore part, tapering, and some compressed posteriorly, of an olive green or yellowish color, flecked in with green. Fins, orange, except the dorsal which is greenish. The skin has no scales, and is covered with mucus which is freely produced from the numerous pores.

Dorsal fin continuous to the tail, with which it is connected by a membrane. The first three rays are spinous.

Pectorals are large and arise near the lower part of the gill openings. Ventrals commence forward of the pectorals. First ray sickle form, covered with thick membrane. The anal fin terminates on a line with the end of the dorsal.

Ray formula—

D. 3, 27 ; P. 16 ; V. 3 ; C. 14.

SYNONYMES.—*Gadus tau*, Linn.
Lophius bufo, Mitch.
Batrachoides vernullus, Lesuer.
B. variegatus, Storer.

FAMILY BLENNIOIDÆ, *Bona.*

Mucous fish.

CHAR. *Body.*—Generally small and offers a great variety of forms from a rounded and subfusiform shape to an elongated, tænoid, and very much compressed one. Scaly in some genera, scaleless in others ; scales either ctenoid or cycloid in shape and structure.

Head. The pseudobranchiæ are gill-like and conspicuous.

Fins. Are as diversified according to the genera in structure, and aspect as the body. Ventrals, when present, are separated from one another, and situate in advance of the base of the pectorals. In some genera these fins are quite rudimentary, whilst in others they are altogether wanting.

Stomach. No pyloric appendages to the intestine.

Air bladder. Absent in a great majority of the genera.

SUB-FAMILY BLENINÆ, *Bona.*

GEN. CHAR. Body elongated, compressed very much. Head small and oblong, with an obtuse snout, and a small mouth. The maxillar teeth are velvet or ard-

13

like, disposed upon one row on the lower jaw and upon a double row on the upper jaw. Velvet-like teeth on the front of the vomer. Palatine bones and tongue occasionally provided with a few prickles. Dorsal fin occupying the whole length of the back, and composed exclusively of spiny rays. Anal fin long and low, provided anteriorly with two spines. Ventrals excessively small, inserted under base of pectorals and often reduced to a single ray. Caudal fin slender, exteriorly rounded and contiguous to the dorsal and anal. Scales very small. Lateral line not perceptible.

Gunnellus mucronatus, Cuv.

Butter fish.

The American butter fish is a beautiful species of the Blenny family, and is so called on account of the thick covering of mucus which envelops its body. It is found among the rocks on the coast from Nova Scotia to New York. It is not unfrequently found at low tides in the shoal water among the stones, and sometimes partly buried in the sand and mud; but sometimes moving slowly and leisurely along, although it is capable of very swift motion. When first taken from the water it is almost semi-transparent, so near it that when held up against a strong light, its back bone (vertebræ) can be very plainly seen.

Characteristics. Greyish with a series of dark oval rings along the sides. Dorsal fins not united to the caudal. In place of the ventrals are two short spines. Length from four to twelve inches. —*DeKay.*

Color. The living fish is of an olive brown with numerous indistinct darker bands upon the sides; about twelve black ocelli along the base of the dorsal fin, each surrounded by a yellow ring. Fins yellow; the anal barred with white. Pupils black; irides golden. Abdomen yellowish. An oblique black band passes from beneath the eye to the throat.—*Storer.*

Description. The head is about one-tenth the length of the body, and blunt at anterior part. Mouth nearly vertical. Jaws equal, but when extended lower one a little the longest. Each jaw has minute sharp teeth—somewhat distant, with a small cluster of them on the vomer. Branchial rays six. Branchial aperture large. The body of this species is elongated, much compressed and without scales.

Fins. The dorsal is single, long, slightly raised above the back, commences above the branchial aperture on a line over the posterior angle of the operculum and extends nearly to the base of the caudal. It contains from seventy-five to seventy-eight sharp spinous rays which are enveloped and nearly or quite concealed in a thick membrane.

Pectorals are situated just below the posterior angle of the operculum, are rounded and weak. Ventrals are wanting, and their place is supplied by two small spines in front of the pectorals.

The anal is nearly equal throughout its length, and extends nearer to the caudal than the dorsal, but is not connected with it. Its first two rays are spinous, the others are soft and flexible. Caudal when spread has a rounded margin.

Ray formula—

D. 75 to 78; P. 11 or 12; V. 1; A. 36 to 40; C. 16 to 18.

Prof. Gill has removed this species into the genus Muraenoides, (*Muraenoides mucronatus*, Gill.)

SYNONYMES.—*Ophidium mucronatum, spinous ophidium*, Mitchill.
Gunnellus mucronatus, Cuv. et Val., DeKay, Storer.
Blennius (*Centronotus*) *gunnellus*, Lin., Rich.
Muræenoides guttata, spotted gunnell, Lacepede, Storer.

GENUS PHOLIS, *Fleming*.

GEN. CHAR. Body elongated, dorsal fin extending along the back, and composed of simple flexible rays. Skin smooth and without scales. Branchial rays six. Ventral fins placed forward of the pectoral and under the throat, and composed apparently of two rows. No cirrhi on the orbits nor any fleshy crests as there are in the Blennius genus.

Pholis subbifurcatus, Storer.

The Radiated Shanny.

This is a very rare species. I have never met with it, and therefore insert it here on the authority of others. It was first brought to notice by Dr. Storer and I copy his specific description of it entire, in addition to DeKay's.

Characteristics. Dorsal fin extending to the tail. Filaments on the nostrils. Three dark bands passing from the eyes. Lateral line sub-bifurcated. Length 5½ inches. *DeKay.*

Specific Description. General color of the body, reddish brown, several lighter colored circular patches along its upper parts, at the base of the dorsal fin; the spaces between the rings darker than the rest of the body presenting the appearance of bars. There is beneath the eye a broad black band, wider at its base, which crosses the operculum obliquely ; two other bands of the same color extend from behind the eye backwards, in nearly a straight line, the distance from one to two lines. Body beneath the lateral lines lighter colored, abdomen yellowish white. Head above brownish ; opercula and preopercula yellow ; numerous black spots upon the dorsal fin. Those upon the first five rays larger. Pectorals light, with some darker shades. Edge of anal dark colored. Small dark colored spots upon caudal.

Description. Length including tail, five inches, five lines ; depth across on a line with the anus, one inch ; body much compressed. Body smooth, scales very minute. Length of head from tip of snout to posterior angle of the operculum, is to the entire length of body, as one to three ; jaws somewhat protracted, armed with prominent sharp teeth ; lips large and fleshy ; over the nostrils is a minute filament one third of a line in length ; circumference of eye two lines.

The lateral line commences just above the angle of the operculum and having extended two lines, sub-bifurcates ; passing down in a gradual curve a little more than a line, it is continued in a straight course to the base of the caudal fin ; while the upper portion abruptly terminates opposite the fourteenth ray of the dorsal fin.

The dorsal fin, commencing on a line with the posterior angle of the operculum, is continued to the caudal fin ; the first five rays of this fin are shorter than the sixth ; the rays become again shorter as they approach the tail.

The pectorals are rounded ; they arise on a line with the posterior angle of the operculum.

The ventrals are situated two lines in front of the pectorals ; the rays are united throughout the greater portion of their extent ; extremities free. The anus is situated two and a half inches from the extremity of the jaws. The anal fin commences just half way between the tip of the snout and the extremity of the tail. The caudal is rounded.

Ray formula—

D. 43 ; P. 13 ; V. 3 ; A. 30 ; C. 14.

Professor Gill has removed this species into the genus *Sticheus* of Rheinhardt and describes it under the name of *Sticheus sub-bifurcatus*, Gill.

SUB-FAMILY ZOARCEINÆ, *Gill.* GENUS ZOARCES, *Cuv.*

GEN. CHAR. Body elongated, and covered with a mucus secretion, in which are imbedded small scales. Dorsal, anal, and caudal united ; no spinous rays in the dorsal, except on its posterior part. Ventrals are jugular and small. Vent with a tubercle. Teeth conical, in two or three rows in front, in a single row on the sides; none on the palate, or tongue. Branchial rays six.

Zoarces anguillaris, Storer.

Eel-shaped Blenny, Thick lipped Eel pout, Ling, Conger Eel.

Early in the spring and first of summer, the fishermen sometimes take, in company with cod, this fish to which, from its general resemblance to the Conger Eel, they frequently give the name of *Conger Eel, and Ling.* It is also caught at other seasons of the year, but not so often. It is much prized by some people as a savory fish. Its common length is from one foot and a half to two feet. Occasionally one is caught from three to four feet long, but those of that size are rare. Weight from 1 to 100 lbs.

Characteristics. " Dark olive brown, varied with dusky blotches. Dorsal and anal fin margined with rufous. *DeKay.*

The living fish is of a bright salmon color, mottled with irregular olive blotches, darker towards the head. The front and top of the head are of light brown ; two indistinct oblique bands upon the operculum, one back, the other in front of the eye and each side of the head. Body beneath white ; neck flesh colored. Dorsal almost white, salmon colored at the edge. Pectorals of a true salmon color lighter at their origin. Ventrals salmon colored. Anal flesh colored at its base, salmon colored at its edge, with seven distinct white blotches in its length. Dorsal, pectoral and anal fins perfectly transparent. In the dead specimen the colors change essentially." *Storer.*

Specific Description. Body much elongated, compressed and tapering to a point, and covered with minute cup like depressions, slimy. The head, though large, is compressed on its sides, flat to the angle of the eyes with a convex forehead. Cheeks full and protuberant. Upper lip very large and fleshy and projected over the under one at the angles. Nostrils somewhat tubular and placed about half way between the snout and the eyes. Its teeth are conical and large, the hindmost ones are sharpest, and there is a slight circular furrow around the base, also some longitudinal furrows.

There are three rows of teeth on the upper jaw in front. Those in the forward row larger than the others. The innermost row, on the lower jaw, has four, and the intermediate row, three teeth. There are also strong pointed teeth on the pharyngeals, but none on the palate, or tongue. Eyes of moderate size. Branchial aperture moderate. Branchiostegal rays six.

Fins. The dorsal fin unites indirectly with the caudal. It commences with a short ray forward of the pectoral and is of moderate height ; it is highest in front, sloping gradually posteriorly until it comes near the tail, when it breaks off, leaving the *stubs* or bases of seventeen or eighteen spinous rays, without any membrane, which continue to the caudal.

The pectorals are broad, and at their extremities rounded ; their inferior rays somewhat scolloped ; they contain twenty rays. Ventrals are mere *nubs* being enveloped in a strong membrane. They contain two small rays.

Anal fin is long, and lower in height than the dorsal, and seems to unite with the caudal. It contains about 100 rays. Caudal pointed.

Ray formula—

D. 118 to 120 ; P. 19 to 20 ; V. 2 ; A. 100 to 105.

SYNONYMES.—*Blennius anguillaris*, Peck. Mem. Am. Acad. Vol. 2.
Blennius labrosus, Mitch.
Le Zoarces a grosse levres, Cuv. et Val.
Zoarces anguillaris, Storer, De Kay.

SUB-FAMILY ANARRHICANINÆ, *Gill*. GENUS ANARRHICAS, *Linn.*

GEN. CHAR.—Head smooth, rounded, muzzle obtuse; body elongated, covered with minute scales; dorsal, and anal fins. Teeth of two kinds, those in front elongated, curved, pointed, (upwards of five in each jaw) the others on the vomer, as also on the jaws truncated or slightly rounded; branchiostegous rays six.

Anarrhicas vomerinus (*Ag*) *Storer.*

American Wolf-fish.

This "*wolf*" among fish, like the wolf among animals, inhabits a broad range, and prefers a cold region to a warmer one. It is most abundant among rocky places, but is not unfrequently found among cod fish, on the banks or shoaler fishing grounds, and the appearance of its "*ugly mug*," as the fishermen say, when they haul it up, is often the signal for some rather *irreverent expressions.* Unlike the land wolf, it is very good food, and the smaller ones make quite a savory dish. When smoked, or dried, it is thought by some, to equal in flavor Salmon, prepared in the same way. It is found largest in size in the more northern regions of its territorial limits. It is a savage among fish, and the expressions of its features, and the snapping of teeth, with which it is well provided, and the ferocity of its actions when taken, give true manifestations of its character. Some of them taken in the high latitudes have measured eight feet in length.

Characteristics. "Leaden grey, with dusky vertical bands on the dorsal fin, extending irregularly over the sides. Length three to five feet." *De Kay.*

Specific Description. The head is arched from the nape of the neck to the point of the snout, but slightly flattened on the top and at the sides. A few rows of pores pass up from the snout to the eye, and beneath it to the back of the head. There is another circular row around the eyes, others are seen on the cheeks and on the lower jaw. These pores produce a thick covering of mucus over it which hide the scales which are discovered when this is removed. Nostrils small. Jaws are well provided with teeth. In the lower jaw are two long stout ones projecting forward, and there are two others of the same size bent backward, and behind these half a dozen more, very sharp and of different sizes. "There are six in the intermaxillaries; many above, larger, and diverging outwards; back of these on each side, are six smaller conical ones, sharply pointed." [*Storer.*]

There are also nine on the vomer with flat tops, increasing in size as you pass back, sometimes forming a solid mass; and in addition to all these there is a double row of molar teeth, some of them having pointed crowns. Eyes of medium size, iris yellow; lips large loose, and fleshy—tongue large and dusky.

The body is cylindrical—somewhat compressed on sides and tapers regularly to the tail.

It is of pinkish brown, or leaden grey color, and has a series of about a dozen blackish bands passing transversely over its back, uniting or running into each other on the sides.

The dorsal fin commences near to the nape of the neck and passes along to near the tail, all the way of a uniform height. It has black rays, while its membrane which is fleshy and tough, is of a slate color.

The pectorals are large, broad with rounded edges, of a lead grey color, as are the other fins.

Ventrals very small "like warts," and has two rays enveloped in tough membrane.

The anal commences about mid way of the body and runs to the caudal.

The caudal is rather small, and short, with a rounded or circular reddish margin.

De Kay says the duodenum is so large as to present the appearance of two stomachs, and the urinary bladder very large.

Ray formula—

D. 118 or 120 ; P. 19 ; V. 2 ; A. 100 ; C. 14.

SYNONYMES.—*Anarrhicas lupus*, Mitch. Storer. De Kay
L'Anarrhique loup, Cuv. et Val.

FAMILY CRYPTOCANTHOIDÆ, *Gill.* SUB-FAMILY CRYPTOCANTHINÆ, *Gill.*

GENUS CRYPTOCONTHODES, *Storer.*

GEN. CHAR.—Body elongated, much compressed, and gradually tapering to the tail. Destitute of scales. Head broad, with no projecting spines; the scapular and humeral spines, and the inferior edge of the preoperculum prominent to the touch. Numerous depressions in frontal, suborbitar, inferior maxillary, and preopercular bones. Banchiostegous rays seven; mouth oblique; a single dorsal fin composed of strong spinous rays enveloped by a common membrane, runs nearly the entire length of the fish, and unites, as does the anal, to the tail. No ventral fins. *Storer.*

Cryptocanthodes maculatus, Storer.

Spotted wry mouth.

This is another one of those very rare fishes first discovered and described by Dr. Storer. Several specimens have been collected and received by him in a range extending from Nova Scotia to Cape Cod. This authorises me to enumerate the species among the Maine fishes, although I have not been able, as yet, to obtain any for examination.

I therefore copy the description of Dr. Storer in part, and hope that some of our fishermen will be successful in taking some of them and supplying the State cabinet with a specimen. I will here state that it has been removed by Prof. Gill from the "Triglidæ" and made the type of a new family the Cryptocanthoidæ, and sub-family Cryptocanthinæ.

Color. Body a dark reddish brown tinged with violet. Abdomen and throat a dirty greyish white. A row or two of moderate sized dark brown blotches above the lateral line ; and another row immediately beneath it extends throughout the greater length of it to the tail. Top and sides of the head, snout and anterior portion of the under side of lower jaw, marked with smaller spots of the same color as those on the sides. Pupils black, irides golden.

Description. Length of head about one sixth the entire length ; greatest breadth about one half the length of the head. On each side of top of head, two prominent long ridges run directly back from posterior angle of eye to occiput. The posterior angles of operculum and preoperculum ; the lower edge of preopercle ; the scapular bones,—all seem like sharp points and edges concealed by the skin. The operculum is large and triangular, covered by the skin, as is also the preoperculum which present to the touch two sensible carinæ.

Eyes circular, deeply sunk in the projecting orbits ; diameter of orbits about equal to distance between the eyes. Nostrils tubular, situated on the side of the prominent snout just at the edge of the intermaxillary bones.

Lips fleshy, lower jaw projecting above the upper, mouth slanting obliquely downwards, numerous teeth in jaws and upon vomer and palatine bones—those in the back part of the jaws recurved, while those in front are smaller and nearly straight. Gape of mouth moderate. Branchiostegal membrane extended along and connected with the sides for a short distance. Lateral line straight and interrupted.

The dorsal fin arises above the posterior half of the pectorals, and is united with the caudal ; all its rays are spinous and strong, concealed by a stout and fleshy membrane ; the first few rays are shortest.

The pectorals arise beneath the membrane of the branchiæ ; they are fleshy, small and rounded.

The anal arises upon the anterior half of the body ; it is similar in its form and the character of its rays and their enveloping membrane to the dorsal, and is also, like the fin connected with the caudal.

The caudal is rounded and appears like the prolongation of the dorsal and anal fins.

Ray formula—

D. 78; P. 15; A. 50; C. 15.

Length three feet. *Storer's History Mass. fishes in Memoirs of Am. Acad. Vol.* 5, *p.* 82.

FAMILY LOPHIOIDÆ, *Bona.*

CHAR. Generally scaleless, some have bony tubercles.

Body in some reduced and tapering; in others subelliptical and compressed.

Head, in most of the genera very large and broad, in others moderate compared with the body; suborbital bone wanting; gills variable in number, according to genera, some two, some three and a half, others two and a half. The two carpal bones are elongated so as to constitute a kind of peduncle at the extremity of which the pectoral fin is articulated (hence this family have sometimes been called *Pectorales pediculati.*)

Branchial apertures open behind the insertion of the pectorals.

Stomach is simple except in "Devil fish," which has a few pyloric appendages.

SUB-FAMILY LOPHINÆ, *Bona.* GENUS LOPHIUS, *Linn.*

GEN. CHAR. Head enormously large, broad and depressed. Mouth large, armed with slender conical teeth on the jaws, palatines, vomer, and pharyngeals. Tongue smooth. Branchial rays six, branchial arches three. Dorsal fins two; the anterior rays distant, detached, forming long filaments supporting fleshy slips.

Lophius Americana, Cuv.

Goose fish, Monk fish, Sea devil, Bellows fish.

If the generic term (*Lophius*) were translated *Loafer* it would give a more expressive name than the many already given to it, as it seems to represent among fishes what the "*loafer*" is among men, a lazy, stupid, gormandizing fellow, careless of itself or how and where it gets a living. It has an enormous mouth enabling it to swallow bodies almost as large as itself. Dr. Storer makes a statement on the authority of Capt. West of Chilmark, that one was caught that had "six coots in its stomach in a fresh condition." It grows oftentimes to a large size, individuals sometimes are taken weighing 60 or 70lbs.

It is never fat, notwithstanding its voracity and the great capacity of its stomach, and, like other loafers, is useless when alive and good for nothing when dead.

It takes the hook readily, and is also taken in nets, generally in the autumnal months. They are so stupid that they frequently commit suicide by running ashore, not knowing enough to turn round into deep water again.

Specific Description. The head is broad and flat, consisting almost wholly of mouth, it having an enormous gape. The top is of a brown color, smooth and scaleless. Lower jaw longest, and fringed around its margin with a row of fleshy barbels, or cirrhi, about an inch long. Similar smaller ones are continued along the sides of the body to the base of the tail. On the top of the upper jaw, about in its center, are two long bristle-pointed fleshy tentacula which the fish has the faculty of raising or depressing at pleasure. The eyes are oval horizontally, pupils black and irides yellowish brown. The lower jaw has a single row of long sharp teeth, curved backwards.

Tongue bony on each side, on which are two rows of teeth also curved backward. The intermaxillaries can be pushed beyond the maxillaries, and have a single row of short teeth on each side and two rows in the middle, these last are larger than the others, curved backward. "Upon the upper jaw, at its tip, is a space of an inch and a half destitute of teeth ; on each side of this space is one quite large tooth, and a second much smaller ; about half an inch outside of this is another single row of eight or ten teeth, the first three or four of which are much the largest ; on each side of the pharynx are three rows of sharp incurved teeth resembling spines ; these rows are arranged directly above each other and are double." *Stor.*

There are several Spines situated upon the head.

The body is flattened, rather globular in front, tapering behind, of a dark brown, with netlike markings, lower part lighter colored.

Two dorsals ; first has three sharp spines, of which the posterior one is shortest, all of them project above the membrane. The second dorsal is more uniform in height, rounded at its posterior margin, and its length twice its height and rises about two inches behind the first.

Pectorals rise by a "strong pedicel" on a line with the front of the first dorsal, expanding broader at their margins where their rays project beyond the membrane.

The ventrals of moderate size ; one spinous ray on their external edge. Anal rises on a line with the commencement of the second dorsal ; has five rays, and its posterior is the highest.

Caudal long, rather narrow and fleshy, and notched on its margin by the projection of the rays from the membrane.

Ray formula—

D. 3–12 ; P. 24 or 25 ; V. 1–5 ; A. 10 ; C. 9.

SYNONYMES.—*Lophius piscator*, *Mitch.*
Lophius piscatorius, *Storer.*

—

This closes the description of those fishes which were enumerated in the synopsis, belonging to the suborder Physoclisti. The succeeding orders embrace some of the most interesting as well as valuable species, whether marine or inland, such as the Gadoids (Codfish family) Clupeoids (Herrings and shad,) Salmonoids (Salmon and trout,) &c., &c. These all enter largely into commercial as well as domestic life.

The taking and carrying of them employ an immense amount of capital and labor, and they are all intimately connected with the pleasures, the comforts and prosperity of the community.

I have collected many valuable facts, and much statistical information in reference to the fisheries of these species, and would willingly record them here, but the time prescribed for publishing this report presses, and will not allow me to continue these descriptions in full any farther at present.

I therefore shall only add descriptions of a few new species of the Salmo genus (trout) which have been recently discovered in Maine and believed to be peculiar to our waters only.

FAMILY SALMONOIDÆ, *Cuv.*

CHAR. Body more or less scaly. Two dorsal fins, the first with soft articulated rays, the second small and adipose, numerous coecal appendages and a swimming (air) bladder. There is great variation in the arrangement of teeth in the jaws. Inhabit salt and fresh water, and mostly ascend rivers periodically.

SUB-FAMILY SALMONINÆ, *Bona.* GENUS SALMO, *Linn.*

GEN. CHAR. Head large; mouth generally deeply cleft and armed with conspicuous teeth. Premaxillary bones short and rather situated upon the sides of the snout than immediately upon its extremity. The maxillaries are attached behind them and composed, each, of a single piece. The lower jaw is strong and terminates oftentimes into a small knob or tubercle, which in some species acquires a very great developement. Strong and conical teeth are inserted in a single row on the dentary; but the teeth vary in different species. Body fusiform in profile, one anterior dorsal fin followed by a small adipose one. Caudal fin well developed, and either truncated posteriorly or slightly emarginated.

Salmo Toma, Hamlin.

Togue.

This trout known among the aborigines as the *Togue, Tuladi, etc.*, has been classed by some observers, as identical with the *Salmo Hucho* of the Danube and of the lakes of Northern Europe; but in these classifications, peculiarities of anatomical structure have been overlooked, and the habits of the two fishes have also been noted as similar, whereas in reality they present great contrasts, for the one, agile and alert, seeks the swift and foaming currents of the clearest streams, and the other sly and sluggish, haunts always the quiet waters of the deepest lakes. It is mentioned by Mr. Gesner in his report upon New Brunswick, and identified with the *Salmo lacustris* of Lake Geneva; a proper examination of the two fishes, however, will satisfy the naturalist that few positive analogies can be drawn; and again it is identified with the *Salmo ferox* of Loch Awe in Scotland, in the descriptive catalogue of fishes of New Brunswick, by Mr. Perley, who identifies from the characters drawn by Sir W. Jardine and Mr. Yarrell, some of which would certainly lead the observer, unless minute, into the same error, for it cannot be denied that great similarities are to be observed, but there are also as many with the *S. erythinus* of Siberia.

There is none among all the Salmonidæ, which resembles it more in form, color, linear markings, etc., than the *S. Siscowet* described by M. Agassiz, and until that eminent naturalist in a momentary examination observed differences, it was regarded as identical with that species.

In shape it is not so elegant as that of some other species of the Salmonidæ, but its whole form indicate great strength and swiftness, although it has the reputation of being slow and sluggish. The female is more perfect in its proportions than the male, not having that gibbous appearance at the nape, where the outlines of the head pass into those of the back, and besides, its general contour is more delicate.

A rich pearly lustre covers the ventral regions, deepening into russet towards the lateral line, above which the color appears of a deep mottled gray, still deepening into blue as it approaches the dorsal summit. The same pearly hues, blended and intermingled with gray, are observed upon the opercula. Spots and markings of a light sienna color appear on the sides; these spots are circular without being ocellate, and appear indistinct and grayish upon the dorsal and upon the commencement of the caudal. All these colors vary according to the seasons and local influences, being brighter at the spawning period than at other times.

Its proportions are quite harmonious. The following are the measurements of a small specimen:

Entire length,				18 inches.
Greatest depth,				$3\frac{7}{8}$
Head,	length	$2\frac{7}{8}$,	with oper.	$4\frac{1}{2}$
Pectoral,	"	$2\frac{7}{8}$		
Ventral,	"	$2\frac{1}{8}$		
Anal,	"	$2\frac{1}{4}$	in width,	$4\frac{1}{2}$
Caudal,	"	$3\frac{1}{4}$	"	$1\frac{1}{8}$
Dorsal,	"	$2\frac{1}{2}$	"	2

Br. 12; P. 12–13; V. 9; A. 11–12; D. 13; C. 19.

Cæcal appendages, 113; Ver. 65.

Scales are small and elliptical. They decrease in size as they approach the thoracic arch. There are 53 in a vertical row anterior to the ventrals, of which 24 are above the lateral line number 123—are long, narrow, with a deep grove passing through them, and strongly attached. They measure on specimens of 18 inches, in length 1-16 in their short diameter, and in their long 3-16.

The lateral line arises from the height of the upper third of the operculum, curves slightly downwards and proceeds with a slight inflection to its caudal insertion. The pectorals are not proportionally so long as those of the *Siscowet,* and they arise much nearer the branchiostegii, leaving a greater distance between their extremities and the plane of the commencement of the dorsal. The ventrals arise vertically beneath the sixth ray of the dorsal, are orange in color, and margined anteriorly with white. Their outer circumference is slightly oval. The anal is not so high as the dorsal by one-quarter, whilst in the *Siscowet* it is of equal height; terminal line obtuse and parallel with the axis of the dorsal. These fins are of an orange hue and tipped with white or light gray. The dorsal arises in the middle of the back, is of a dark gray color and spotted in the form of transverse bands—terminal line obtuse. Caudal long and much furcated, much more so than with the *Siscowet*, nor does age change much the acuteness of its terminal line.

The branchiostegal rays are 12 in number, and are of a pure white except the last, which is irregularly spotted with gray. Eye large and circular, with irides of a golden yellow, and pupil angulated towards the snout, which is obtuse. The upper maxillaries are longest, and at their union show in both sexes a singular depression, into which is received the curve of the lower maxillaries.

The maxillaries, intermaxillaries and palatines, have each a row of conical and inflected teeth. Those upon the lower maxillaries are large and strong; those of the intermaxillaries are next in size; upon maxillary and palatines next, and those upon the vomer smallest, numbering only three or four, and not confined to the anterior extremity, but extending a good way backwards. The tongue is deeply grooved and furnished with inflected teeth, arranged in lateral rows.

The opercular apparatus is somewhat concealed by the thick skin which envelopes it, but the outer lines of the operculum are quite distinctly marked. The operculum is quadrilateral, of greater height than breadth, well rounded in its posterior free margin, denticulated in its lower and nearly square in its upper, the anterior angle of which is characterized by a strong and prominent process. Suboperculum is nearly one-third smaller than the operculum, is triangular in its upper portions, elliptical in its lower borders, and terminates at its articulation in the form of a fish hook. The interoperculum has, as usual, the form of a long square, but square on the posterior side, and forming an acute angle, with its lower margin; slightly rounded on the anterior side. Finally, the preoperculum is long, slender, crescentic and almost vertical in its position ; it is thick and furnished with a prominent ridge and three foramina upon its anterior surface.

This trout inhabits many of the great lakes and deep mountain tarns of Maine and New Brunswick, but it is believed not to exist in those of Eastern New Brunswick, which singular hiatus in its distribution, perhaps may be explained by the absence of deep waters in that country. It haunts the deepest waters, where the cold or the repose to which it leads, favors that development and conservation of fat which is indeed a characteristic, and it steals forth in quiet at the approach of twilight or at early morn, to the shoals and the shores in quest of its prey, which consists, for the most part, of the *Lota* and *Cyprinidæ*, but its baffled voracity often contents itself with substances entirely foreign, as its stomach presents sometimes a heterogeneous mass of bones, leaves, twigs, and fragments of decayed wood.

Its habits vary in some localities ; in certain lakes they are bold, and ranging near the surface, at times may be taken by trolling, but never rising to the fly, whilst in other lakes they are timid and seek the obscurest recesses ; thus, for instance, their existence in the Tunk Lakes, was unknown for more than half a century to the inhabitants living near their shores.

Its mysterious nature has furnished the all-observing Indian with some proper idioms, and it appears again in the vague mythology and wild legends of that almost extinct race. Its names are various among the different tribes, and if the present are not of the half-breed Canadian date, they are perhaps of recent origin, since the few remaining dialects have changed greatly within a century past. Considering then, the uncertainty of its ancient name and the diversity of its synonym, I propose my friend Toma of the Openangos.—*Copied from a brochure on the Togue, published by A. C. Hamlin, M. D., Bangor.*

Salmo sebago, Girard.

Sebago lake trout, Salmon trout. (?)

The following is a description of a species of trout taken in Sebago lake, Cumberland county, in this State, by Dr. Girard, and published by him in the proceedings of the American Academy of Natural Sciences, Penn., Aug. 16th, 1853.

I am inclined to think that this species is identical with that called *Salmon* trout caught in the Schoodic lakes, in the eastern part of the State (Washington county,) but I have as yet had no opportunity of comparing the two together side by side.

Dr. Girard considered it a new species, and in his description observes—"Its large scales and fusiform body recall to mind the salmon, but on a more close examination the general shape and outline are far more elegant than in the salmon, preserving altogether better proportions between the different regions of the body. The head forms about a fourth of the entire length, whilst in the salmon it is about the sixth only. The eyes are of medium size and sub-circular in shape, their diameter being contained about seven times in the length of the head. The posterior half of the maxillary, which is regularly and most decidedly curved downwards, gives to the shape of the mouth a peculiar aspect. The anterior margin of the dorsal fin is equidistant between the tip of the snout and the base of the caudal. The posterior margin of the latter is regularly crescent-shaped. The adipose fin is elongated, club-shaped, and situated opposite the posterior half of the anal. The ventrals are inserted under the middle of the dorsal, somewhat nearer the anal than the pectorals. The scales are remarkably large, contrasting greatly when compared with those of *Salmo erythrogaster*, (*red-bellied trout*,) and *S. fontinalis*, or *S. Namaycush* or *amethystus*. There are about 115 of them in the lateral line. The color in the female is uniform silver-grey, darker on the back and head. Sub-quadrangular or sub-circular black spots are observed upon the sides of the head behind the eyes, along the back, and the half of the flanks, also on the dorsal and caudal fins, where the red is sometimes but faintly indicated. The name of *Salmo sebago* is proposed for this species which inhabits the southwestern part of the State of Maine."

Salmo oquassa, Girard.

Blue back trout.

A species of trout known by the name of "Blue backs," is found in the lakes at the head of the Androscoggin river, in Franklin County. This name is given them on account of the peculiar blue color of the back and upper parts of their bodies. Dr. Girard took opportunity to visit the lakes and make an examination of the distinctive characters of this trout, and subsequently published the results of his investigations in the Proceedings of the Boston Society of Natural History (Vol. IV, p. 262,) of which the following is a copy:

" He had often been told by Anglers that the trouts of those waters (upper lakes of the Androscoggin) *Salmo erythrogaster* and *S. fontinalis* are subject to considerable variations, making it probable, in their opinion, that there were more than two species.

" Visiting the locality he had an opportunity of seeing and comparing large numbers of individuals. He satisfied himself that all the varieties spoken of are mere varieties of color, and all referable to either *Salmo fontinalis* or *S. erythrogaster*. He was told however, that about the 10th of October another trout, smaller in size than the common brook trout, and inhabiting the deep waters of Moosillamaguntic lake, would make its appearance near shore and ascend in large numbers the eastern inlet called Kennabago. This actually took place, and the trout on examination proving to be very different from *Salmo fontinalis* and the other species of *Salmo*, he named it *Salmo oquassa*,* Girard, and gives the following

"*Specific Description* of it. It is from eight to ten inches in total length. The body is subfusiform, slender, and the most graceful of the trout family. The head is proportionally small, conical, coregonoid in shape.

" The mouth is smaller than in *Salmo fontinalis*. Differences are likewise observed in the structure of the opercular apparatus. The fins have the same relative position as in the brook trout, but are proportionally more developed, with the exception of the adipose which is considerably smaller. Their shape is alike except that of the caudal, the crescentic margin of which is undulated instead of being rectilinear. The scales are somewhat larger, although they present the same general appearance as those of the brook trout.

" The lateral line is similar in both of these species. A bluish tint extends all along the back from the head to the tail, so that

*The Dr. gave it this name undoubtedly from the Indian name (*oquassa*) of the lake in which he found it—now Rangely lake. I object to his mode of spelling it. Matalluck, an Indian of the St. Francis, who for a long time lived on those lakes, and who used to be considered the guardian genius of that locality used to pronounce the name of this lake, Argwas-suc, making a slight pause between the second and third syllables.

when seen from above, the fish appears entirely blue; hence the name *Blue back* given to it by the settlers of that neighborhood.

"The sides and abdomen are silvery white in the female, and of a deep reddish orange in the male, spotted in both sexes with orange of the same hue as the abdomen. The dorsal and caudal fins are brownish blue, bordered with pale orange in the male, the pectorals, ventrals and anal of a fiery orange, blackish blue at their base, with their margin of the purest white.

"When just taken out of the water it is impossible to imagine any thing more beautiful and more delicate in the way of coloration in fishes of the temperate zone.

"The abode of the Blue back is, as stated above, the Moosillamaguntic Lake, in which it is concealed during the greatest part of the year, but about the 10th of October, it comes near shore and ascends in schools the Kennebago for the purpose of spawning. Half a mile above its mouth the Kennebago receives the outlet of Oquassa (Rangely Lake,) the trout there leaves the Kennebago to the left and runs toward Oquassa Lake where its voyage comes to a close.* After the middle of November it goes back into Moosillamaguntic Lake and is seen no more until next year.

"The flesh of this fish is highly flavored, and more delicate than that of the brook trouts in Europe and America. It resembles that of *Salmo umbla* of the Swiss lakes, both in the peculiarity of its habits and its delicacy.

"*Salmo umbla* is a lake trout, an inhabitant of the deep, making its appearance near shores in January and February, to spawn, and never ascending the brooks or rivers, tributaries of these lakes."

Salmo Gloveri, Girard.

Union River Trout.

The following description of a species of trout, considered by Dr. Girard as a new one, caught in Union river in this State, is copied from the Proceedings of the Philadelphia Academy of Arts and Sciences for 1855, page 55:

Body of the male is subfusiform and rather slender, particularly

*This is not correct. Prof. Hitchcock who was in that region last fall, informs me that they go through Rangely Lake and up the Sandy river some distance.

in the caudal region ; the head being regularly subernial, contained five times in the total length.

The maxillaries are gently curved, extending backwards to about the posterior margin of the orbit. The female is stouter, with peduncle of the tail shorter ; the head has the same general shape, but is not contained five times in the total length. The maxillaries are less curved, but extend as far backwards as in the male.

The eye is very large, its diameter being contained nearly five times in the length of side of head. The caudal is deeply emarginated posteriorly, giving to it a more forked appearance than either in *Salmo oquassa* or *Salmo sebago.* The adipose fin in the male is situated opposite the anterior margin of the anal, whilst in the female it corresponds to the posterior margin of the same fin.

The scales are well developed, being somewhat smaller, however, than in *Salmo sebago*, and considerably larger than either in *Salmo oquassa* or *Salmo erythrogaster.* On the dorsal and ventral regions they are considerably smaller than upon the sides, and along the peduncle of the tail. They extend, diminishing in size, over nearly the half of the length of the middle rays of the caudal fin.

The lateral line takes an almost straight course along the middle region of the flanks. The following is our approximate formula of the rays of the fins :

D. 2, 12 ; A. 7, 9 ; C. 81, 8, 9 ; I. 5 ; V. 1, 9 ; P. 14.

There are two anterior rudimentary rays to the dorsal, one or two to the anal, one to the ventral, eight or ten to the upper lobe of the caudal, to five or six to the lower lobe. The upper surface of the head and dorsal region is blackish brown ; the sides are silvery white, the belly yellowish, the region above the lateral line is densely spread all over with black, irregular spots, some of which are confluent. A few scattered ones may be seen beneath that line upon the middle of the abdomen. Four to six of these spots, well defined, are always observed on the operculum, one of which may occasionally reach the preoperculum.

A few reddish orange dots, individually situated in the middle of a black spot, are occasionally observed along the middle of the upper part of the flanks. Whether these dots are peculiar to the female, or proper to both sexes, I am not prepared to say, from want of sufficient information upon that point. This species was first brought to my notice by Mr. Townsend Glover, of Fishkill

Landing, Dutchess county, N. Y., who caught it in the upper affluent of Union river, Me., during the middle of September. I propose the name of *Salmo Gloveri* as a token of gratitude.

Salmo namatus, Cuv.

This species of trout, according to the observations of Prof. Agassiz, is found in the eastern waters of New Hampshire, (and probably the adjacent waters of Maine.) In the proceedings of the Boston Society of Natural History, Vol. VI, p. 518, "Prof. Agassiz remarked that of the European species of Salmo, the *Salmo salar*, Linn, (common salmon,) is found on both sides of the Atlantic, while the *Salmo eryox*, Linn, called *Salmo namatus* by Cuvier, remarkable for the hook in the lower jaw of the male, and differing from the other in color and shape, has been hitherto considered as confined to Europe. But, on the 29th of October a fish of the latter species was caught in the Merrimac river and examined by him—another example of Arctic species coming down on the American as well as the European coast."

It may not be improper here to state that no part of the world affords finer trout fishing, or a greater variety of trouts than Maine. At all times of the year, except in April and May, there are capital opportunities for the angler to exercise his skill and gratify his taste in the "gentle art." During those two months, probably on account of the breaking up of the ice and the consequent disturbance in the waters, they do not bite freely; but in summer and autumn, either on lake or by stream and brooklet, or during mid-winter, in deep lake water, through the ice, they can be caught in great abundance. In our large rivers, as in the upper Penobscot for instance, in warm weather, they abound near the mouths of the cool water spring brooks, and can be caught in unlimited numbers. In summer no better sport of the kind can be found than that afforded to the amateur by fly-fishing for the salmon trout on the Schoodic Lakes, in Washington county, or on Sebago Lake in Cumberland county, as well as in many other parts of Maine.

In October, the streams which flow into our numerous lakes in every part of the State, and especially those on the frontier, are crowded with trouts of the different species which, impelled by their natural instincts, are hurrying up to the shoal waters to spawn. Barrels of them are then caught and preserved by the provident settler for domestic use during winter.

BIRDS OF MAINE—(ADDENDA.)

In addition to the list of birds published in last year's report, I am enabled by the kindness of our zealous ornithological friend, Geo. A. Boardman, Esq., of Milltown, to enumerate the following species as having been obtained by him in his vicinity during the past season:

INSESSORES.

Black and Yellow Warbler. Dendroica maculosa, *Baird.*
Sylvia magnolia, *Wilson.*

SCANSORES.

Banded three-toed Wood-pecker. Picoides hirsutus, *Gray.*
Picus hirsutus, *Vieitt.*

GRALLATORES.

Northern Phalarope. Phaleropus hyperboreus, *Temm.*
Tringa hyperborea, *Linn.*

NATATORES.

Burgomaster Gull. Larus glaucus, *Brunnich.*

Through the politeness of Prof. C. E. Hamlin of Waterville College, I have been furnished with the following additional species found by him last summer in that vicinity:

INSESSORES.

Least Fly-catcher. Empidonax minimus, *Baird.*
Tyrannula minima, *Baird.*
Traills Fly-catcher. Empidonax Traillii, *Baird.*
Muscicapa Traillii, *Aud.*
Yellow-bellied Fly-catcher. Empidonax flaviventris, *Baird.*
Tyrannula flaviventris, *Baird.*
Olive-sided Fly-catcher. Contopus borealis, *Baird.*
Muscicapa Cooperi, *Nutt.*
White-crowned Sparrow. Zonotrychia leucophrys, *Sw'n.*
Emberiza leucophrys, *Fors.*
Chestnut-sided Warbler. Dendroica Pennsylvanica, *Baird.*
Sylvia Pennsylvanica, *Latham.*
Indigo Bird. Cyanospiza cyanea, *Baird.*
Fringilla cyanea, *Wilson.*
Maryland Yellow-throat. Geothlypis trichas, *Cabanis.*
Sylvia Marilandica, *Wilson.*
Mourning Warbler. Geothlypis Philadelphia, *Baird.*
Sylvia Philadelphia, *Wilson.*

MAMMALS.

I have also been informed by Prof. Hamlin that he has found a very rare species of Sorex in his neighborhood, (Waterville) the *Sorex Thompsoni.* Only one specimen of this Sorex has been found in Maine besides this, and that was in Norway, discovered by A. E. Verrill.

SUB-ORDER INSECTIVORÆ. FAMILY SORICIDÆ. SUB-FAMILY SORICINÆ.

GENUS SOREX, *Linn.*

GEN. CHAR. Ears large, valvular concha directed backward, partly furred on both surfaces, tail about as long as the body (exclusive of the head,) or longer, its hairs of equal length, except at the tip; feet moderate, not fringed; skull slender anteriorly and elongated, upper anterior incisor with a second basal hook, and a small angular process on the inner side, near the point; two anterior lateral teeth somewhat larger than the next.

Sorex Thompsoni, Baird.

Thompson's Shrew.

This species was named by Prof. Baird as a mark of respect to the late Prof. Zadoc Thompson of Vermont, who first discovered it in that State.

Prof. B. gives the specific characters thus:

Very small and slender. Ears large, about as long as the fur, which measures 1½ lines. Feet very small and slender, hinder ones barely exceeding four lines. Tail shorter than the body, exclusive of head; terminated by a pencil.

Only four lateral teeth above, the third in contact with the first molar. Anterior upper incisors with a serrated internal lobe near the point.

Color above, dark olive brown, slightly hoary, paler on sides. Beneath ashy white, no tinge of chesnut or reddish brown.

Length 2 inches. Tail, 1¼.

We give the above description with the hope that it will lead others to watch for more specimens of the kind.

Respectfully submitted,

EZEKIEL HOLMES.

BOTANICAL REPORT.

E. Holmes, M. D., *Naturalist to the Scientific Survey of Maine:*

Sir :—During the present season, I have prepared a catalogue of the Flowering Plants of Maine. A portion of it, extending from the genus Atragene to the genus Ostrya, has been already printed in the first number of the Proceedings of the Portland Society of Natural History. The remainder, embracing the names of all other species known to exist in the State, will probably be published in a subsequent number of the Proceedings of the same Society, and will form, with the list already printed, a complete catalogue of the plants represented in the Herbarium at Portland. Owing to some oversight, the specific localities of many plants have been omitted, but, in most cases, reference to the Botanical report of last year will enable any one to distinguish localities, and they can be marked in the catalogue by marginal notes. Any additions, either to localities or species, will be most gratefully received by the Botanist or any member of the Survey.

Many species peculiar to the North of Maine will be found noticed in the catalogue. During this summer, we have been quite fortunate in securing fair specimens of these plants, which, with others already upon the shelves of the Herbarium, form abundant material for exchange. I sincerely trust, Sir, that you may devise some method, by which these specimens may be of the greatest service to botanical students throughout the State. Packages of preserved plants, designed for delivery to teachers of those institutions in which Botany is taught, either have lain uncalled for upon the shelves of the Society, or have been bestowed, unasked for, upon those manifesting the slightest interest in botanical pursuits. But a parcel of dried plants thus thrust upon those who do not care enough about the specimens to even ask for them as a gratuity or by way of exchange, is as worthless in such hands, as a handful of hay. Holding this opinion, I have endeavored to be not too forward in offering to give away specimens from the State collection; but proposals to exchange have been regarded, in all cases, as worthy our most prompt and careful attention. We now have, as I have already stated, abundant material in the Portland collection for many exchanges, and I commend, Sir, to your most serious consideration, this subject of facilitating its useful and judicious distribution.

I have been requested to give, in this report, some plain directions for collecting and preserving plants. Most excellent rules are laid down in Dr. Gray's "Lessons" and "Structural and Systematic Botany," also in other text-books upon the same science; and, for this reason, it seems to me to be entirely superfluous to present extended directions in regard to the matter of plant collection. Nevertheless, since it is desired that some guiding rules should be given in this report, I will offer a few brief directions to those wishing to commence collecting plants, prefacing them by the remark that Botanists in Maine will find Dr. Gray's "Manual" and "Structural Botany," indispensable to a proper understanding of the Flora. Although the "Manual" has a very extensive geographical range it does not, in embracing so much, slight any part of its survey. This remark, which may be thought to have too personal a character for a report of this kind, is prompted solely by a desire to call the attention of Maine teachers and students to the absolute necessity of using the most thorough, precise and useful manuals of botany.

Plain directions for collecting and preserving plants:

1. Gather plants upon a dry day, if possible, and shield them from sunlight and wind.

2. Plants should be dried between sheets of even, thick bibulous paper to which considerable pressure is applied. This mechanical pressure can be obtained by means of heavy weights or, more conveniently, by a botanical press.

3. A press, portable, easily adjusted, and in every way satisfactory, is constructed of three boards and two strong leathern straps. The boards should be at least fifteen by twelve inches in size, and be kept from warping by means of firm oaken splints secured to the ends. The middle board serves to equalize the pressure.

4. Between the boards should be placed eight or ten quires of thick, unsized, but smooth paper. The plants, as soon as convenient, must be laid evenly between these sheets of paper, having

16

ten or a dozen thicknesses of paper to absorb the moisture of the fresh plant.

5. Change the papers each day, till each plant is perfectly dry; then place the dried specimens in folios of white sized paper, with the name of the species, if possible, the date of collecting and the locality. The plants are now ready for the Herbarium.

6. Herbarium specimens are best preserved by being moistened with a solution of corrosive sublimate in diluted alcohol. Having been thus poisoned they are to be fastened by hot glue to single sheets of thick paper. Specimens illustrating one species may be attached to a single leaf; the generic and specific name being written on a separate slip of paper and fastened to the right-hand lower corner of the sheet. The several species of a genus are usually contained in a folio of stiff paper of a color different from the single sheets.

It is advisable to collect all plants which have not been previously placed in the Herbarium, whether the names are known or not. Unknown plants become, in a little while, far more interesting to the Botanical student than those with which he is perfectly familiar. Please send duplicates of all plants which the collector is unable to determine, to the State Collection at Portland, where they will, if possible, be gladly studied and named.

These brief directions may aid many young botanists in Maine in commencing to form Herbaria of much importance, and materially advance the knowledge of the plants of our State. By the members of the Survey, the Phænogamia have been studied as thoroughly as time would allow: but much remains still to be done.

As the names of those plants new to Maine, which have been detected this season, have been already published in the Proceedings of the Portland Society of Natural History, I shall not enumerate them in this place. The facts of botanical and agricultural interest, which I noticed during the tours in the wild-lands, will be given in the detailed account of those journeys through the valleys of the St. John and Schoodic Rivers. There are, however, one or two facts of some interest which may be as well alluded to now. I refer to the occurrence of several rare plants in Western Maine. While assisting in running a Geological section from Mount Desert to Canada, in July, I observed in a swamp two miles North-west of the hotel at Parlin Pond, and on the South side of the highway, many fine specimens of the interesting species of Juncus, J. Stygius, *L.* This European species was first detected in the United States, by Dr. A. Gray. He discovered it on the borders of Perch pond in Northern New York, but I have been informed that it is not at all plenty in that locality. Therefore Botanists will be interested to learn of a new and easily accessible locality, which I have, for this reason, been particular in describing. All my specimens of J. Stygius, *L.*, differ, from the specific description, in having a sheathing, filiform leaf clasping the middle of the stem, instead of being "naked above."

Eriophorum vaginatum, *L.* common along the "Canada road," so called. Nardosmia palmata, *Hook.* very abundant in swamps near the Canada line, upon the same thoroughfare.

Arnica mollis, *Hooker.* This showy plant is found sparingly, near Moxie Falls, a few miles from the Forks of the Kennebec. It occurs in great beauty and profusion in the vicinity of the cataract of Parlin Pond Stream, where its orange flowers are sprinkled by the spray of the falling water. The iridescence of the flowers as they were bathed in the sunlight and the spray, was a spectacle of much beauty, the orange of the blossoms here and there, overpowering the rainbow coloring of the drops of water.

The Vegetation of Aroostook County.

The local distribution of plants is a matter of much interest to the Botanist and Agriculturist. It is not my purpose, however, at this time, to express my crudely formed opinions concerning the laws which have regulated the distribution of plants in Maine, nor to advance any theory in regard to the occurrence of certain species, but to present some facts which have come to my notice during the surveys of 1861 and 1862. Many of the facts were presented in the report of last year, but it is necessary to repeat them in this connection in order to give a clearer understanding of the Flora of the north of the State. It will be seen that I have ventured to divide the upper portion of our State into Botanical districts of considerable extent. That I may place myself right in regard to this subject, let me preface the descriptions of the districts by some remarks upon plant distribution throughout New England.

Notwithstanding hundreds of our Phænogamia are common throughout the length and breadth of New England, I think no one will be unwilling to adopt the belief that there are certain well-marked districts of plants in the six States. The maritime plants may be readily classed together, and, in like manner, the Alpine and sub-Alpine Flora of the White Mountains and Katahdin. But there are others, more or less confluent perhaps, which all botanists will readily recognize. Thus the West Connecticut district, so thoroughly examined by Drs. Barratt and Ives, contains very many herbs, shrubs, and even forest trees, not known to exist in other districts.

Other well marked districts are Western Massachusetts; South-eastern Massachusetts, including Rhode Island; Northern Vermont, etc. As a general rule the characteristic species of each district are few in number, but the individuals of each species occur abundantly. Let us notice, for example, the difference between the plants of Hampshire county, Mass., and the plants of Southern New Hampshire. These districts are one hundred miles apart—a distance hardly great enough to account on any climatic hypothesis for so great a difference as exists. In the former we find occurring frequently Lygodium palmatum, Ophioglossum vulgatum, Camptosorus rhizophyllus, Allosorus atropurpureus and gracilis, Asplenium Ruta-muraria and angustifolium, Carex squarrosa, Verbena angustifolia, Pedicularis lanceolata, Pterospora Andromedea, Azalea nudiflora, etc., etc. I have searched very carefully among Oakes' lists of plants of Southern New Hampshire, and among many catalogues and Herbaria, and a large portion of the district itself, for any of these species, but in vain. And since English botanists make smaller and less important districts than these two are, I think there can be no reason for not accepting them. The question now arises, how can one know the points along the dividing line of any two districts? Since we have chosen these two as an illustration, it may be advisable to give the results of protracted herborizing, along the borders, on either side of what is accepted as the provisional line of demarcation. None of the above plants have as yet been found farther northeast than a line running from Athol to North Brookfield, Mass., and the New Hampshire plants are not found, plentifully, west of the same line.

It will be perhaps noticed that these two districts have, for their centres, Concord, Mass., and Greenfield, Mass., these nearly cöinciding with those marked out in the Mass. catalogue of plants appended to the final Report upon the Geology, in which Boston and Amherst are made two central points.

If, then, it is admitted that districts of limited extent do exist, is it unreasonable to believe that several may be distinguished in Maine, a State fully as large as all the rest of New England? I am aware that severe, extremely careful and faithful study is needed to define the limits of such districts with anything like accuracy, and I long hesitated about marking out such lines upon the map of Maine. But I am sure of two facts; one, that the effort, although perhaps premature, may induce Maine botanists to explore Aroostook county in order to satisfy themselves of the truthfulness of these conclusions, and secondly, that such a map will elicit charitable but impartial criticism as to the expediency of defining such limited districts of vegetation. Although the members of the survey have been able to sketch roughly the limits of various districts in the State, it is thought expedient to defer presenting these lines of definition in this report, exhibiting now only two districts of Northern Maine. The upper fourth of the State is now, owing to the scattered and scanty population, all embraced within one county, Aroostook. It is of this county that I wish to speak particularly at the present time.

The country lying along the river St. John, from Boundary branch to Grand Falls, is marked by the very frequent occurrence of certain Northwestern plants. And the district comprised by the curved northern limit of Maine and a line drawn from Grand Falls to a point between Baker Lake and Boundary branch will be found to be nearly the range of these plants in our State. This district is so entirely distinct botanically from any other portion of Maine, that its limits can be said with confidence to be clearly defined. The following list of plants may be considered as comprising the most characteristic species of the St John district:

Anemone parviflora, *Michx.*—Abundant along the main river, in the disintegrating slates.

Astragalus alpinus, *L.*—Common in rocky, damp woods.

Astragalus sp. ign.—Much resembling A. Robbinsii, *Gray*, but appearing to possess specific differences. Very frequent along the shore.

Oxytropis sp. ign.—Dr. Gray has examined specimens of this plant, but considered them too mature for proper identification. It agrees pretty well with O. Uralensis, *L.*, var. b. Quite abundant in moist woods.

Artemisia borealis, *Mx.*—Common in clefts of rocks along the shore, particularly near falls.

A. Canadensis, *Mx.*—With the last.

Tanacetum Huronense, *Nuttall.*—Plentiful in rocky soil, and very thrifty.

Vilfa cuspidata, *Torrey in Hooker's Flor. Bor. Am.*—Not infrequent along the shore of the main St. John.

Besides the foregoing species we also find in this district many which occur in other peculiar localities in New England; for instance, at the singular precipice at Willoughby Lake, or some cold maritime swamp.

Astragalus Robbinsii, *Gray. Phaca, Oakes.*—Very abundant on the shores of the river St. John.

Hedysarum boreale, *Nutt.*—Quite common and flourishing throughout the district.

Primula Mistassinica, *Michaux.*—So abundant is this delicate plant at some points along the River St. John, that the shore assumes a faint red or purple tinge when viewed at a little distance.

Solidago Virga-aurea, var. alpina, *Bigelow.*—With the last.

Nabalus racemosus, *Hooker.*

Tofieldia glutinosa, *Willdenow.*—Wet grounds along the river.

These plants occur on the shores of many tributaries of the upper St. John and in the neighboring woods. The whole region through which these plants are distributed is covered by a thick growth of coniferous trees, most of which are of good size, and are considered valuable for "tun timber" and "deal."

Immediately south and east of the lower limit of this district we come into a different vegetation. The St. John plants have entirely disappeared, except along the river banks, to which they have been floated by the spring freshets. One, perhaps two, of the compositæ are detected high above the usual line of freshets, but it will be remembered that each plumed seed of the species of this vast order is wafted on its own wings far beyond the ordinary limits of the dissemination of other plants. With this exception, no plants of the St. John district were discovered outside the natural reach of water communication. This second region, which we can distinguish by the appellation of *Aroostook district,* is characterized by the occurrence of a different flora. Instead of conifers, we find a prevalence of "hard-wood" trees. Maples, Beeches, Oaks and Amentaceæ form the forests. Under such trees we see flourishing Dicentras, Claytonias, Adlumia, Aralia quinquefolia, Solidago odora;

on the shores of the rivers and their tributaries, Lobelia Kalmii, Anemone Pennsylvanica, and two species of Vitis, V. labrusca and V. cordifolia. Even a third species of the Vine is said to be found near Woodstock, but a protracted search failed to detect it. It will be remembered, perhaps, that it was stated in the report of last year that a section of great fertility was noticed on the east branch of the Penobscot, near the mouth of the Wassataquoik and Sebois. Our limits which we have assigned to the Aroostook belt embrace this portion of the county as well as much of the west branch valley beyond Katahdin. The adaptability of this valley to farming purposes will be shown, at length, in the report upon the "Wild Lands." In order to better exhibit the limits of the two sections, I have made the following map, upon which are represented the St. John and Aroostook districts.

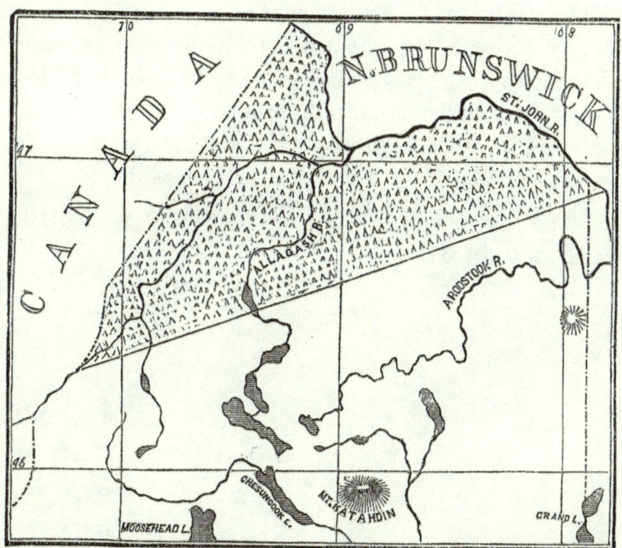

Those who have at hand the Geological map prepared by Mr. Hitchcock will see that the more fertile belt corresponds remarkably to the defining lines of the great formation of calciferous slates and slates of Devonian age. Of course the lower limit of the Aroostook section must, for the present, be considered entirely provisional, because we have not been able to devote sufficient study to this portion of the subject. The southern part of the Aroostook district and the northern portion of the Somerset section are therefore regarded now as having vegetation which is, so to speak, confluent. It is my opinion, however, that it will be found that the "wheat-growing lands," as the farmers call them, are much better north of Weston, on the eastern boundary, than south of the same town. In Washington county, the granite comes in to modify the fertility in a marked degree, and it is very likely that the southern line of the Aroostook vegetation will be best traced westerly from the town of Weston just referred to. The whole matter is one of interest alike to the botanist and the farmer, and deserves greater study than the members of the survey corps have been able, amid other more pressing duties, to bestow upon it.

I must be permitted to acknowledge many favors received from botanists and other gentlemen in Maine during the present season. Very much has been due to Rev J. Blake, of New Hampshire, who has contributed to the Herbarium of the State pretty full sets of certain difficult genera. As his specimens were largely collected in Maine, the value of the gift can scarcely be over estimated. To him and the many others who have assisted me in botanical study this summer I am under great obligations.

With high respect, I am, sir,

Your obedient servant,

GEORGE L. GOODALE.

Portland, Oct. 29, 1862.

REPORT ON MARINE ZOOLOGY.

To Ezekiel Holmes, M. D.,
and C. H. Hitchcock, A. M.,
Directors of the Scientific Survey of Maine:

Pursuant to your instructions, I hereby submit the following brief and partial report of my labors in the department of Marine Zoology during the months of July and August, 1862. It is impossible to furnish anything more than a general statement of the progress of the work, and of the portions of the coast visited, as the proper identification and classification of the specimens collected will require several months' additional labor. The time allotted for my labors being but two months, and this in an advanced stage of the season suitable for work upon the sea-shore, I determined to commence at that point where the excessive fluctuations of the tides were more favorable to an abundant growth and a larger variety of species of marine life than some other portions of the coast.

Accordingly, on the eighth day of July, accompanied by Mr. A. S. Packard, Jr., of Brunswick, I arrived at Eastport. Thence I took my small skiff and selected Treat's Island as the most central point of the work in that region, it furnishing the best advantages on account of the extreme rise and fall of the tides, and its proximity to deep water. From this place I made numerous excursions by water to different localities, dredging in all practicable places and in depths of water varying from fifteen to twenty-five fathoms.

Among the localities visited the following are the principal, and the result of the dredging most interesting.

At Treat's Island, between the high and low water marks, the species of the fauna of the coast of Maine, mentioned below, occur.

Sertularia polyzonias, S. argentea ; Ophiolepis robusta, Ophiopholis scolopendrica (plenty ;) Asteracanthion rubens, A. littoralis, Solaster endeca, S. papposa ; Echinus granulatus ; Pentacta frondosa, Chirodota lævis ; Ascidia callosa, Cynthia pyriformis ; Boltenia reniformis (rare ;) Pecten islandicus (rare;) Modiolaria discors ; Mya arenaria, M. truncata ; Saxicava distorta ; Tectura testudinalis ; Margarita helicina, M. cinerea, M. undulata ; Littorina littorea, L. rudis ; Purpura lapillus, Buccinum undatum ; Fusus decemcostatus, F. islandicus ; Dendronotus arborescens.

Nearly one half of the above are found at much lower points in the western part of the State, and there, generally, not above the laminarian region.

In the vicinity of the same locality, in from ten to twenty-five fathoms of water, the dredge brought up—Alcyonium carneum ; Actinia obtruncata ; Astrophyton Agassizii ; Ohiopholis scolopendrica ; Cribella oculata ; Solaster papposa, S. endeca ; Echinus granulatus ; Pentacta frondosa ; Gemellaria dumosa ; Ascidia callosa ; Cynthia pyriformis ; Boltenia reniformis (?) ; Terebratulina septentrionalis ; Pecten tenuicostatus (rare ;) Nucula delphinodonta; Yoldia sapotilla, Y. myalis ; Modiolaria discors, M. corrugata ; Cryptodon Gouldii, Astarte semisulcata, Cardita borealis, Lyonsia hyalina, Pandora trilineata ; Chiton marmoreus, Chiton albus ; Entalis striolata ; Crucibulum striatum ; Cemoria noachina ; Scalaria groenlandica ; Lunatia triseriata ; Nassa trivittata ; Buccinum undatum ; Fusus pygmæus, F. decemcostatus, F. islandicus ; Sipunculus Bernhardus ; Sternaspis fossor.

After exploring this region as thoroughly as the time and weather wonld permit, we made a hurried visit to Cobscook river. At Pembroke Point we collected numerous fine specimens of fossils of marine animals, but in consequence of the strong tide which here flows with great velocity, it was impossible to use the dredge.

Proceeding thence to Perry, we visited, on the way, the fossil deposits on Upper Treat's Island, where we procured some specimens, mostly Lingulæ.

At Little river in Perry, a small collection of Devonian fossils rewarded our labors.

Returning to Treat's Island we resumed dredging, for a few days, with satisfactory results, and thence, Mr. Packard having left me, I turned my attention to the exploration of the St. Croix river, dredging between Devil's Head and Robbinston. In the vicinity of St. Croix Island the dredging in seventeen fathoms of water was productive of good results. The following are some of the specimens collected here :

Alcyoneum carneum; Pecten tenuicostatus (not plenty but probably more abundant further up the river;) Trochus occidentalis (rare;) Natica pusilla, N. clausa. At Devil's Head, Chirodota lævis (plenty); Mya arenaria (large and more plenty than at Eastport); Chiton marmoreus; Tectura testudinalis; Margarita undulata; Buccinum undatum; Homarus Americanus (abundant); Cancer irroratus (very rare.)

Before leaving this section of the State I occupied some days in examining the bottom in the neighborhood of Eastport, where, at Shackford's Head I found some of the finest specimens which I had taken during the exploration, consisting of Actinia obtruncata, Boltenia reniformis (very abundant), Velutina haliotoides, Hyas coarctata and a number of crustaceans.

A little below Todd's Head at Eastport, we found the Corymorpha nutans so abundant that they were attached to nearly every mesh of the dredge net.

At a short distance from this point, in from fifteen to forty fathoms of water I found Astrophyton Agassizii; Solaster papposa, S. endeca (plenty); Yoldia myalis (rare); Callista convexa; Anatina papyracia; Cochlodesma Leana; Thracia truncata; Trochus occidentalis; Trophon clathratus; Fasciolaria ligata.

Dredging from Lubec to West Quoddy Head, on a hard nullipore bottom, I took large numbers of Chiton marmoreus, finding them attached to almost every pebble.

I observed in this section the absence of the eel grass (Zostera marina) so common in the western part of the State, its place seeming to be supplied by the Chorda filum. The shores here are well lined with the fuci, while in the Passamaquoddy bays the Laminaria are quite small.

My time having now nearly expired, I concluded to devote the remainder to a preliminary reconnoisance of other sections of the work. I accordingly embarked upon a vessel bound up the coast, and visited, among other places, Machias bay, where I found, in from four to seven fathoms, Actinia obtruncata, (?) (plenty); A. sipunculoides, (1 specimen); Astarte semisulcata; Cardita borealis; Sternaspis fossor, (very large,) and some fine forms of Polyzoa;—Narraguagus bay, securing the Nucula proxima in great plenty;—Rockland harbor, taking, in five fathoms, nullipore bottom, Cuvieria Fabricii, plentifully;—Pemaquid light, near which, in thirty-five fathoms, I dredged fine specimens of Corymorpha nutans; Entalis striolata; Yoldia thraciæformis. At Pemaquid outer and inner harbor I found in abundance, in from three to seven fathoms, Echinerachnius atlanticus, on sandy, nullipore bottom.

I had now arrived at Casco bay, and here I spent some days visiting and exploring the numerous islands, bays, inlets and channels of this beautiful locality. At Jewell's Island, at the lowest tide mark, I found the Pholas crispata imbedded in peat and logs, usually covered with from two to three feet of water at ordinary low tides. At this place, beneath the surface of the water and extending to some distance from the land, is a submerged tract thickly covered with the remains of a forest visible in smooth times of water and in the early morning. It was with considerable difficulty that I was able to procure specimens of the Pholas.

On the Brown Cow and the Green Islands, which are nearly destitute of vegetation, I found the following land shells in abundance, viz: Helix nemoralis, (wood snail,) Helix albolabris, (white-lipped snail,) and Succinia obliqua. On Eagle Island, a short distance from the above, Helix alternata occurs in great numbers.

Almost every island in the bay has a mollusk peculiar to itself, and coincident with its soil or flora; Eagle Island bearing spruce and fir, producing Helix alternata, while one of the Goose Islands, with a hard wood growth, produces Helix albolabris abundantly.

This completed the labors in this branch of the survey for the season. The specimens collected were carefully preserved, and are deposited for the purpose of identification and arrangement in the hall of the Portland Society of Natural History. Glass jars and alcohol will be required for their permanent preservation, and should be provided at the earliest opportunity.

In consequence of the limited time during which my work was performed, you will readily perceive that this report must be quite imperfect; in fact the work was in its nature hardly more than preliminary. The process of collecting specimens and making explorations in this department of science is attended with so many difficulties arising from the state of the weather and of the tides, together with the necessary exposure and labor, that a complete examination of so extensive a range of coast as that of Maine, and a thorough classification of its fauna is unavoidably slow. I can, however, assure you that what has been done has been done well, and that the specimens enumerated above are but a very small proportion of those actually taken and preserved. The collection comprises animals belonging to almost every class in Marine Zoology known upon our shores, and probably contains several new and rare forms.

In conclusion, I desire to acknowledge the valuable assistance rendered me by Messrs. A. S. Packard, Jr., C. A. Shurtleff, George Hayes, U. S. Treat, L. M. Barbour and Wm. I. Beals.

Respectfully,

C. B. FULLER.

NOTES UPON CERTAIN MAMMALS IN MAINE.

Quite recently our attention was called to descriptions of the characteristics and habits of some of the mammals inhabiting the wilds of Maine, by J. G. Rich, Esq., of Upton.* These articles appeared first in the columns of the *Bethel Courier* and the *Oxford Democrat.* Impressed by their value, we requested some of them for publication; and Mr. Rich has kindly handed to us the following sketches for this purpose:

NORTHERN HARE, OR RABBIT, *Lepus Americanus*, *Erxl.*

We have but one variety of this little animal in our good State of Maine, although in the neighboring State of Massachusetts there exists a distinct variety—being much smaller than the kind we are here to treat upon, and having quite different habits.

Authors of Natural History, in several instances, confound the rabbit and hare; and although resembling each other exactly to the common observer, yet there exists a distinct difference, especially in the young leveret.

In the first place, let me here observe that in no instance has the Creator shown His care and provision for animal nature more conspicuously than by providing so bountifully this species of animal. They cover the earth to the full extent of its capability of support for them; are extensive breeders, more than any other, and have no weapon of defence except flight,—and on which depend for sustenance, in a great measure, the bear, lynx, fox, fisher, mink, weasel, ermine, among the animal tribes, and owls, hawks, eagles, and many others among the birds,—in fact, it may truly be said the rabbit is the substantial food of a large class of animal nature.

Our hare is about two feet from the nose to the hind feet, short head, full eye, receding forehead, large, open, long ears, short tail, long hind legs with four toes, and short forward legs with five toes; feet well covered with coarse hair, making a good brush, when dry, for many purposes; loose, long hair on the body, always whiter under the body, and in winter white all over; but in summer, of a yellowish brown, varying to a rufous brown. Next to the body is a soft, loose fur, of a silky texture, of a lead color on the back, and which fur, I think, keeps its constant color, and only the long hair on the upper parts of its body changes color.

The weight of this animal is from five to seven pounds; and so constantly

* We are glad to ascertain that Mr. Rich proposes to publish all his sketches of mammals, birds and fishes in a book, together with apropos descriptions of the best fishing grounds in the northwest part of the State; of life in the woods, and some sketches of his own experience of sixteen years as a hunter, trapper and guide. Persons as familiar as he is with the natural inhabitants of the forest, confer a favor upon the public when they present to them such experiences.

out of condition, that it is proverbial among hunters, when visited by our young city sportsmen, to enlarge upon the great and excellent qualities of shaving soap made of "rabbit's grease" and "cedar ashes."

The meat of the hare is quite flavorless and light colored, and depends much upon the condiments in dressing for its savory taste,—what epicures think to the contrary notwithstanding. Still, it is not void of a good share of nutrition, and makes a very simple diet for invalids. Indeed, I have not unfrequently been obliged to resort to this sort of food entirely, after getting short of provisions, in many a hunting tour in the wilderness; and we can almost always safely depend upon taking by trap, snare, or gun, a sufficient supply of hare venison almost anywhere in our forests of Maine.

The female hare is capable of bearing young before they are one year old,—say those born in August and September multiply the following spring; and in this State, I am of opinion that they quite regularly have two litters each season of from three to five each.

They do not burrow as some of the genus do, but make a nest under a brush heap or the thick foliage of a small tree, of leaves and soft moss. They go with young about five weeks, and nurse them for three weeks, when they gradually leave the original nest and mother and take care of themselves. I have often picked them up in the woods, and when you find one you may be almost sure the rest of the family are within ten rods of you. They are born with a good dress of brown hair, and eyes open, and teeth well cut through; and what is quite peculiar to this animal by a curious formation of their genitals are often found to have a superfœtation.

The food of our hare is chiefly browse of small and tender bushes, and they especially love the buds of yellow birch. I have often baited up a score of them in two nights by chopping down a birch tree, and limbing it down, and among the branches setting my traps.

Hunters depend almost entirely on the hare and muskrat for bait for their traps to take larger game.

I have noticed that the higher up among the mountains I go the larger are the hare, and I have no doubt that the atmosphere and even the soil have great influence on the native animals.

The noise of the hare when frightened or hurt is a high note, cut short at very frequent intervals, and the voice kept up, and very shrill and very plaintive, sounding like filing a mill saw. They also have a peculiar grunting noise that is used when near each other, and in their families to make known their wishes to each other. They also stamp with their feet like the domestic sheep; and often resemble, when jumping around the camp, the step of a heavy animal.

I have often had them come into my open camp, when I had nicely bivouacked for the night, and, in several instances, jump upon my body, causing a sudden fright and leap which sent "Master Fatty" (as he is familiarly called,) away in a hurry. They are always attracted by a campfire in the night, and we can always, of a clear night, shoot them by keeping awake after all is quiet.

I have seen the fisher follow and take the hare. One on a certain occasion followed down the Richardson Lake about one mile, after they came on, when the hare commenced to circle, and the fisher, who was but a short distance behind, also continued the chase, but kept inside the circle of the hare, and by so doing made quite a gain of the game, and in a very short time was able to overcome the hare, of which he ate a portion, and dragged the remainder on shore and buried it for future use.

THE WEASEL, *Putorius pusillus*, *Aud. and Bach.*

This little animal is very common, and doubtless well understood by observing people; yet there may be simple facts about its habits worthy of a passing notice.

The length of its body is about eight inches, with short legs, long neck, large, open ears, small eyes set in the head nearer the nose than the ears. Color, in the summer, brown above and white under the limbs, while in the winter it is white over every part of the body and limbs, with the exception of the end of the tail, which is always black for about an inch. The length of the tail will always reach exactly to the foot of the hind leg if both are drawn out straight.

Its home or common retreat is in piles of loose stone, wood, or other material, or in hollow trees, where it brings forth its young of two or sometimes even three litters in a year, of from four to five at a time, commonly the latter number, and which it will defend with indomitable courage. They are remarkable for their perseverence and courage, and have been known to attack men when in companies of three or four.

Their food is commonly mice, but they will devour most all kinds of birds and eggs, or at least the brains and blood of birds, which they like better than the fleshy part.

The color of the weasel when changing from brown to white, or *vice versa*, is very prettily variegated. Their voice, or noise, is quite peculiar, and sounds like a serpent—a sharp, shrill, compressed sound, quickly repeated two or three times.

We find the weasel everywhere, in the woods as plenty, and perhaps more so, than in the farm yard and house. They are uncommon mousers, being much more expert than the house cat, and so much smaller in body that they can follow their prey in many places where the cat cannot enter. It can run up a smooth perpendicular board with facility.

The enemies of the weasel are chiefly the owl and hawk, and they are often obliged to come down after seizing and rising into the air with it, by his opening a vein under the wing.

A larger species of Mustelidæ, which is often confounded with the weasel, and is called in history the "Stoat," Mustela erminea, is very plenty on high mountains in the forests in this State. I have often caught them while sable hunting. I caught one the past winter. They are full twice the length and bigness of the weasel, and subject to the same change of color; and when in the white state are called "Ermine," and much used in foreign countries for lining and trimming fur garments. Of their habits but little is known, except that they prey on larger game than the weasel—such as the hare, partridge, &c.

The Canada Lynx, *Lynx Canadensis, Raf.*

This animal is the largest of the wild cat species in this State. Arranged in the group by naturalists called "Feline" (Felidæ,) and in the order "Carnivora."

There are about eight kinds of lynx described by naturalists, but they are so confused that their histories are of little account. They describe the "Canada Lynx" for the Boreal Lynx and the Caracal, and so mix them up that it is about impossible to distinguish them by existing descriptions.

The Lynx is one of those animals of which the ancients told so many fables. That they could see through opaque bodies, and even through stone walls, and that their urine often contained a valuable stone called "*lapis lincurius*," so the old maxim "Lynx eyed," &c.—be this as it may, our Lynx has a very sharp, large, round eye, and is capable of staring at you for a great length of time. Such is however the natural ferocity of this animal that it is believed to be impossible to perfectly subdue it.

This Lynx is about three feet in the length of its body and stands about twenty-one inches high; it has a round head like our domestic cat, but much larger. Its fur is long and soft, and changes its color twice in the year. In the hot months of summer it is dingy reddish grey and when prime, in the coldest part of the season, it is a beautiful stone grey along its sides and mixed along its back with long hairs tipped with black, while underneath it is white, beautifully mottled with black spots. Its tail is about four inches long, tipped with black. Its ears stand erect and are quite conspicuous, being tipped with a tuft of black hair, and on either side of its lower jaw is quite a bunch of grey hair mixed with long black hair.

Dr. Richardson states that the early French writers on Canada, who ascribed to this animal the habit of dropping from the limbs of trees, on to the backs of *deer* and destroying them by tearing their throats and drinking their blood, gave them the name of *Loup Cervier*, or wolf stag. This animal has very long, owl shaped, retractile claws, and four toes on each foot, and the bottoms of their feet are covered with fur. Their legs are very muscular. And its whole contour denotes great activity.

Virgil calls the Lynces of Bacchus, "*varica.*" and in another place alludes to the skin of the spotted Lynx, "Maculasca Lynces."

I think the "Boreal Lynx" and our "*Canadensis*" are the same, and one animal. The former ranges in northern Europe, and the latter in North America. The Hudson Bay Co., a few years ago, used to export to Europe from seven to nine thousand pelts of the Lynx. The most beautiful skins of the Lynx are from Siberia, and belong to the "*lupus cervarius.*" Buffon says that all animals of America are smaller than the same kinds in Europe, and that the Lynx in Siberia is compared to the wolf, but in America, to the wild cat.

I have been informed of another kind of Lynx in this State, but have never seen one of that kind. It is said they live in the open cultivated regions, and have no fur on the bottoms of their feet, and are not so thick furred, neither so handsome; but I cannot describe them from personal knowledge, and therefore will let them pass.

Our Lynces breed once a year, and bring forth sometimes two and sometimes three at a birth, and like most other carnivorous animals will defend their young with their lives. They are not, however, so speedy as most people think, and can easily be treed with hounds. When they run they leap and strike all their feet together. When this country was first settled, this animal was quite troublesome among the sheep and lambs. They went in droves and were more bold than when single. The State of New Hampshire now pays a bounty of one dollar on their heads.

When I first moved to the Mollychunkemunk Lake I trapped and killed forty-nine of these animals in one hunting season, and since that time have killed a great many each year, but have not kept an account of the number. I have seen them swimming the lake and they appeared to be good swimmers. Will often swim two miles at a time, and about as fast as we can paddle a boat. They can be easily trapped when it is good travelling for them on the snow, but when the snow is deep and soft they keep in the thick swamps and do not travel much. They live chiefly on the rabbit, and when we trap them we have to bait them with fresh meat or scent of Assafoetida, or beaver castor, of which they are very fond. Their teeth are feline, very long, and extremely sharp. They take their prey principally by watching and creeping upon it. I frequently see their tracks where they are creeping along very slow, and with steps not more than three inches apart, although, when they they leap they go from seven to ten feet.

Capt. John M. Wilson informs me that he once saw one lying stretched out on the limb of a tree directly over his head, but as he moved quietly along, the cat did not seem inclined to attack him.

I have a preserved specimen of the Lynx before me now, but I think the eye a little too yellow, as the natural color always appeared to me in the living animal to be like bright silver; and after eyeing the creature for a short time, and giving it time to grow mad, it increased in size and brightened to something like livid fire; and would seem to challenge the nerves of a strong minded hunter to look them out of countenance.

It is my impression that this animal seldom, if ever, attacks a man, when enjoying the freedom of nature, but if cornered up, and unable to escape, no doubt would defend itself to the last.

Mr. Z. F. Durkee, of Magalloway, killed one of these animals in the Richardson farm house, some years ago, in this wise :—The cat was in the woodshed and Mr. Durkee went into the house through the shed and the cat ran in ahead of him in hopes of escaping, but he cornered it in the dining room and the creature jumped up against the window and before it could recover, he struck it down with a club he had in his hand.

Mr. A. P. Gould, of Boston has had several hand to hand fights with this animal, and in every instance, I think, has succeeded in conquering the beast; although, at one time, he nearly surrendered to an old male, which had the advantage of position, and had it not been for his knowledge of the manner of the animal's attack, and his own superior skill with the knife, he would certainly have been overcome. I am not at liberty to give the full details now, of this encounter.

Mr. Robert Torrey, of Cambrideport, Mass., came near being attacked by one of these animals in the summer of 1858, in an old barn between the Mollychunkemunk and Mooseluckmaguntic lakes, where he and a few friends composing his party were camping for the night. The cat appeared to have been concealed somewhere on the beams of the building, and was not noticed by our party until we had got quietly stowed away for the night among the old refuse hay on the upper floor. The animal then descended to where we lay, not with a bound and a growl, but softly, as if to make sure that we were fit for immediate use without cooking. He commenced his operations creeping along the floor near our feet, and making a rustling noise in the old hay, that set poor Torrey in high dudgeon, and he immediately drew his dirk knife and prepared for a close combat; but owing to the darkness, and the proximity of our bodies, he did not commence offensive operations, and the cat escaped.

The hunter who taught me to hunt when I first came to this country (Wm. H. Leverett, who has since removed to Marquette County, Wisconsin,) frequently accompanied me to my traps, and I remember at one time, when we were following along on our line of traps many miles from human habitation, we heard one of these animals screech, and knew by the peculiar sound what it was, and this being the second one I had caught, I was highly excited and elated. Leverett, on the contrary, was perfectly cool and collected, and told me while we were hurrying along to where the trap was set, that he would show me how to kill a Lucivee, (as we called them.) When we came up, behold we had a large specimen of the Lynx in a trap! Leverett immediately broke him a stick of dry alder and said: "It takes but a small stick to kill one of these creatures." But, said I, "do take a sound stick, for the one you have there is rotten." Said he, "it is sound enough." So he walked directly up to the cat and struck him over the head, and his stick broke in two, and the cat leaped toward him and he jumped backward and at the same time caught his foot in some bushes and fell flat on his back, when the cat leaped upon him, trap and all, and but for my assistance with the axe, he would at least have been badly scratched, if not entirely done up.

I think it very foolish for hunters and others to dally with these and other wild animals, because they have them in traps and security. I have known many instances of hunters losing their game in this way, and even getting into bad scrapes. I have lost bear and moose and some other game by not immediately killing them at the first opportunity.

I will leave this animal by relating my first adventure with one, the first one I ever saw. I had been hunting but a few days when I came one evening just at sunset, to where I had set a trap a few days before—this was in township No. 4, Range 2d. My trap was gone and I followed on the trail by the marks in the moss the traps had made. I followed on at a rapid pace for it was nearly night and I could but just see the signs in the thick woods. I had gone perhaps one fourth of a mile when, coming quickly out of a thick place into an open spot and leaning forward on a dog trot, I was completely knocked over by the cat, trap, and all, coming against me with a rush, and a spit, and a growl, as quick as thought. I did shoot the beast, and ever after followed on their signs more cautious, but the fright it gave me was never forgotten.

THE BEAVER, *Castor Canadensis*, *Kuhl.*

This almost amphibious animal is now chiefly confined to this continent, although formerly it was found in many parts of northern Europe and Asia, but has become nearly extinct in those countries. Its body, from the end of the snout to the insertion of the tail, is almost three feet long; the tail, or caudal paddle, is about one foot long, one inch thick, and five or six inches wide. Its teeth are two incisors in each jaw, and a vacant space between them and the eight upper and under molars. These incisors are no doubt what they chop their wood with, and as they wear shorter by continual use, the teeth grow out—for they are from two to three inches long and arching in form, and protrude about one-half to three-fourths of an inch outward. The feet have five toes, exterior and posterior; the forward toes are short and close, and the hinder ones long and palmated; being the only animal of this description known in nature. Its body is covered with two kinds of hair, one outward coating of hair two inches long and glossy black—the inner coat quite thick and downy, like fine silk. Its tail is covered with scales and looks like a fish. The meat of the forward parts of the body tastes like land animals, and that of the hinder parts and tail like fish.

The largest beavers weigh about sixty pounds. The beaver attains its full growth at the age of three years, and is supposed to live from ten to fifteen years. It breeds once a year, and has from two to six at a birth, and even breeds before it attains its own maturity. It goes with young four months, and brings forth about the close of winter. Its food is chiefly vegetable, and it will not taste of meat, raw or cooked. In winter it subsists mostly upon the bark of green wood which it has laid up in store for this purpose, and it is also quite fond of fish. The color of the body is a cream brown, intermixed with black hair.

This animal differs from all other animals in several particulars; first, it has a pair of glandular sacks between the hind legs, (which, I believe, have no connection with the organs of generation,) containing a substance quite odorous, called castor, (*castoreum*,) which hunters call "barkstone." It is used extensively in medicine, and is quite valuable; but the castor which comes from Europe is esteemed by medical men as by far the most virtuous. This substance is worth about one dollar for each sack, and the fur of the beaver now commands two and one-half dollars for each skin. Not many years ago the skins were sold by weight, and brought one dollar an ounce. Since that time the beaver hat gives place to the silk hat, hence the difference.

The savages are said to use the oil of the tail for many diseases, and we know they esteem it dressed and cooked, as far above any other food you can offer them. The second toe of their hind feet is armed with double nails. They make a very unequal gait in walking on the land, owing to the sloping of the body behind the centre down under towards their feet; and of their hind feet standing out, which is an advantage to their progress in the water. They are also remarkable for the great appearance of reason they seem to possess. They are more shy than the fox, and more knowing than the elephant. They are very acute in their scent and hearing, and naturally timid, with no way of defence except in flight. They live in the enjoyment of all the blessings of life, peacefully; and no doubt have a way of communing one with the other. Their houses and dams have been recently so well described that I will not now go into the particulars of them, but only add some new items or ideas that I have myself noticed.

In the unincorporated Townships of this county, (Oxford,) you cannot cross a meadow or follow up a valley without discovering the old signs of beaver. Their dams on flat land, where there is no current, are always made of grass and mud, but where there is a current, sticks and drift wood are used. And where the current of the stream is quite strong, the dam is made bowing, and presenting a convexity to the tide. Their houses are situated immediately above the dam, and are constructed of sticks and branches of willow and other wood, with mud and stones, all mixed together without any particular method, except to leave a hole to live in. They are built very strong and are generally two stories hight, and have a window, or hole facing the water up stream, apparently to admit fresh air. They have holes through the chamber floor to escape through, when alarmed. And at such times they always slap the water powerfully with their tails, which arouses the whole colony. I think this habit of slapping the water, is not for any particular purpose, but merely a habit, notwithstanding others have said that they do this while building, to put a finish to their work. They always lay with their hind quarters under water, and on their belly, and some suppose this gives their tail the fishy taste which it is known to have.

When swimming, they keep almost an upright position, their heads and shoulders being above water. They generally congregate in the month of July, and choose their mates and live together until April or May. They live from six to ten in a house, and their houses, if joined together by one partition, never have a passage through. They sink their wood for winter food, by means of stones and mud, and not by any mysterious way, as some have supposed. They always choose a locality for colonizing where the water is deep enough not to freeze clear through. In the spring, the males leave their habitations first, and leave the females to take care of their young a few weeks, when they all roam about during the summer months, regardless of their home. They can be shot between sunset and dark, by lying in wait for them, as they take that time to swim out for relaxation. They work chiefly in the night.

They seem to know how high the water will be the coming spring, and build their houses accordingly; whether this be reason or instinct, it never fails. They only seem to enjoy their full powers of reason or intellect, when they are living in society, and families, for when they get strayed apart, from any cause, and wander off alone, they seem to be as dumpish as a musk rat, or any other animal. If a colony gets scattered, it is the opinion of the hunters that they never collect to live together again.

In our upper rivers there are now many small families of beaver. I saw new work on the Diamond river last year, and a hunter informed me of beaver, last fall, on the Cupsuptic river, and on the Beamus stream, seven miles from the upper dam, there were a few, and I met a hunter lost in the woods, trying to make his way to hunt them.

I think beaver now choose very small streams in the deepest recesses of the wilderness to build their dams and houses, since they have been so often disturbed. They remind one of the native Indians. The white man is not contented to let either live in peace on the possessions given them by our common Father.

The beaver seems to be the last of the genus "Castor," and they are fast dying out, and will soon become extinct. In the year 1745, there were imported into London and Rouchelle nearly one hundred and fifty thousand skins. And in the year 1827, the importation had fallen away to about fifty thousand, although four times the ground was hunted over to get them. And now there are very few beaver sent to Europe.

This animal is truly a wonder in the great economy of creation, and presents to our minds the connecting link between instinct and reason, indeed the knowledge of the beaver seems sometimes to be far in advance of the lower order of the human family.

LIST OF REPTILES AND AMPHIBIANS FOUND IN THE STATE OF MAINE.

At our request Dr. B. F. Fogg, Curator of Herpetology in the Portland Society of Natural History, prepared a catalogue of all the Reptiles and Amphibians found in the State, which was published originally in the Proceedings of the Society. It is now reproduced with notes upon their distribution, and a revision of the nomenclature of the Frogs, Toads and Salamanders, by Mr. A. E. Verrill, of Cambridge, approved by Dr. Fogg. Specimens verifying all the species may be found in the collections of the Society. The catalogue contains several species not heretofore credited to the State.

REPTILES.

Turtles.

Nanemys guttata, *Ag.* The yellow-spotted Turtle. Common in the southwest part of the State.

Chrysemys picta, *Gray.* The Painted Turtle. Common.

Ozotheca odorata, *Ag.* The Mud Turtle. Quite common.

Chelydra serpentina, *Schmig.* The Snapping Turtle. Common.

Glyptemys insculpta, *Ag.* The Wood Turtle. Common except in extreme eastern part of the State.

Cistudo Virginea, *Ag.* The Box Turtle. Only one specimen found.

Savria.

Scincus fasciatus, *Linn.* The blue-tailed Lizard. Rare. *C. B. Adams.*

Serpents.

Eutainia sirtalis, *B. & G.* The Striped Snake. Common.

E. saurita, *B. & G.* The Riband Snake. Common.

Storeria Dekayi, *B. & G.* The little Brown Snake. Common.

Chlorosoma vernalis, *B. & G.* The Green Snake. Common.

Crotalus durissus, *Linn.* The Banded Rattlesnake. Rare. In the southwest parts of the State only.

Bascanion constrictor, *B. & G.* The Black Snake. Rare. In the southwest parts of the State only.

Nerodia sipedon, *B. & G.* The Water Snake. Common.

Storeria occipito-maculata, *B. & G.* The Spotted-neck Snake. Common.

Diadophis punctatus, *B. & G.* The ring-necked Snake. Not common.

Ophibolus eximius, *B. & G.* The Milk Snake. Common.

AMPHIBIANS.

Frogs and Toads.

Rana Catesbianus, *Shaw*. The Bull Frog. Common.

R. clamitans, *Daud*. The yellow-throated Green Frog. Common.

Rana palustris, *Leconte*. The Pickerel Frog.

R. sylvatica, *Leconte*. The Wood Frog.

R. halecina, *Kalm*. The Leopard Frog.

Hylodes Pickeringii, *Hobb*. Pickering's Hylodes. Common.

Hyla versicolor, *Leconte*. The Tree Toad. Common.

Bufo Americanus, *Lan*. The Common Toad. Common.

Salamanders.

Plethodon erythronotus, *Baird*. The Red-Backed Salamander. Common.

P. glutinosus, *Tesch*. The blue-spotted Salamander. Common.

Salamandra opaca, *Gray*. The Banded Salamander. Quite common.

S. punctata, *Lac*. The violet colored Salamander. Common.

S. maculata, *Green*. The brown-spotted Salamander.

S. granulata, *Hol*. The granulated Salamander. Rare.

Desmognathus fuscus, *Baird*. The Painted Salamander. Rare.

Spelerpes bilineata, *Baird*. The Striped Back Salamander. Common.

Pseudotriton salmoneus, *Baird*. The salmon colored Salamander.

Notophthalmus viridescens, *Baird*. The crimson-spotted Triton. Common.

N. miniatus, *Raf*. The symmetrical Salamander. Common.

ENTOMOLOGICAL REPORT.

BRUNSWICK, December 28, 1862.

To the Gentlemen in charge of the Scientific Survey:

I transmit herewith some instructions about collecting and observing the insects of our State, which will, I hope, lead to an extended cooperation in furthering the knowledge of the habits and forms of our noxious and beneficial insects.

Very respectfully,

Your obedient servant,

A. S. PACKARD, JR.

Dr. E. HOLMES,
Prof. C. H. HITCHCOCK.

HOW TO OBSERVE AND COLLECT INSECTS.

INSECTS IN GENERAL.

That branch of the Animal Kingdom, known as *Articulata*, is so called from having the body composed of rings, like short cylinders, which are placed successively one behind the other. In the class of *Worms* these rings or segments, are arranged in a continuous row, and their number is indefinite. The organs of locomotion consist of branches of cilia and bristles placed in a row, one on each side of the body; while on the first ring there are slender feelers directed forwards and placed around the mouth-opening. In the class of *Crustacea* this continuity of rings is broken; and there is a definite number, (21) which are gathered into two regions; the head-thorax and abdomen. The number of jointed legs is also indefinite, the number varying from ten to fourteen. In the class of *Insects*, the number of rings is still more limited, (14,) the head is distinctly separated from the thorax, thus forming, with the abdomen or hind-body, three distinct regions.

In the Insects again, there are three modes of disposing the rings, and their appendages:

1. Where the number of segments is indefinite, and much like each other in form, supporting both thoracic and abdominal legs; as in the order of *Myriapoda.*

2. Where the head and thorax are closely united; and there are eight pairs of legs attached to the thorax alone, as in the *Arachnida.*

3. Where there are three distinct regions to the body; the head, thorax and abdomen, as in the *Insecta.* Moreover the true insects have three pairs of legs attached to the thorax; and are winged.

The Myriapods grow by the addition of rings, after hatching from the egg; the Arachnids by frequent moultings of the skin; while the winged insects pass through a distinct metamorphosis. The young insect after being hatched from the egg is called the *larva,* from the Latin term meaning a *mask,* since it was the ancient belief that it concealed beneath its skin the form of the perfect insect. When full-fed, the pupa-skin rapidly forms beneath the tegument, and the insect in that form escapes through a slit in the back of the larva. The perfect insect is often called the *imago.* The larval state of insects which resembles worms, has also an analogous form to the Myriapods; so spiders are analogous to Crustacea, while reminding us of the pupa state of the winged insects.

Moreover, worms and crustacea are, generally speaking, aquatic, breathing by gills, while insects are terrestrial and breathe by pores in the side of the body which communicates with a complex system of air tubes, including tubular blood vessels.

The order of winged-insects is subdivided into seven divisions, occupying an intermediate rank between orders and families, and called by naturalists *suborders.* Of these the *Hymenoptera* seem to be highest in the scale, and the *Neuroptera* the lowest.

Before characterizing these suborders, a few explanations will be necessary to understand the terms applied to the different parts. In insects as in the higher animals, the parts are repeated on either side of the middle of the body, with the exception of the single intestinal canal, and the dorsal vessel, which performs the functions of a heart.

In this head of a bee here figured we have all the parts connected with the function of sensation, and those adapted for seizing and

Fig. 4.

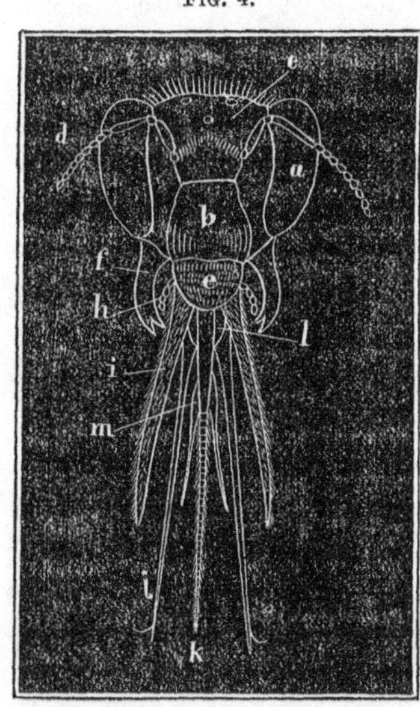

chewing the food. Two large *eyes* (*a*) composed of numerous facets, and three small simple *eyelets* (*c*) arranged in a triangle on the top of the head, and the *antennae* or feelers, (*d*) composed of numerous joints, are the most important sensory organs.—A pair of *mandibles* (*f*) for grasping, often toothed for tearing the food; two *maxillae* (*i*) for collecting and manipulating the food, on the base of which is a pair of *palpi*, (*h*) or touchers, which are used in conjunction with the antenna, as feelers; together with another pair articulated on to the *labium* (*l*) or so-called under lip, corresponding to the *labrum* or upper lip, which is attached to the *clypeus* (*b*); and the labium which is prolonged into the *lingua* (*k*) or tongue having a pair of rude palpi-like organs called the *paraglossæ* (*m*), form the organs for seizing and chewing the food.

Of the three rings of the thorax, the first (*prothorax*) is specialized to support the head; the second (*meso-thorax*) carries the first pair of wings (primaries;) the third (*meta-thorax*) carries the second pair (secondaries) To each of these three rings is articulated a pair of five-jointed legs, of which the last joint or *tarsus* is divided into five smaller joints, the last terminating in two claws. The abdomen contains the viscera, and the organs of reproduction, surrounded, externally, by several pairs of sheath-like pieces in the male, which are in the female united into the ovipositor and its sheath-pieces. All these parts exist in a rudimentary state in the larva and pupa.

Hymenoptera, (Bees, wasps, &c.) are known by their hard compact bodies, distinct head and thorax, the small narrow wings irregularly veined, and by the possession of a hard ovipositor, often forming a poisonous sting. Their transformations are the most complete of all insects. The larva being most generally a white footless, helpless grub, partly curved, and rapidly tapering at each end. The pupa has the limbs free, contained in a thin silken cocoon. The species are all terrestrial.

Lepidoptera, (Butterflies and moths,) have the mandibles obsolete, the maxillae greatly prolonged and rolled up between the labial palpi; and soft bodies covered with scales; and broad, regularly veined wings, also covered with dust-like scales. Their transformations are complete. The active larvae assume a worm-like form with several pairs (1–5) of fleshy false legs besides the thoracic ones; they spin silken cocoons before changing to pupae (chrysalids, nymphs,) with the exception of the butterflies. The limbs of the chrysalids are soldered together, and the abdomen is movable upon the head and thorax. Some of the lower families are somewhat aquatic, feeding on water plants.

Diptera, (flies) have the mouth parts formed into a kind of proboscis; the second pair of wings are undeveloped, being reduced to a pair of pedicelled knobs serving as *balancers* or *poisers*. Their transformations are complete, the larva being maggots or elongated worm-like embryos. The pupæ often change within the skin of the larvæ, which serves as a cocoon. The limbs are free. Many of the species are aquatic. Here we first find wingless parasites.

Coleoptera, or beetles, are known by their hard bodies, free and well developed mouth parts, and by the first pair of wings being hardened into sheaths (*elytra*) for the protection of the second pair. The larvæ called *grubs*, often have a terminal prop-leg besides the thoracic or true jointed legs, and pass by a complete metamorphosis to the imago state. The pupæ are often protected by a cocoon, and have their limbs free. Some of the species are aquatic. One family is parasitic but is winged.

Hemiptera (bugs,) have the mouth parts formed into a sucking tube. The first pair are often thickened at the base and laid flat upon the abdomen, are thin, somewhat net veined, and inclined over the hind body. The transformations are incomplete, as in the orthoptera. The species are largely aquatic. Some lower groups are true wingless parasites.

Orthoptera (grasshoppers,) have free mouth parts, and the organs of nutrition very highly developed. The first pair of wings are still partly hardened to protect the broad net-veined hind pair which fold up like a fan upon the abdomen. The transformations are not complete, the larvæ and pupæ resembling closely the imago, both being active. All the species are terrestrial.

Neuroptera have the mouth parts free again, the wings large and net-veined, the hind pair being often larger than the primaries.

Their bodies are more elongated than those of other insects. The metamorphosis is incomplete, the lavæ and pupæ closely resemble the imagines, and are both active, and with few exceptions they are all aquatic. The different species present strong analogies to all of the other suborders. The wingless lower genera present more analogies than other insects to the Myriapods.

Insects differ sexually in that the female generally has one abdominal ring less, and in being larger, fuller and duller colored than the males, while the males have often marked differences in the sculpture and ornamentation. In collecting, whenever the two sexes are found united they should be pinned upon the same pin, the male being placed highest. When we take one sex alone, we may feel sure that the other is somewhere in the vicinity ; perhaps while one is flying about so as to be easily captured, the other is hidden under some leaf, or resting on the trunk of some tree near by, when every bush must be vigorously beaten by the net. Many species rare in most places have a *metropolis* when they occur in great abundance. There are also *insect years* like apple years, when a species is more abundant than for three or four years succeeding. The collector should then lay up a store, against years of scarcity.

In different seasons of the year insects are found in different stages ; thus there are spring and fall insects, and summer species alone. Few insects hybernate in the perfect state, the species is more often represented in winter by the egg, or larva, or pupa. At no time of the year need the entomologist rest from his labors. In the winter, under the bark of trees and in moss he can find many species, or on trees, &c. detect their eggs, which he can mark for spring observation when they hatch out.

He need not relax his endeavors day or night. Mothing is night employment. Skunks and toads entomologize at night. Early in the morning, at sunrise, when the dew is still on the leaves, insects are sluggish and easily taken with the hand ; so at dusk when many species are found flying ; and in the night, when many species fly that hide themselves by day, and many caterpillars leave their retreats to come out and feed, and the lantern can be used with success to draw them out, the collector will be rewarded with many rarities.

There are species frequenting gardens, lawns, fields and deep woods, and swamps and pools, that are not met with away from those localities. So there are insects frequenting mountains that are not found in the vallies below. More lepidoptera inhabit the summits of high mountains than beetles and other insects. In Maine there are found species which inhabit the sea coast alone, others that abound most on the sandy plains that run back from the sea to meet the hills of the interior, and some of the most productive places are those towns situated on the border of the low lands and hilly regions of the State. Other species are found only in the thick forests of the wild lands.

Moreover it has been found that two assemblages of insects called *faunae*, people the surface of the State. The one called the *Canadian fauna* comprises a large mass of species that inhabit British North America, the Great Lakes, and the lakes and forests of northern Maine, including Eastport and the coast towards Mt. Desert. The other assemblage called the *Alleghanian fauna*, is that which covers the southern half of the State, besides New-England generally, (except the White Mountain region which belongs to the Canadian fauna,) and sweeps down the Alleghany range towards the southern States. The plants of the summit of Mount Katahdin belong to a more arctic region still than the Canadian flora; whether the insects partake of the subarctic character remains to be determined.

Hymenoptera.

In studying this suborder we must remember that *every part of the body* varies in form in the different genera, forming admirable and plain distinctions to characterize the genera. To the form of the head and its appendages, that of the thorax and its appendages, the wings in the venation of which we can perceive at a glance those characters which separate genera, and in the legs especially of the fossorial families, where there are found to be great differences, the student must look closely. The best specific characters lie in the sculpturing and color, but the spots and markings are apt to vary greatly. The great differences in the sexes are liable to mislead the student, and hence large collections are indispensible to their proper study. The Hymenoptera are the most numerous in species of all the suborders except the Coleoptera. They have been less studied in this country than almost any other suborder, though so deserving from their interesting habits. Especial attention should be paid to collecting the smaller species, and to the families of the *Chalcididae*, the *Cynipidae* and the *Crabronidae*.

They should be pinned through the hard thorax, high up on the pin, and numbers should be preserved in alcohol.

Their habits should be studied long and patiently, and attention be given to rear in the same way as given for Lepidoptera, the saw-flies, the gall-flies, &c. The Eurytomae can be found in wheat fields, &c., after harvest; the galls in autumn.

Apidae, (honey bees, bumble bees, &c.) They are known from other families by their bodies being densely hirsute, the mouth parts lengthened and partially united to form a kind of proboscis that can be folded up out of sight under the head; and in their broad, flattened hirsute hind legs, adapted for collecting and carrying pollen. They are social, and the species often consists of *males,* or drones; *females,* or *queens;* and *imperfect females,* or workers, improperly called neuters, which are much smaller than the others. *Apis mellifica* is the honey bee, whose complex œconomy and hives are well known. Siebold, a German physiologist, has ascertained that the queen and neuters are hatched from fertilized eggs, while the drones come from eggs that are unfertilized. There is one queen to a colony or swarm. The workers sometimes lay eggs producing males, and there is a difference between them in other respects. The humble bees (*Bombus*) contain many species, which build hemispherical nests of moss under ground in pastures. The cells are large, oval and partially separate. There are from fifty to seventy in a swarm. The nests are built by the females, of which there are several in the spring which survive the winter; they then lay their eggs, which hatch out the workers late in the summer; soon after another brood of males and females alone, and in the fall, still later, a few more of both sexes appear. There are two kinds of females; the earlier born differing in size and producing male eggs only; so also there are two kinds of workers. The remaining species are solitary, and consist of males and females only.

Xylocopa, the Carpenter-bee, has black wings; it forms a tube a foot long, in which it lays its eggs, arranged in successive layers in masses of pollen.

Megachile, the Leaf-cutter, cuts circular pieces out of leaves, with which it makes a honey-tight cartridge-like cell, which it builds in holes excavated in trees and rotten wood.

Osmia, the Mason-bee, is blueish, and has a circular abdomen. It constructs its nest of sand, large enough to hold three to eight cells, in crevices in fences. Other species burrow in the sunny side of cliffs or sand banks, or in rotten trees; while others occupy dead snail shells.

Coelioxys, with a sharp triangular abdomen, is parasitic, laying its eggs in the nests of Megachile, Osmia and other bees.

Nomada is not hirsute, and in its slender form and gay colors resembles wasps, It enters the nests of *Andrena* and feeds on its food, hence it is called the Cuckoo-bee.

Andrena resembles very much the hive bee, though it is smaller. The Andrenae "are all burrowers in the ground, some species preferring banks of light earth, others hard trodden pathways, &c.; their burrows differ in depth, but are seldom less than about six, whilst others excavate to nine or ten inches; at the bottom of each burrow is formed a small oval cell or chamber, in which the industrious female lays up a small pellet of pollen mixed with honey; these little balls are usually about the size of a garden-pea, varying somewhat in size in different species." *Smith.*

Halictus is one of the smallest of the family. *Angochlora* embraces bees whose bodies are slightly hirsute, and of a rich shiny green.

Mr. Fred. Smith, an English entomologist, says of the economy of this genus, that "it is so remarkably different from that of all other solitary bees, except of those belonging to the genus *Sphecodes*, that I am surprised it had escaped the researches of my predecessors, who, like myself, 'have loved to hear the wild bees' hum.' It will be observed that the females of *Halictus* and *Sphecodes* make their appearance in June, and are to be found from that time until late in autumn; but no males of these genera will be observed until long after the appearance of the females: my observations on a colony of *H. morio* will serve as the history of the whole genus, making allowance for the different periods of their appearance. 'Early in April the females appeared, and continued in numbers up to the end of June; not a single male was to be found at any time; during the month of July scarcely an individual was to be found; a solitary female now and then might be seen, but the spring bees had almost disappeared; about the middle of August the males began to come forth, and by the end of the month abounded; the females succeeded the males in their appearance about ten or twelve days: these industrious creatures began the tasks assigned to them, burrowing and forming their nests; one of their little tunnels had usually others running into it, so that a single common entrance served as a passage to several cells, in each of which a little ball of pollen was formed, and a single egg deposited thereon; the larvae were usually ten or twelve days consuming it, by which time they were fully fed; in this state they lay until they changed to the pupa-state, when they very shortly became matured.' I have reared individuals of *H. rubicundus* from the egg to the perfect insect; on the 15th of July I procured cells containing the pollen balls, with an egg on each; in twelve days the larvae were full-fed; the change to the pupa-state took place about the 25th of August, and during the first week of September the perfect state was acquired. The history of *Halictus*, therefore, is as follows: the males and females appear in the autumn; the latter being impregnated, pass the winter in the perfect state, appearing during the following season to perform their economy, as detailed above in the case of *H. morio*."

All these females of solitary species are found in spring on the blossoms of fruit trees, of wild cherries and about flowers.

Vespidae, (wasps, yellow-jackets.) The hornet is the *Vespa crabro* of Europe. The group is characterized by the folding of the wings, longitudinally. Vespa lives in colonies of three kinds of individuals, constructing complex nests either under ground or attached to the branches of trees, consisting of several galleries of hexagonal cells, with their mouths downward, connected and supported by pedicels, and surrounded by an outer papery envelop. The females which have survived the winter begin in the spring to form their small nests, consisting of a single tier of a few cells, in which they lay their eggs and feed the young workers. The males and females do not appear until autumn. Reaumur has observed that there are two sizes among the males.

"Notwithstanding the powerful sting of the wasp, it is liable to the attacks of other insects. *Rhipiphorus paradoxus* and the larva of a *Volucella* infests its nests, devouring the larva; as does also *Anomalon vesparum*, and another species of Ichneumon. Dr. Leach also mentions that wasps are much infested by *Lebia linearis*. I have also observed a spider sucking a wasp, which it had killed." —*Westwood.*

Wasps should, if possible, be collected by the whole colony, when the individual variation of the three kinds of individuals—the males, females and workers—can be studied. For this purpose visit the nests by night, plug up the hole with a sponge saturated with ether or chloroform, and the inhabitants can thus be stifled. Or stand by the nest and net the insects as they go to or from the hole. Little or nothing is known about these interesting insects in this country, and persons who will spend the time can find out much that is new to their history. The following genera have no workers:

The common brown *Polistes* builds an exposed nest, consisting of few or many cells arranged in one tier, and attached to leaves and twigs by a short pedicel.

The solitary wasps, *Odynerus* and *Eumenes*, build nests of sand glued together and hidden in cavities, hollow branches, &c., and they store them with great numbers of caterpillars, flies, larvae of beetles, and spiders. Thus it seems that the larvae of the social wasps are daily fed with sweets by the workers, while those of the solitary species, which have no workers, have a store of insect-food laid up for them by the female.

The following families are truly fossorial sand-diggers, making their holes in sunny paths, &c., of which the ants are the most familiar examples. Their ovipositor is adapted for stinging, and by the poison conveyed into the wound, for benumbing their victims, which live for a long time half alive, for the larvae to feed upon.

"Although there is much general similarity in the habits of the truly fossorial species, there is considerable diversity in the details of their proceedings: thus, whilst *Oxybelus* conveys its prey by means of its hind legs, *Pompilus* and *Ammophila* walk backwards, dragging it with their mandibles. '*Astata, Tachytes, Psen, Crabro, Mellinus* and *Cerceris* fly bodily and directly forward with it in their mandibles, assisted by their fore legs.' *Shuckard.* From my own observations each species appears ordinarily to confine itself to its own particular prey. Instances are on record, however, in which considerable diversity in the prey of the same species has been observed; this probably arose from the female not being able to discover her legitimate prey; thus Serville and Saint Fargeau state that *Bembex rostrata* indifferently collects the species of *Eristalis, Stratiomys,* and the larger Muscidae; but it may be regarded as the ordinary rule that each species confines itself to its peculiar prey: thus, numbers of the *same* species of fly or larva are found in the *same* cell, although this must sometimes be a matter of difficulty." * * * "The prey is moreover very various, comprising insects of nearly every order; the Coleoptera, Hemiptera, Lepidoptera, Hymenoptera, Diptera and spiders, contributing to the support of this tribe; and insects in the larva, pupa and imago state are employed for this purpose. The number of individuals enclosed in each cell varies according to the size of the species, and of the progeny for whose support it is buried: thus, whilst Ammophila sabulosa buries a single lepidopterous larva, as many as fifty or sixty Aphides are shut up in a single cell by other species." *Westwood.*

Crabronidae, sand-wasps. It is this family that many of the Syrphus-flies resemble so closely. They have cuboidal heads, a somewhat flattened, spherical thorax, and a flattened abdomen, rarely pedicelled. The fore legs are broad, adapted for digging, and they often have a broad, banner-like expansion, to use perhaps as a shovel, while the hind and middle legs are spined for retaining the prey the sand-wasp carries off. The insects are of moderate size; they are found resting on leaves in the sunshine. They occur generally rarely, and little is known of the extent or habits of the family in this country. Crabro (Fig. 5) has slender legs, and digs into rotten posts, fences, stumps, where it makes its nest, provisioning it with caterpillars, flies, &c. *Gorytes* has been seen protruding her sting into the frothy secretion of Tettigoniae on grass, and carrying off the insect. *Oxybelus* is a small, stout black genus, "its prey consists of Diptera, which it has a peculiar mode of carrying by the hind legs the while it either opens the aperture of its burrow or else forms a new one with its anterior pair. Its flight is low, and in skips; it is very active." *Trypoxylon* has a long, club-shaped abdomen, and is black throughout. "Mr. Johnson has detected it frequenting the holes of a post pre-occupied by a species of *Odynerus*, and into which it conveyed a small round ball, or pellet, containing about fifty individuals of a species of Aphis; this the Odynerus, upon her return, invariably turned out, flying out with it, held by her legs, to the distance of about a foot from the aperture of her cell, where she hovered a moment, and then let it fall; and this was constantly the case till the *Trypoxylon* had sufficient time to mortar up the orifice of the hole, and the *Odynerus* was then entirely excluded; for although she would return to the spot repeatedly, she never endeavored to force the entrance, but flew off to seek another hole elsewhere."

FIG. 5.

Shuckard. *Cerceris* has a long abdomen, with convex rings. It is gaily marked with golden yellow. It has not been known to use its sting upon its captors. It lays up stores of young grasshoppers and *Curculionidae.* "*Philanthus* burrows in hot sandy situations, and provisions its nests with hive-bees; a single individual of which, after being stung, is deposited with an egg; and as each deposits five or six eggs, the number of bees destroyed must be at least equal to that, if not more considerable, which is most probable; and Latreille counted as many as fifty or sixty females occupied in making their burrows in a space of ground one hundred and twenty feet long." *Westwood.*

This is a most difficult family to study. The two sexes differ greatly, and are apt to be mistaken for distinct species, and the collector is fortunate if he comes upon a "metropolis" of a species. In limiting the species, more value must be placed upon the size and sculpture than the coloration, which varies greatly.

Larridae. This is rather a small group, having a sessile conical flattened abdomen, and with the legs of the females very hirsute. They are generally dark in color. They are caught about sandbanks. *Larra* provisions its nests with the caterpillars of small moths.

Bembecidae. We have but two genera, *Bembex* and *Monedula,* which have large heads and flattened bodies, bearing a strong resemblance to syrphus flies from their similar coloration. The labrum is very large and long, triangular, like a beak. The species are very active, flying rapidly about flowers with a loud hum. "The female Bembex burrows in sand to a considerable depth, burying various species of Diptera (Syrphidae, Muscidae, &c.,) and depositing her eggs at the same time in company with them, upon which the larvae, when hatched, subsist. When a sufficient store has been collected, the parent closes the mouth of the cell with earth." "An anonymous correspondent in the Ent. Mag. states that *B. rostrata* constructs its nests in the soft light sea-sands in the Ionian Islands, and appears to catch its prey (consisting of such flies as frequent the sand; amongst others, a bottle-green fly,) whilst on the wing. He describes the mode in which the female, with astonishing swiftness, scratches its hole with its fore legs like a dog. *Bembex starsata,* according to Latreille, provisions its nests with *Bombylii.*" *Westwood.*

Sphegidæ. The mud-wasps are known by their long antennæ, long hind legs and pedicelled abdomen. They are of large size, and are colored black and red, brown and red, or wholly blue or black. They are very active, restless in their movements, and have a powerful sting. *Ammophila* is long, slender, with a long, pedicelled abdomen, the tip of which is red, and flies and runs on sunny paths and about pumps. "The species inhabit sandy districts, in which *A. sabulosa* forms its burrow, using its jaws in burrowing; and when they are loaded, it ascends backwards to the mouth, turns quickly round, flies to about a foot's distance, gives a sudden turn, throwing the sand in a complete shower to about six inche's distance, and again alights at the mouth of its burrow.

"Latreille states that this species provisions its cells with caterpillars, but Mr. Shuckard states that he has observed the female dragging a very large inflated spider up the nearly perpendicular side of a sand-bank, at least twenty feet high, and that whilst burrowing it makes a loud whirring buzz; and in the Trans. Ent. Soc. he states that he had detected both A. sabulosa and hirsuta dragging along large spiders. Mr. Curtis observed it bury the caterpillars of a Noctua and Geometra. St. Fargeau, however, states that A. sabulosa collects caterpillars of large size, especially those of Noctuae, with a surprising perseverance, whereas A. arenaria, forming a distinct section in the genus, collects spiders." *Westwood.* Pelopaeus, which is the true mud-wasp, builds in length a row of parallel adjoining cells an inch or more long, and enveloped in an outer case of mud or clay, in the corners of rooms, on rafters, &c., enclosing in each cell some insect. *Pelopaeus coeruleus* is our common shiny blue "sand-dauber." *Pompilus* has a short pedicel to its abdomen, and very long hind legs. They run very swiftly in grass and over sandy places, looking like winged black spiders, on which they prey.

Scoliidae. This group has long, rather narrow hirsute bodies, with short, spiny fossorial legs, sessile abdomen, with two prominent terminal spines in the males, and often lunate eyes. They are black, often with bright yellow spots along the sides of the hind body. The genus *Scolia* is very large, often two inches long, marked with yellow. It is found in the hottest places about strongly scented flowers. It makes deep burrows in sand-banks, provisioning its cells with grasshoppers, &c. Other species are sluggish, crowding on stems of grass. *Sapyga,* known by its unusually narrow body and long, club-shaped antennæ, is said to be parasitic on bees of the genus *Osmia,* in whose nests it lays its eggs.

Mutillidae. This interesting family is characterized by the females being wingless, by the want of the three ocelli on the top of the head that other hymenoptera possess, while the form of the body resembles the Scoliidæ, though more hirsute. They are deep red and black, and are solitary in their habits. They belong more to the Southern and Middle States,—one species only being found in Massachusetts. The females run in hot places, and hide themselves quickly when disturbed, while the males frequent flowers. They take flies by surprising them. The sting of *Mutilla coccinea* in this country is said to be very powerful. This family, in its wingless females and structural features generally, leads to the ants, where we have three kinds of individuals, as has been noticed in the bees, but differing in the workers being wingless.

Formicidæ. Ants have a triangular head, round eyes, long elbowed antennæ and slender legs. Some species have a sting like the other fossorial families. The males are much smaller than the females, and the wingless workers are a little smaller than the males. The mandibles in those species that do not themselves labor, but enslave the workers of other species, are slender and smooth, though they are generally stout and toothed. As in the bees, there have been found in some species two sets of workers, (a few being of larger size than usual, with very large heads,) which are said to make honey in cells, like worker-bees.

The habits of our ants in America have not been recorded. The little yellow ant that digs its holes in paths; the pismires that excavate their galleries in stumps; the ferocious red and brown species that raises its hills of sand in woods, or of clay in clayey places, and the large Pennsylvania ant nearly an inch long, whose colonies we find under boards, &c., are but little known. In collecting them they should be caught when swarming, that is when the winged sexes come out of their holes and fill the air in countless hosts. The little yellow ants swarm thus in the second week of September on a hot day that we generally have at that time. Hundreds of them should be pinned, or better, thrown into alcohol, keeping the colonies separate. So also their eggs, with the larvæ and pupæ, should be taken in large numbers.

Unlike the bees, ants are represented in winter by the workers alone, the winged sexes only appearing in the summer. After swarming, the females lay their minute eggs, and Gould, an English observer, says that those destined to hatch the future females, males, and workers, are deposited at three different periods. The larvæ are like those of hymenoptera in general, being footless grubs, short, thick and white. How the larvæ are fed and the pupæ are cared for by the neuters, and the habits of ants generally, are found in all the books. Sometimes the pupæ are naked, but generally they are enclosed in thin cocoons.

Chrysidæ. These insects are very different from the ants in their oblong compact form, their nearly sessile oblong abdomen, having only three to five rings visible, the remaining ones being drawn within, forming a long, large jointed sting-like ovipositor which can be thrust out like a telescope. The abdomen beneath is concave, and the insect can roll itself into a ball on being disturbed. They are green or black. The sting has no poison-bag, and in this respect, besides more fundamental characters, the Chrysis approaches the Ichneumon family. They best merit the name of "Cuckoo-flies," as they fly and run briskly in hot sun-shine, on posts and trees, &c., darting their ovipositor into holes in search of other hymenoptera, &c. in which to lay their eggs. Their larvæ are the first to hatch and devour the food stored up by other fossorial bees and wasps. "St. Fargeau, however, who has more carefully examined the economy of these insects, states that the eggs of the Chrysis does not hatch until the legitimate inhabitant has attained the greater part of its growth as a larva, when the larva of the Chrysis fastens on its back, sucks it, and in a very short time attains its full size, destroying its victim. It does not form a cocoon, but remains a long time in the pupa state."

"In the Ent. Mag. has been noticed the discovery of *Hedychrum bidentulum*, which appears to be parasitic upon *Psen caliginosus*; the latter insect had formed its cells in the straws of a thatched arbor, as many as ten or twelve cells being placed in some of the straws. Some of the straws, perhaps about one in ten, contained one or rarely two, of the Hedychrum, placed indiscriminately amongst the others. Walkenaer, in his Memoirs upon *Halictus*, informs us that Hedychrum lucidulum waits at the mouth of the burrows of these bees, in order to deposit its eggs therein; and that when its design is perceived by the bees, they congregate together and drive it away. "St. Fargeau states that the females of Hedychrum sometimes deposit their eggs in galls, while H. regium oviposits in the nest of Megachile muraria; and he mentions an instance in which the bee, returning to its nearly finished cell, laden with pollen paste, found the Hedychrum in its nest, which it attacked with its jaws; the parasite immediately, however, rolled itself into a ball, so that the Megachile was unable to hurt it; it however bit off its four wings which were exposed, rolled it to the ground and then deposited its load in the cell and flew away, whereupon the Hedychrum, now being wingless, had the persevering instinct to crawl up the wall to the nest, and there quietly deposit its egg which it placed between the pollen paste and the wall of the cell which prevented the Megachile from seeing it."—*Westwood.*

Proctotrupidæ, Egg-parasites. In this family are placed very minute species of parasitic Ichneumon-like hymenopters which have rather long and slender bodies, with antennæ of various lengths, often haired on the joints, while the wings are covered with minute hairs and most of the nervures are absent. Here the ovipositor has its true function, and its puncture conveys no pain; this may be said of the remaining families of the hymenoptera. These minute insects which can scarcely be distinguished by the naked eye unless specially trained, are black or brown, and very active in their habits. They may be swept off grass and herbage, from aquatic plants, or from hot sand banks. They prey on the wheat-flies by inserting their eggs in their larvæ, in gall-midges, and gall-cynips, and in fungus-eating flies, in which places they should be sought. In Europe species of *Teleas* lay their eggs in those of other insects, especially butterflies and moths and hemipters where they feed on the juices of the growing larvæ and pupæ within the egg, coming out as perfect Ichneumons.

"*Mymar ovulorum* oviposits in the eggs of other insects from which the tiny parasite emerges only in the perfect state, a single butterfly's egg often nourishing the transformation of many individuals." A species of *Platygaster*, a short broad genus, lays its eggs in those of the Canker-worm moth just after their deposition. It is one twenty-fifth of an inch long. Another species infests the eggs of the Hessian fly. *Ceraphron destructor*, which is a larva-parasite of the Hessian fly, is a tenth of an inch long.

We must have many species of these insects in this country. They occur in great numbers where they are found at all. They are almost too small to pin, and if transfixed would be unfit for study, and should therefore be put into homeopathic vials of alcohol.

Chalcididæ. This is also a group of great extent, and like the preceeding, the species are of small size; but they are of shiny colors, as the name implies, being often bronzen, or metallic. They have also elbowed antennæ, and the wings are often deficient in nervures. In some genera, including *Chalcis* the hind thighs are thickened for leaping. The differences between the sexes, generally very marked in hymenoptera, are here especially so. The male of Eurytoma has the joints of the antennæ swelled and furnished with long hairs above. Some of the species, such as those of *Pteromalus*, are wingless, and closely resemble ants.

They infest eggs and larvæ. Some species prey upon the Aphides, others lay their eggs in the nests of wasps and bees. One species is known in Europe to consume the intestines of the common House Fly. Others consume the larvæ of the Hessian fly, and those *Cecidomyiæ* that produce galls, and also the true gall flies (*Cynips.*) Some are parasites on other Ichneumon parasites, as there are species preying on the genus *Aphidius*, which is a parasite on the Aphis. So also in Illinois a species of *Hockeria* and of *Glyphe* are parasitic on a *Microgaster*, which preys upon the Army worm; and *Chalcis albifrons*, Walsh, was bred from the cocoons of *Pezomachus*, an Ichneumon parasite of the same caterpillar.

The genus *Leucospis* is of large size and known by having the ovipositor laid upon the upper surface of the abdomen, and by its resemblance to wasps. *Eurytoma hordei* (fig. 6,) is found in gall-like swellings of wheat stalks. The pupæ of this family have often the limbs and wings soldered together as in *lepidoptera*, and the larvæ seldom spin a silken compact cocoon as in the succeeding family. We have probab y in this country a thousand species of these small parasites, nearly twelve hundred having been named and described in England alone. They are generally large enough to be pinned or stuck upon cards; some individuals should be preserved in this way, others, as wet specimens.

FIG. 6.

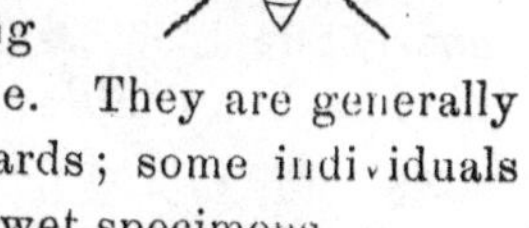

Ichneumonidæ. The Ichneumon-fly (Fig. 7,) represents the most extensive family of the suborder as regards numbers. They are long and narrow bodied, with long and straight antennæ; the ovipositor is generally long and protected by two sheath-pieces of the same length. In those genera that have the ovipositor short, the eggs are placed in exposed larvæ, while those provided with longer ones, such as in the figure, are adapted for penetrating into

FIG. 7.

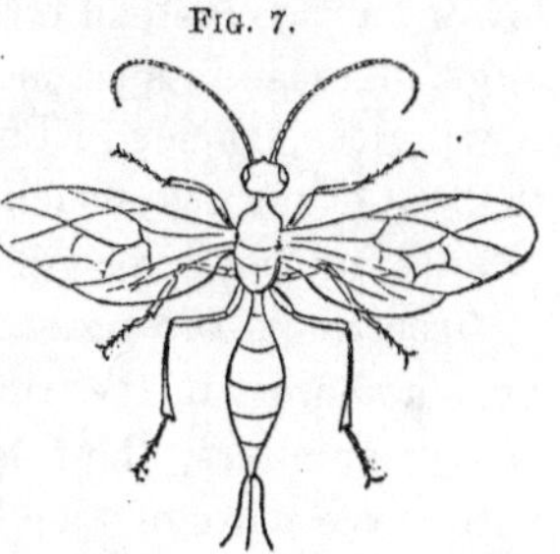

holes under bark, or in crevices, &c., and for this purpose they are often two or three times the length of the body. There are scarcely any insects which do not suffer from the attacks of these parasites. They are the best friends of agriculturists. The eggs are either laid on the surface of the larvæ, when they eat their way inwards; or the egg being placed within the body of the victim, it hatches out and feeds on the fatty issues of the larvæ, gradually consuming its life until the parasite turns to pupa, when it dies. There may be one large Ichneumon thus feeding within, or numbers of them. Thus the caterpillar of *Acronycta*, found on the alder in October, is often seen adhering to the leaf, preserving the semblance of life, while the inside of the body is packed with little cocoons placed vertically next to one another.

Of course Ichneumons abound most in summer when larvæ are most plenty, when they are found in great numbers on umbelliferous flowers. But many species appear in April. The species of *Ophion*, with compressed arched yellow bodies, come to light in summer. In Europe nearly 2000 species of this family have been described. *Evania*, with its very short abdomen, *Pelecinus*, with a very long one, which is abundant in summer, represent a small family, the *Evaniidœ*, which lead to the

Cynipidœ, or Gall-flies. The species are of small size, with short broad heads, a globular thorax, and short compressed abdomen. With their long slender ovipositor they insert their eggs into leaves, &c. which causes by the irritation a hollow swelling on the leaves, buds or stalks of plants. Those large swellings on oaks imported from the East, known as *galls*, have given the name to these productions. Galls are of various forms and sizes, and differ with the species of gall-fly that produces them. They may contain one or several grubs, which are small, fleshy and footless, with tubercles in the lower surface instead of feet, to move by. The eggs increase in size as the gall itself enlarges. A wingless species in England makes its galls at the foot of the oak, beach, &c. *Cynip dichlocerus*, (Fig. 8,) forms long galls in the stem of rose bushes.

FIG. 8.

Uroceridœ, Boring-sawflies. These rather rare insects pass their lives as borers in the trunks of trees. Unlike the previous families of hymenopters, their larvæ are long, cylindrical, and furnished with three pairs of true legs. The saw-flies are likewise cylindrical and long, and the sides of the body continuous, not being insected as usual, while the abdomen is blunt, and a large saw-like ovipositor projects from beneath. They are among the largest of hymenoptera. *Tremex columba* bores into beech trees. *Urocerus albicornis* and another species is found on pines. *Oryssus* is a much smaller genus with a slender ovipositor. There are but few species, and they are found in August flying about with a loud buzz.

Tenthredinidæ, Saw-flies. We now come to a family whose affinities are closest to the Lepidoptera. In their bodies the three divisions are less marked than usual, they only fly in the warmest days, their larvæ have 18–22 legs, and closely resemble caterpillars, and like them feed exposed on leaves. The flies are sluggish; their heads are transversely oblong, and the antennæ are simple, club-shaped or feathered. Their wings are folded at rest upon the body, overlapping each other somewhat. Their ovipositors are toothed like little saws, with which they bore into the stems and leaves of plants to deposit their eggs. The larvæ spin compact cylindrical oval cocoons. They are found in companies on the leaves of the alder and birch, holding on by their true legs while the rest of the body is suspended and curved curiously upwards; or they occur as slimy slugs on the leaves of the pear and rose, while others feed on the stems of plants, or construct cases of bits of leaves to hide in, like Tineids, or roll up a leaf like the Phyganidæ. The large solitary larva of *Cimbex Americana* is found partially rolled up, on the elm and birch. *Lophyrus abietis* (Fig. 9, female,) feeds on the fir. *Selandria vitis* and *rosæ* feed upon the vine and rose, and can be taken when those plants have leaved out. Many can be taken early in summer about alders and willows.

Fig. 9.

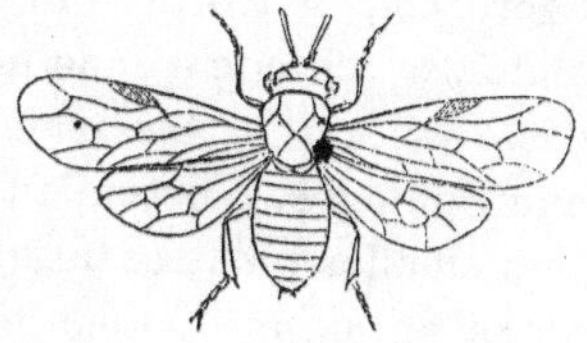

Lepidoptera.

Butterflies are easily distinguished from the other groups by their knobbed antennæ; in the *Sphinges* and their allies the feelers are thickened in the middle; in the *Moths* they are filiform and often pectinated like feathers. Lepidoptera have also been divided into three large groups, called Diurnal, Crepuscular and Nocturnal, since butterflies fly in the sunshine alone, most Sphinges in the twilight, (many of them fly in the hottest sunshine,) and the moths are generally night-fliers, though many of them fly in the day time, thus showing that the distinctions are somewhat artificial.

In studying these insects the best generic characters will be found in the antennæ, the shape of the head parts, the neuration and proportions of the wings. Very slight changes in these parts separate genera. Size and coloration, which are very constant, afford good specific characters.

The caterpillars, chrysalids and perfect insects, besides being preserved dry, should be collected largely in alcohol. In collecting them to pin dry we must remember that the least touch will remove some of the scales from the wings and bodies, thus injuring them for study and spoiling their looks. The collector should have the *ring net,* the *beating net,* plenty of pill boxes, a large box lined with cork to pin his captures into, which should have pinned in the bottom a sponge saturated with benzine, (which is the cheapest poison,) and though after frequent airing the box loses the strong odor, yet there is enough left to keep the specimens from fluttering, until more can be applied at home. This box should be small enough to slip into the coat pocket, and with the cover made to open easily with one hand. A bottle of alcohol is needed about the person for the reception of duplicates, larvæ, &c. Pins of various sizes should be carried in a cushion suspended from the neck or from a button-hole. The best insect pin is that of German make. The different sizes can be had of F. W. Christern, 763 Broadway, and Theodor Schreckel, 14 North William street, New York. Two sizes, No. 2 and 5, which come done up in square packages of five hundred pins each, will do for the majority of insects, the larger for butterflies and Sphinges, Noctuæ and Geometræ, while for the micro-lepidoptera smaller pins are needed, which will be mentioned further on when speaking of them more specially. The *net* most convenient is a sugar-loaf-shaped bag of silken gauze (which can be bought as cheaply as muslin or musquito-netting, and does much better,) fastened to a margin of cloth sewed previously onto the ring. The net should be made a foot and a half deep, attached to a frame of stout brass wire twelve inches across, which should be soldered on to a tube half an inch or so in diameter, into which a slender stick six feet long can be thrust. A light net like this can be rapidly turned upon the insect with one hand. The *beating net* is stouter and made of thick muslin, and fastened on to a short stick. It is used for beating bushes and herbage for moths and their larvæ. It can be also used for collecting all other insects. In this connection should be mentioned the *water net,* (Fig. 10,) which may be round, or of the figure indicated. The ring should be of brass, and the shallow net be made of grass-cloth or coarse millinet. Small aquatic species can be fished up in mud which will strain through the net, leaving them to be picked up and pinned. When the insect is taken in the bag-net by a dexterous twist of the handle, which throws the bottom over the mouth, it should be confined with the other hand with great care, and then pinned through the thorax, when in the net. The pin can be drawn through the meshes upon opening the net. The pin should be thrust through the thorax so that three-fourths of it should be below the insect; care should be taken to preserve some uniformity of height from the cork in the different specimens. After being pinned the specimen should be handled with a pair of curved pincers, whose jaws should be roughened to retain the pin, and kept apart by their opposite ends being united, as in the surgeon's dissecting forceps; or the handles may be large, and a special spring introduced between to keep the branches apart. These pincers are indispensable in handling specimens, especially those on slender pins.

FIG. 10.

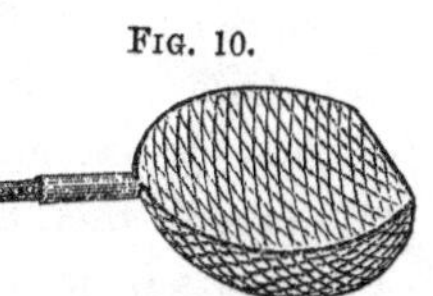

Some specimens should be preserved as they look when at rest. To *set* specimens a number of *setting-boards* will be necessary. These should be made of soft wood, with grooves or cracks of different sizes, in the bottom of which strips of cork, or corn-stalk, or pasteboard should be fastened, into which the insect's body can be received, while the pins stick through beneath. The surface of the board should incline a little towards the groove, as the wings often fall down a little after the specimen is dry. The wings can be arranged with a needle stuck into a handle of wood, the wings set horizontally, and the front margin of the primaries drawn a little forward of a line perpendicular to the body, so as to free the inner margin of the secondaries from the abdomen, that their form may be clearly seen, as in the figure (11.) When thus arranged they can be confined by pieces of card, as indicated, or by square pieces of glass laid upon them. Several days are requisite for them to dry thoroughly. Several of these setting-boards can be made to slide into a frame covered with gauze-wire, to keep them from devouring insects, while the air may at the same time have constant access to them.

FIG. 11.

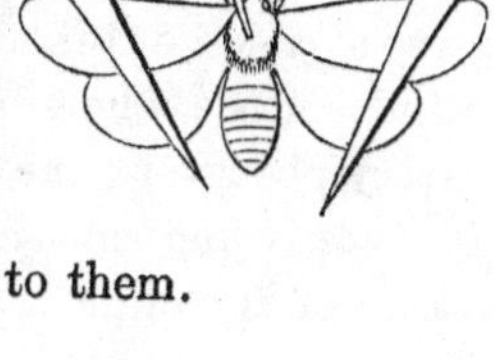

Rearing Caterpillars. The larvæ of butterflies are rare; those of moths occur more frequently, while their imagines may be scarce. In some years many larvæ, usually rare, at other times occur in abundance, when they should be reared in numbers. In hunting for caterpillars bushes should be shaken and beaten over newspapers or sheets, herbage should be swept carefully, and trees examined carefully for leaf-rollers and miners. The best specimens of moths and butterflies are obtained by rearing them from the egg if possible, or from the larvæ or pupæ. In confinement the food should be kept fresh, and the box well ventilated. Tumblers covered with gauze, pasteboard boxes, pierced with holes and fitted with glass in the covers, or large glass jars, are very convenient to use as cages. The bottom of such vessels may be covered with moist sand, in which the food plant of the larva may be stuck and kept fresh for several days. Larger and more airy boxes, a foot square, with the sides of gauze, and fitted with a door, through which a bottle of water may be introduced, serve well. The object is to keep the food plant fresh, the air cool, the larva out of the sun, and in fact everything in such a state of equilibrium that the larva would not feel the change of circumstances when kept in confinement. Most caterpillars change to pupæ in the fall; then they should be covered with earth, kept damp by wet moss, and placed in the cellar until the following summer. The collector in seeking for larvæ should carry a good number of pill-boxes, and especially a close tin box, in which the leaves may be kept fresh for a long time. The different forms and markings of caterpillars should be noted especially, and they should be drawn carefully, on a leaf of the food plant, and the drawings and pupa skins, and perfect insect, be numbered in the same way. Descriptions of caterpillars cannot be too carefully made or too long. The relative size of the head, its ornamentation, the stripes and spots of the body, and the position and number of tubercles, and the hairs, or fascicles of hairs, or spines and spinules, which arise from them, should be noted, besides the general form of the body. The lines along the body are called *dorsal,* if in the middle of the back, *subdorsal* if upon one side, *lateral* and *ventral* when on the sides and under surface, or *stigmatal* if including the *stigmata* or breathing pores, which are generally parti-colored. Indeed, the whole biography of an insect should be ascertained by every observer ; the points to be noted are :—

1. Date, when and how the *egg* is laid; and number, size and marking of the eggs.

2. Date of hatching, the appearance, food-plant of *larva*, and number of days between each moulting; the changes the larva undergoes, which are often remarkable, especially before the last moulting, with drawings illustrative of these; the habits of the larva, whether solitary or gregarious, whether a day or night feeder: the Ichneumon parasites, and their mode of attack. Specimens of larvæ in the different moultings should be preserved in alcohol. The appearance of the larvæ when full-fed, the date, number of days before pupating, the formation and description of the cocoon, the duration of larvæ in cocoon before pupation, their appearance just before changing, their appearance while changing, and alcoholic specimens of larvæ in the act, and drawings illustrative—all these should be studied and noted.

3. Date of pupation; description of the pupa or chrysalis; duration of the pupa state, habits, &c.; together with alcoholic specimens, or pinned dry ones. Pupæ should be looked for late in the summer or in the fall and spring, about the roots of trees, and kept moist in mould until the imago appears.

4. Date of the insect's escaping from the pupa, and method of escape; duration of life of the imago; and the number of broods in a season. Labels for alcohol may be written in pencil on paper, or in ink on parchment.

Papilionidæ. The Swallow tails are at once known as being our largest butterflies, and by their having the hind wings produced into a tail-like appendage. The yellow *Papilio Turnus* flies in June and July, through woods and about lilacs. Its larva feeds on the apple, and wild thorn. It is green, with two eye-like spots on the thorax. *P. asterias,* the Parsnip Papilio, flies in August about wild parsnip, which grows by river sides; and is found upon the cultivated species. It is dark blue. The larva is yellow, striped and spotted with black. When sailing free on their wings it is almost impossible to capture them. The larvæ when irritated, push out a V-shaped yellow organ from the head.

Pieridæ. (White or Sulphur Butterflies.) *Pieris oleracea,* is white with rounded secondaries or hind wings. It feeds on cabbages and turnips. Its larvæ are hirsute, green, tapering towards each end of the body, and feed on grass. Those of *Colias Philodice* are green and smooth. This is our common "Sulphur Yellow," abounding in roads.

Nymphalidæ. *Argynnis* is known by the under side of the wings being covered with silvery spots, while the larvæ are spined, as are those of *Vanessa* and *Grapta*, whose species are the earliest to appear in spring. *V. antiope* is the large purple species that flies from March to October; its gregarious larvæ feed on the willow and elm. *Grapta progne* with notched red and brown wings, is common in May and September, in woods and about houses; its solitary larvæ feed on the currant. *G. comma* inhabits the northern part of the State. All the species have silvery comma-or semicolon-like markings on the under surface of the secondaries.

Satyrus has the wings broad and rounded, with eye-like spots near the outer margins, and it is of a soft brown color. It is seen as it flies, rising and falling gracefully over fields and through woods. *S. eurythris* inhabits pine woods. It flies towards the last of June, and is the first species of the genus to appear. The others are August species. *S. alope* flies in fields about clumps of golden rod, *S. canthus* by rivers and in low places. *Neonympha semidea* is found only on the summit of Mt. Washington. It must be looked for upon Mt. Katahdin. The larvæ are smooth green, often striped, with forked tails, and feed on grass. They are rarely found and should be especially sought for. By their larval forms and skipping flight these wood Satyrs lead to the small sized—

Lycaenidæ. Lycaena Americana is our common little copper butterfly. Its larva is green, oval, flattened, and feeds on sorrel. The pupa is short and thick, and is fastened by a loop to the under surface of stones.

The Azure butterflies *Polyommatus pseudargiolus*, and *P. comyntas* and *P. lucia*, are pretty species which occur frequently in May, and sometimes in April, on sunny days. *Comyntas* is an August species and has not been found in Maine yet. *Thecla* contains coppery brown species with a slight tail to the secondaries, which fly early in forests. *T. mopsus* and *niphon* are our two common ones; they may be easily captured when alighted in paths. Our largest one is *T. falacer*, which has an orange colored spot on the inner angle of the secondaries, and two unequal tails. It is rare and found in August.

We come now to butterflies with stout bodies, and large heads, whese antennæ have the knob as if untwisted and bent to one side, approaching the form of the antennæ of the Sphinges. Moreover their flight is swift and strong, while they generally skip with a jerking flight. Their colors are a soft rich brown, with yellow square spots. Such are the—

Hesperiadæ or Skippers. The green caterpillars have large heads, and taper rapidly towards either end. They are solitary, feeding within rolled up leaves, as the Tortrices, or exposed on the surface. "Their chrysalis are generally conical, or tapering at one end, and rounded, or more rarely pointed, at the other, never angular or ornamented with golden spots, but most often covered with a bluish white powder or bloom. They are mostly fastened by the tail and a few transverse threads, within some folded leaves, which are connected together by a loose internal web of threads, forming a kind of imperfect cocoon." *Harris.*

Eudamus bathyllus, is a very common species. It is of a darker brown than usual, with a few small white spots. It is common in June and July in paths, and easy to capture.

Sphingidae. (Hawk moths, Humming-bird moths.) These are the largest bodied of the lepidoptera. They have narrow thick wings which enable them to fly with great rapidity, as they frequent flowers at dark or before sunrise in the morning, inserting their long maxillae into the flowers like humming-birds, which they are often mistaken for. They are found about Lilacs, pinks and honey suckles in June and July. *Sesia diffinis* and *Thysbe* are smaller clear winged moths with flattened bodies and have spreading tufts like the tails of humming-birds on the tip of their hind body; they fly in the hottest sunshine, about the flowers of the orchard, of the Rhodora, Kalmia, Lilac and Pink, &c. Our smallest and rarest species is the *Ellema Harrisii*, which lives on pine trees, and is taken in their vicinity at flowers. The large *Macrosila carolina* is not found in Maine. Our largest species is *S. cinerea*; next to that the *S. drupiferarum* which feeds on the plum; *S. gordius* is our most common species in Maine, and feeds on the apple. The larvae are large green caterpillars with a terminal horn, and have the queer habit of elevating the head and front part of the body, (as in Figure 12,) in a Sphinx-like attitude.

FIG. 12.

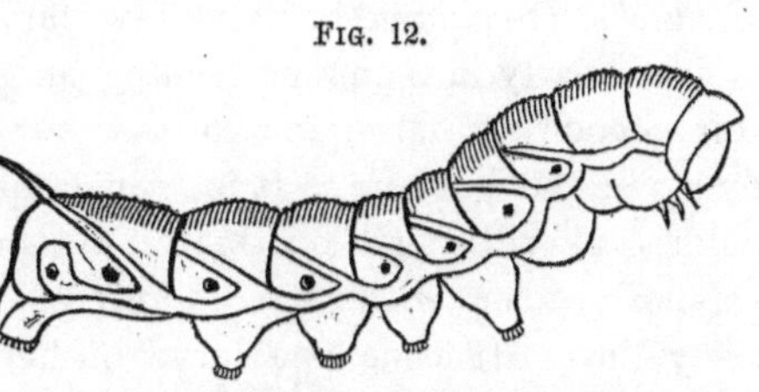

Ceratomia quadricornis has a larva with four short fleshy horns on its thorax. It feeds on the elm. By thrusting a pin dipped into oxalic acid into the body of the moths they can be quickly killed, as also by the fumes of benzine. The larva are found crawling about in September. They descend into the ground and make a rough earthern cocoon before pupating. The chrysalis has the tongue case detached.

Smerinthus has notched wings, and the secondaries are ocellated. The larvae have triangular heads. *S. geminatus* feeds on the apple. *Deilephila* feeds on the Willow herb.

Ægeriadae. These are small species whose larvae are borers. The moths have delicate transparent wings and slender bodies, elegant and gaily colored. *Ægeria cucurbitae* feeds upon the squash. *Æ. exitiosa* bores in peach trees. These two species have the sexes very distinct. The species bear a close resemblance to some hymenoptera. *Trochilium tipuliforme* is a slender blue species. It bores in the stems of the current, and by splitting the stems open in the fall and spring, we shall find the larvae. Towards the last of May they turn to pupae. In the middle of July they appear, often abundantly, flying with great rapidity about the leaves, like certain hymenopters. They are easily caught with the net. The species are rarely met with.

Zygaenidae. The members of this family which contains but a few New England species, fly in the middle of warm sunny days. They are generally blue, with pectinated or nearly simple antennae, slender bodies, and rather narrow wings, and they are covered with fine powdery scales. *Procris americana* is a slender bodied species, of a deep blue color, and saffron-yellow collar, and spreading anal tuft, which feeds on the vine or common woodbine (Ampelopsis.) Its larva is short and thick, yellow, with tufts of short black hairs across the rings. Those caterpillars of genera which approach more to the Lithosians have the body more elongated, and thickly covered with whorls of thick set hairs. *Ctenucha latreillana* has a yellow larva of this description, which is found early in summer feeding on grass. In June it makes a thin cocoon of hairs, and in the last of July appears in fields, flying in the hot sun. It is our largest species, of a dark blue color, and with well pectinated antennae. *Glaucopis Pholus* is a smaller species, with serrated antennae, and the base of the wings are yellow. It feeds as a larva on lichens, and flies about stone walls.

Bombyces. Spinners. This handsome family comprises species of the largest and most regal moths. Their thick heavy bodies and small sunken heads, and often obsolete mouth parts, pectinated antennae, broad wings, and sluggish habits, notwithstanding the numerous exceptions, afford good characters for distinguishing them. Likewise the thick hairy larvae, which spin silken thick cocoons, and change to short thick pupae, separate this family. There are several well marked minor groups, of which the Lithosians, with their slender bodies and wings, simple antennae, and slender verticillated larvae, head the group. They are also day fliers. Most of the group have narrow wings, such as *Deiopeia bella* which has bands of white enclosing dark spots on the fore wings, and scarlet hind wings, edged without irregularly with black. The species of *Crocota* of uniform pale red, look like Geometrids, and *Nudaria* has broad, nearly transparent wings, with square thinner spots.

The *Arctians* have thick bodies, and simple or feathered antennae. Their larvae have whorls of long spinulose hairs, as in the "yellow bears," the young of *A. isabella,* the buff brown species, which is yellow and black, and curls up and lies on its side when disturbed. The common yellow caterpillar is the young of *Spilosoma virginica,* a white species found in gardens, in August. *S. acraea* has a partly buff body, its larva is the Salt Marsh Caterpillar. *Halesidota* has a short thick larva, with raised middle tufts. The moths are yellowish with cross bands of spots, often partially transparent. They lead to the *Dasychirae* or tussock caterpillars, which have long pencils of hairs projecting before and behind the body. The pretty larva of *Orgyia* is variously tufted and colored, and feeds on garden vegetables. The moths fly in the sunshine in September, and resemble Geometrids. The thick and wooly bodied, pale yellow crinkled-haired genus *Lagoa,* leads to the *Cochlidiæ,* a most interesting and anomalous group, when we consider the slug-like, footless larvae, which are either hemispherical, boat-shaped, or oblong with large fleshy spines. The moths are small, thick bodied, and with antennae pectinated two thirds of their length, or they are slender bodied with simple feelers, and resemble closely some of the Tortrices. They are very difficult to raise, as they generally die in confinement.

The *Notodontians* have larvæ singularly humped, with naked or slightly hairy bodies, having the last pair of prop-legs often prolonged and not often used in locomotion, being when at rest elevated over the back. The moths resemble very closely *Noctuæ.* They may be distinguished by their small, sunken heads, feathered antennæ, and often by the tufted inner margin of the primaries.

The *Platypterycidæ* have broad falcated wings, closely resembling the Geometrids, and the larvæ have the last pair of prop-legs united and greatly prolonged. The *Bomycidae* include the *Bombyx mori* or the silk-worm. The *Atticidae* are the central group of the family. The very large, eyed wings and broad doubly pectinated antennæ of this kingly assemblage of moths, and the large, thick, fleshy larvæ with angulated wings, surmounted by scattered tubercles, giving rise to a few short hairs, are represented by *Samia cecropia* and *promethea*, which have the discal spots triangular; *Tropaea Luna* is the immense, tailed, green species, while *Telea Polyphemus* is brown and has large transparent eye-like discal spots. The *Ceratocampadae*, include *Citheronia regalis* and *Eacles imperialis*, which are of gigantic size, and the smaller *Saturnia Maia* and *Hyperchiria Io*, which have triangular subfalcate primaries. The larvæ are cylindrical and armed with hair bearing tubercles; or, as in *Dryocampa*, they have smooth bodies, with a pair of slender horns just behind the head.

The two species of *Clisiocampa*, of which *Americana* and its larva are here figured, (13 and 14,) represent another small group. The leaf caterpillars are most injurious to orchards. The moths fly at light in July.

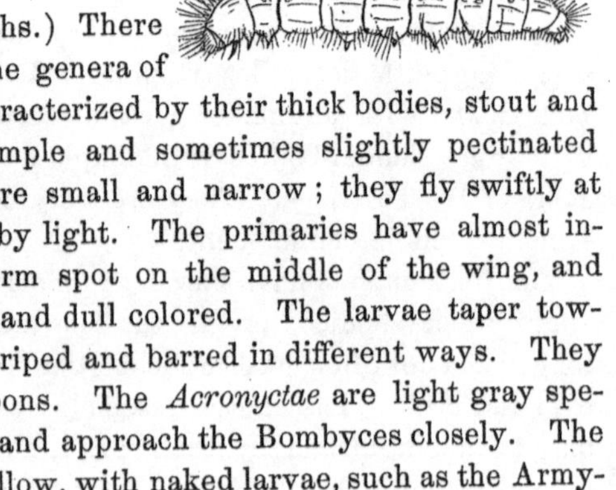
FIG. 13.

FIG. 14.

The *Hepialidae* have long, narrow wings, with both pairs much alike. Their larvæ live in the roots and stems of plants. The moths come to light in July and August, and are rare. *Xyleutes robiniæ* is stout bodied, and bores in the locust tree.

Noctuidae. (Owlet moths.) There is great uniformity in the genera of this family, which are characterized by their thick bodies, stout and well developed palpi, simple and sometimes slightly pectinated antennae. The wings are small and narrow; they fly swiftly at night, and are attracted by light. The primaries have almost invariably a dot and reniform spot on the middle of the wing, and they are generally dark and dull colored. The larvae taper towards each end, and are striped and barred in different ways. They make thin, earthen cocoons. The *Acronyctae* are light gray species, with haired larvae, and approach the Bombyces closely. The *Leucaniae* are whitish yellow, with naked larvae, such as the Armyworm. The *Agrotes* or Dart moths have broad tips to the palpi, their antennae pectinated, and the spaces between the dot and reniform spots dark and conspicuous. The larvae are the noxious *cuts worms*. They lead to the *Mamestrae,* which usually have a W in the middle of the outer line of the primaries; they have rather broad wings, and are of large size. The larvae are long, cylindrical and naked. *Gortyna,* the spindle-worm and its allies, have somewhat falcated wings. The *Acontians* are small, slender bodied, often white species, which fly in the day time about flowers. *Xylina* and *Cucullia* are large dart moths, with tufted fronts. *Plusia* is marked with silver spots.

The previous groups of genera have stout, blunt palpi, and narrow wings; but the *Catocalae* have broad wings, filiform antennae, and long, slender palpi, which reach often beyond the top of the head. Moreover, the larvae are elongated, and have fourteen legs, and a semi-looping gait, approaching closely to Geometrids. *Catocala* is very large, with gray fore wings, and beautifully scarlet, vermillion, or black striped secondaries. *Erebus odora* is a gigantic species, dark as night and faintly banded. *Homoptera lunata* and allies are similar but much smaller broad winged Noctuae.

Noctuae can be taken at dusk flying about flowers, and they enter open windows in the evening and night in large numbers, attracted by the light within. When lighted on the table under the lamp a slight tap with a ruler will kill them without injuring the specimens. In warm, foggy evenings they come in in great numbers. July and August are the best months for this family, but many species occur only in autumn, while others hybernate and are taken early in the spring. "Moths are extremely susceptible of any keenness in the air; a north or east wind is very likely to keep them from venturing abroad. Different species have different hours of flight. Thus, on a mild and dark November evening *Pœcilocampa populi* will occupy from seven to ten o'clock, after which it will make way for *Petasia cassinea,* which will fly till one or two o'clock in the morning. I have, for experiment's sake, sat up in the summer till three o'clock, when the whole heavens were bright with the rising sun, and moths of various kinds have never ceased arriving in succession till that time. Some of these must come from a considerable distance: *Scotophila porphyrea,* being a heath-moth, must come nearly a mile." *Bird.*

"In April the willows come into bloom. In the day time they

are very attractive to bees, *Bombi*, *Andraenæ*, &c., and a few beetles also. At and after dusk the flowers are the resort of several species of moths, (*Noctuina*,) some of which have hybernated, and others have just left their pupa state. It is now some fifteen years since the collectors first took moths in this way, that were likely long to have remained deficient in the collections but for the discovery, by Mr. H. Doubleday, of the attractive powers of the sallow blossoms. I believe it was the same gentleman who found out about the same time that a mixture of sugar and beer, [or rum and sugar or molasses, &c.,] mixed to a consistence somewhat thinner than treacle, is a most attractive bait to all the *Noctuina*. The revolution wrought in our collections, and our knowledge of species since its use, is wonderful."

"The mixture is taken to the woods, put upon the trunks of trees in patches or stripes, just at dusk. Before it is dark some moths arrive, and a succession of comers continues all night through, until the first dawn of day warns the revellers to depart. The collector goes, soon after dark, with a bull's-eye lantern, a ring net, and a lot of large pill-boxes. He turns his light full on the wetted place, at the same time placing his net underneath it, in order to catch any moth that may fall.

"The sugar bait may be used from March to October with success, not only in woods, but in lanes, gardens, and whenever a tree or post can be found to put it upon. The best nights will be those that are warm, dark and wet; cold, moonlight, or bright, clear and dry nights are always found to be unproductive. It is also of no avail to use sugar in the vicinity of attractive flowers, such as those of willow, lime or ivy. Sometimes one of the Geometrina or Tineina comes, and occasionally a good beetle." The Virgins' Bower, when in blossom, is a favorite resort of Noctuæ. Many can be taken by carrying a kerosene lamp into the woods and watching for whatever is attracted by its light.

Geometridæ. (Geometers, Measuring-worms, Span-worms.)—This is a large group of slender-bodied, broad winged moths, with feathered antennae, which at rest have the wings nearly expanding, hardly overlapping each other. The larvae have but ten legs, walking with a looping gait. At rest they often hold themselves out straight and stiff by the muscles of the anal prolegs. *Ennomos* and allies have stout, rather wooly bodies, and angulated wings. They are generally yellow, dusted with ochreons, and the larvae are large, tuberculated, and spin rather thick cocoons among leaves. *Boarmia* has wings crossed by numerous bands of dark irregular dots. The *Macariæ* have falcated primaries, and are of smaller size than the foregoing groups. The allies of *Abraxas* have wings rounded at the apex. A species that is pale buff with smoky spots, inhabits the currant, whose larva is golden yellow with white and dark spots. Some genera have wingless females ; such as *Hybernia* which appears in October, and whose wingless female is ornamented with a double row of square black spots along the back ; and the *canker-worm* (*Anisopteryx vernata* Fig. 15, larva, Fig. 16, moth,) which is rarely found now in Maine, but will probably be abundant before many years.

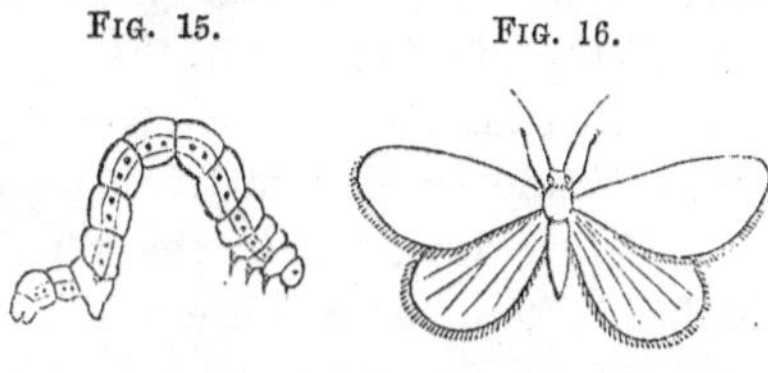
Fig. 15. Fig. 16.

Acidalia is a very delicate slender bodied genus, of large extent, whose wings are banded much as in the Boarmiæ. The genus *Geometra* which is large and green, we do not have here ; but some smaller species belonging to the genus *Racheospila,* whose abdomens are scarlet spotted above are frequent. The smallest speeies are found in the *Eupithiciæ,* which have long triangular wings. Nearly all the species can be taken in June and in July, in damp shady woods, or in open fields. *Larentia* and *Cidaria* come at light with Noctuids in July and August.

Pyralidæ. (Delta moths.) The species have the habit of placing the wings in the form of a triangle, when at rest, since they do not overlap each other. Their bodies are slender, the antennæ nearly always simple, while the palpi are greatly enlarged, so as sometimes to be thrown back over the head. *Hypena* and its allies are of large size ; the fore legs are frequently curiously tufted. They are found in company with Geometrids. *Hydrocampa,* as a larva feeds on aquatic plants, constructing a case like the Phryganeids, whieh it carries about with it. *Pyrausta* is generally red, striped with dark. *Botys* (Fig. 17,) is of a pale straw color with transparent spots, and long slender body and legs.

Fig. 17.

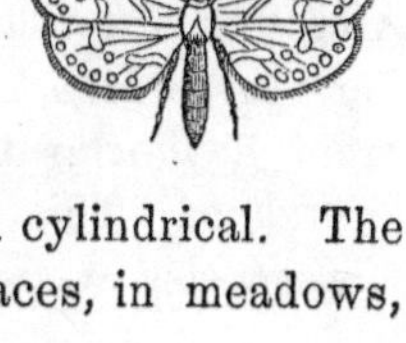

Aglossa is found about houses, and feeds on fatty substances. Some of the larvae are half-loopers, while those of the smaller species are naked, or with a few scattered hairs, slender and cylindrical. The smaller species are nearly all taken in damp places, in meadows, grass lanes, or by rivers and pools in summer. Some of the species are day fliers.

Tortricidæ. (Leaf-rollers.) These are small, broad-winged moths, which at rest fold their wings, roof-like, over their bodies, in the form of a triangle, (as in Fig. 18.) They are abundant in June and July, in low bushes, herbage, or on leaves of trees, where they can be swept by the net. The larvae are rather thick greenish caterpillars, which roll up leaves; their work can thus be easily detected. When disturbed they wriggle out of the other end of their domicile, and let themselves down by a silken thread. Others feed on buds and flowers, such as *Loxotænia;* while another tortrix *Carpocapsa pomonella,* the "Codling moth," lays its eggs in the plumage of the young apple, and in the fall is found as a white fleshy grub in the core.

FIG. 18.

Tineidæ. These are the smallest of moths, and are known not only by their minute size, but by their narrow wings, often falcate, or pointed acutely in both pairs, and edged with a long fringe of exceeding delicacy. The maxillary palpi are greatly developed, while the labial palpi are of their usual size, and are sometimes recurved as in the Pyralidae.

Crambus and its allies have long palpi and oblong wings, generally white and buff yellow, sometimes ornamented with golden spots. They fly in grass in great abundance, resting on the spears with their heads down. To this group belongs the *Bee moth,* (*Galleria cereana*) which as a larva eats wax. There are two broods in a season.

Hyponomeuta has long maxillary palpi, and very long antennae, closely resembling some of the smaller Phryganids. The Tineids, generally, are moths of rare beauty. The family is one of great extent, and the species are very destructive to vegetation, and have innumerable modes of attack. Thus, *Tinea vestianella,* the clothes-moth, and allied species, construct a case of the fibers they eat, and bear it about for their protection. In June the moth appears and lays its eggs. *Tinea granella* make a silken web of the grains they devour. Another species, still more destructive in granaries is the Angoumois moth, (*Anacampsis cerealella,* Fig. 19,) which secretes itself within the grain, devouring the mealy substance.

FIG. 19.

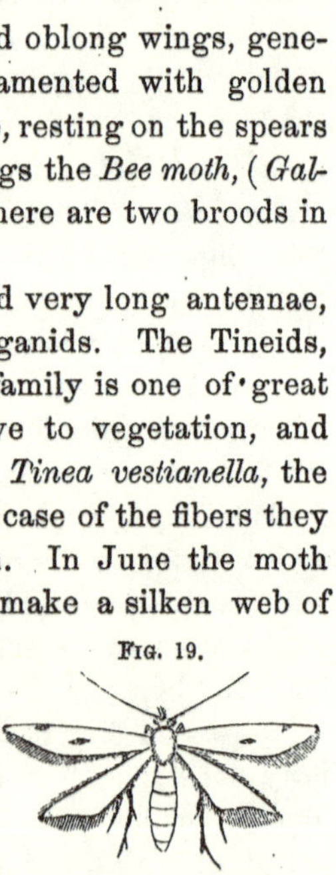

Alucitæ. This is a family comprising a few species whose wings are divided into numerous delicately fringed branches. They are found in July and August, in herbage. *Pterophorus marginidactylus* is a common species, and flies in at light in July and August.

For collecting and preserving these minute and delicate moths, which are called by collectors, *micro-lepidoptera,* especial instructions are necessary. When the moth is taken in the net, it can be blown by the breath into the bottom. "Then by elevating the hand through the ring, or on a level with it, a *common cupping glass* of about two inches in diameter, or a *wine glass* carried in the pocket, is placed on the top of the left hand over the constricted portion, the grasp relaxed, and the insect permitted to escape through the opening into its interior. The glass is then closed below by the left hand on the outside of the net, and may be transferred to the top of the collecting box, when it can be quieted by chloroform." Clemens. Or the moth may be collected in pill boxes, and then carried home and opened into a larger box filled with the fumes of ether or benzine. In pinching any moths on the thorax, the form of that region is invariably distorted, and many of the scales removed. In searching for "*Micros*" we must look carefully on the *lee side* of trees, fences, hedges, and undulations in the ground, for they avoid the wind.

In seeking for the larvae we must remember that most of them are *leaf miners,* and their burrows are detected by the waved brown withered lines on the surface of leaves and their "*frass*" or excrement thrown out at one end. Some are found between united leaves, of which the upper is curved. Others construct portable cases which they draw about the trunks of trees, fences, &c. Others burrow in the stems of grass, or in fungi, toadstools, in the pith of currant or raspberry bushes. Most are solitary, a few gregarious. A bush stripped of leaves and covered with webs, if not done by *Clisiocampa,* (the American Tent Caterpillar,) will witness the work of a Tineid. Buds of unfolded herbs suffer from their attacks, such as the heads of composite flowers which are drawn together and consumed by their larvæ.

After some practice in rearing larvæ it will be found easier and more profitable to search for the leaf miners, and rear the perfect fresh and uninjured moths from them. In this way many species never found in the perfect state can be secured.

In raising *micro* larvae it is essential that the leaf in which they they mine be preserved fresh for a long time. Thus a glass jar, tumbler or jam-pot, the top of which has been ground to receive an air-tight glass cover, the bottom of which has been covered with moist white sand, will keep a leaf fresh for a week, and thus a larva in the summer will have to be fed but two or three times before it changes; and the moth can be seen through the glass without taking off the cover. Or a glass cylinder can be placed over a plant placed in wet sand, having the top covered with gauze. The pupae easily dry up; they should be kept moist, in tubes of glass closed at either end, through which the moth can be seen when disclosed. Instead of benzine, powdered and crushed *laurel* or *kalmia* leaves, which contain prussic acid, is often used instead of ether, chloroform or benzine.

How to set micro-lepidoptera: "If the insect is *very small* I hold it by its legs between the thumb and finger of the left hand, whilst I pierce it with the pin held between the thumb and finger of the right hand; if the insect is not very small I use a rough surface, as a piece of blotting-paper, or piece of cloth, for it to lie upon and prevent its slipping about, and then cautiously insert the point of the pin in the middle of the thorax, as nearly as possible in a vertical direction. As soon as the pin is fairly through the insect, remove it to a piece of soft cork, and by pressing it in, push the insect as far up the pin as is required.

"For setting the insects I find nothing answers as well as a piece of soft cork, papered with smooth paper, and with grooves cut to admit the bodies. The wings are placed in the required position by the setting needle, and are then retained in their places by a wedge-shaped thin paper brace, placed over them till a square brace of smooth card-board is placed over the ends of the wings." *Stainton.* Pieces of plate glass are often used instead of card-braces. Small slender insects pins No. 19 and 20, are made by Edleston & Williams, Crown Court, Cheapside, London.

Diptera.

North American flies have been but little studied, though so interesting and numerous. The different parts of the body vary greatly, and often give easy characters for discrimination. Thus the parts of the head, the form and disposition of the nervures and the intespaces of the wings, give good generie and specific differences. Their habits are very variable. Fresh water aquaria, consisting of glass jars with a few vegetables to oxygenate the water, are necessary for the maintenance of aquatic larvae. If quantities of swamp mud and moss with decaying matter is kept in boxes and jars, multitudes of small flies will be hatched out. Leaf mining species can be treated as micro-lepidoptera, and earth inhabiting larvae, like ordinary caterpillars. Dung, mould in hollow trees, stems of plants and toad stools contain numerous larvae or *maggots*, as the young of flies are called, which must be kept in damp boxes.

Flies can be pinned alive, without killing them by pressure, which destroys their form ; and numbers may be killed at once by moistening the bottom of the collecting box with creosote, benzine or ether. Minute species can be pinned with minute No. 19 or 20 pins, or pieces of fine silver wire, and stuck into pieces of pith, which can be placed high up on a large pin. In this way the specimen can be handled without danger of breaking. Small moths can be treated in this way. In pinning long legged, slender species, run a piece of card or paper up under their bodies upon which their legs may expand, and thus prevent their loss by breakage.

Of these insects, as with all others, duplicates in all the stages of growth, should be preserved in alcohol, as the minute species often dry up unless put in homeopathic vials.

Culicidæ. Mosquitoes, Gnats, have the mouth parts produced into a proboscis half as long as the insects themselves, which they can push into the skin. The females lay their eggs in a boat-shaped mass, which floats on the surface of the water, and in the spring the larvae are seen in pools by thousands, jerking themselves up and down in the water, after protruding a star-like respiratory organ above the surface to obtain a supply of fresh air. The pupae are club-shaped, with very large heads, to which two respiratory feeler-like organs are attached. There are several generations in a season. A large four-spotted species, (*Anopheles quadrimaculatus*) is abundant very early in spring and late in the fall. There are several genera and species of this family.

Tipulidæ. (Daddy-long-legs. Crane-flies.) The long palpi and antennae, slender body and very long legs of the members of this family, make them well known. The smaller species belong to the genus *Chironomus,* which is musquito-like, with feathered antennae, and abounds in swarms in early spring. Their larvae are worm-like, of a blood-red color, and are found in the bottom of ponds.

Ceratopogon, like the musquito, is a blood-sucker, has the male antennæ partially feathered, and the larvæ live in mushrooms, or under the bark of decaying trees.

Cecidomyia. Gall flies have slender bodies and verticillate antennæ, their wings have few nervures, and are placed roof-like over the body. The female lays her eggs in the stalks of cereals, and of stems and leaves, which produce galls inhabited by maggots. The Hessian fly does not, however, produce an enlargement of the stalk, as is usual. Those species injurious to wheat, &c., can be collected by sweeping the fields in the spring, at evening, when they are laying their eggs.

Psychoda is a minute genus, with white, broad, oval wings, which is found flying about and into, little pools, and in great numbers on windows. The larvæ live in dung.

The *Mycetophilidæ* are of small size, and very active, leaping little flies, which are found in damp places. The larvæ are long, nearly round, white or yellowish; they are gregarious, living in decaying vegetable matter, fungi, or in dung, one species forming a gall. *Rhyphus* is common on windows; it has rather broad, spotted wings, and the larva lives in cow dung. The large *Tipulæ*, which fly all the summer and form a numerous group, live as larvæ in the mould of gardens, at the roots of willows, and in rotten wood. The pupa have the margins of the rings spiny, to be able to push themselves along, as do many other Diptera. Other species are aquatic, and should be raised in aquaria. *Chionea*, the snow-gnat, looks like a spider, being wingless, and is found in March on the snow.

The *Bibionidæ* comprise species very injurious from their feeding on the roots of grass; the thorax is stout, and the legs are short. *B. albipennis* flies in swarms in June and October, alighting slowly on the passer-by. *Simulium*, the black fly, has a stout body and short legs, often silvery in color. It is aquatic, its larva living on the stems of plants.

Stratiomyidæ. The insects of the following families have broad bodies, and short, bristle-like antennae, the basal joints being enlarged. Those of this family are found among herbage in damp places. The larvae live in the water, in decaying subtances or dung.

Tabanidæ, Horse-flies. The parts of the mouth are here again converted into a proboscis. The horse-flies are of large size, and troublesome from their formidable bite. Their eyes are very large, and the thorax large and oblong-square. They abound in pastures and woods. Their larvae live in the earth. The species of *Chrysops*, the golden-eyed breeze fly, are very troublesome, as they fly about one's head unceasingly, striving to alight and draw the blood. The following genera represent families of small extent. *Anthrax* is rather broad and flat bodied, with a round head, gaily colored with yellow and black, the wings often partially black; it frequents sunny paths, flying with great swiftness. *Bombylius* has the body covered with long hairs, which gives the genus an oval outline, with slender legs. They are exceedingly swift on the wing, and are found in sunny paths and glades early in the spring, and can only be taken when lighted on the ground.

Leptis has large palpi, a fleshy proboscis and elongated form. Their bodies are often spotted, and the wings also spotted or banded. They are found resting on flowers and shaded sides of trees. The larvae are footless grubs, which widen gradually towards the terminal segment, which has two short appendages. The larva of an European species entraps other insects in holes in the sand, like the ant-lion, and is three years in coming to the perfect state.

Midas is a genus of large size, *M. clavatus* being banded with orange, and expanding over two inches. It flies in July and August. The larva, according to Harris, is white, cylindrical, tapering before and almost rounded behind. Two breathing pores are situated in the last ring but one. The pupa is brown, nearly cylindrical, and provided with a forked tail. It lives in decayed logs and stumps.

Asilus comprises several species, which have long, slender bodies, a rather stout thorax, and are covered with short, stiff hairs, variously colored. They are rapacious, seizing other insects and flying off with them like the sand-wasps. *Proctacanthus philadelphicus* is a very large, dark species, which frequents sunny places. The larva of *A. sericeus* lives on the roots of the rhubard plant. It is three-quarters of an inch long, cylindrical nearly, and tapering at either end. Their pupa cases, with forked tails, are found sticking out of the ground and the roots.

Laphria is one of the same family, though the body is much stouter and more densely covered with yellow and black hairs. Indeed, in their loud buzz, swift flight and appearance they closely resemble bumble bees. They are found in sunny places, preying upon other insects.

Syrphidæ. These gaily colored flies, so useful to agriculture from their habit of feeding upon plant lice, are very like the hymenoptera in form and coloration, having hemispherical heads, rather flattened bodies, ornamented with yellow bands and spots; they hover in the hot sun over and about flowers, resting upon them to feed upon their sweets. The eggs are laid among a group of plant lice, which hatch out footless, eyeless, flattened grubs, having extensile bodies to reach up and grasp the Aphis by their jaws, which are peculiarly modified for seizing their prey. They do great damage among these enemies of vegetation. The species of *Eristalis* which flies abundantly in May about the blossoms of gooseberries and currants, live in the water during their larval state, and are called rat-tailed maggots. The abdomen of *Conops* is pedunculated, whilethe thorax is globular like Eumenes, a genus of wasps.

FIG. 20.

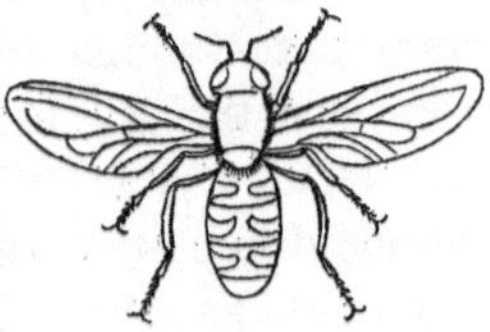

Empis represents a small group of species that are allied in form to the Asilidæ. They are active flies, and very rapacious, seizing upon other insects and sucking out their juices. They often assemble in swarms.

Dolichopus and allies have long legs, and are generally green colored, and occur solitary in leaves or in damp situations, or in numbers flying and running on the surface of pools and running brooks, appearing very early in spring.

Œstridae, (Bot-flies.) In these flies, which are of large size, the mouth parts are nearly obsolete, the flies themselves having thick bodies, covered thickly with hairs. The fly lays her eggs upon that part of the animal from which the larvae as they hatch out may find their way by some means to burrow in the back or stomach of the animal which they infest. From thence, when full grown, they escape and pass through their remaining changes in the earth. These grubs are very thick and soft, being broad oval, with rows of minute spines along the wings of the body to aid in locomotion. The Horse bot-fly larva is provided with hooks which are modified maxillae, to enable it to maintain its position in the stomach of that animal. The Sheep bot-fly larva lives in the frontal sinus; and that of *Œstrus bovis* in the back of cattle, forming large open tumors.

Muscidae. The common house fly, the blue bottle fly, and the flesh fly, at once recall the appearace of this family, one of great extent, and much subdivided by entomologists. "The larvae are in general footless, soft vermiform, ringed grubs, of a cylindrical-conic form, attenuated in front, and thickened and obtuse behind, with a head of variable form, furnished with two retractile hooks; the terminal segment of the body in many, and also that immediately succeeding the head, furnished with two spiracles, in some species inserted upon horn-like appendages. The pupa, which is very unmature in its form, with a swollen head, is enclosed within the contracted and indurated skin of the larva, which sometimes assumes the form of an oval, horny exasticulate mass, but in other species retains more of its former appearance." *Westwood.*

Tachina is parasitic upon caterpillars, and destroys great numbers in the same way as Ichneumons. Some of them are parasitic in the nests of bees. *Sarcophaga*, the flesh fly, is viviparous, the larvae being placed upon the meat by the parent fly.

Musca Caesar, the blue bottle, and *vomitoria*, the flesh fly, lay their eggs also upon meat and decaying animal matter, the larvae developing with great rapidity. The larvae of the House fly live in dung. *Anthomyia raphani* is the grub that attacks the radish roots. Other species live in onions, turnips, and the pulpy parts of leaves, and in rotten substances and dung. The species are very numerous; they are rather small and fly feebly.

Ortalis and allies produce galls in plants, or lay their eggs in fruit, such as raspberries, &c. They are found in shady places; their wings are generally spotted. *Tephritis asteris* causes the large swellings in the stems of tall asters. *Oscinis*, in Europe, does great damage to cereals by laying its eggs in the flowers of grain, the larva afterwards consuming the grain itself. Thus by collecting heads of wheat and composite flowers and keeping them in boxes, &c., these flies may be reared, and much light thrown upon their history and modes of attack. Many of these small flies, like the micro-lepidoptera, are leaf-miners, and can scarcely be distinguished from them when in the larva state.

Hippoboscidae, (Spider-flies.) These are small, flat-bodied flies, of disgusting appearance and habits, which by their large clawed legs run over the surface of quadrupeds and birds with great agility, burying themselves in the fur or feathers.

Nycteribia, or Bat-tick, is a wingless genus, with long legs and a spider-like body, and has similar habits to the Hippoboscidae. *Mellophagus ovis* is the Sheep-tick. "These singular creatures are not produced from eggs, in the usual way among insects, but are brought forth in the pupa state, enclosed in the egg-shaped skin of the larva, which is nearly as large as the body of the parent insect. This egg-like body is soft and white at first, but soon becomes hard and brown. It is notched at one end, and out of this notched part the enclosed insect makes its way, when it arrives at maturity." This species is probably viviparous, and the larvae are hatched within the body of the parent.

Pulicidae. Fleas are but wingless flies, with hard, compressed bodies, a long, sucker-like arrangement of their mouth-parts, and large hind legs, formed for leaping. Their metamorphosis is complete, the larvae hatched from eggs laid upon hairs, being worm-like, as in flies. They come to maturity in a few days; spin a sort of cocoon, and change to pupae, when the perfect insects appear in about ten days. Thus a generation may be produced in a month. Different species inhabit man, cats, dogs, &c. Those infesting the lower animals do not pass from one species to another.

Coleoptera.

Beetles have been studied much more than other insects; in this country there have been described some 8,000 species, but from the difficulty of finding their larvae and carrying them through their successive stages of growth, the immature forms of but few native species are known. The family forms are easy to distinguish and characterize, the genera are based upon marked changes in the different parts of the body, which vary greatly, and some of the best characters lie in the relative size of the head pieces and those pieces that make up the flanks of the three thoracic rings, and the basal joints of the legs. The relative size and the sculpture of the body and of the elytra; and lastly, the coloration, which varies much among the individuals, afford good specific characters.

The most productive places for the occurrence of beetles are alluvial loams, covered with woods, or with rank vegetation, where at the roots of plants or upon their flowers, under leaves, logs and stones, under the bark of decaying trees, and in ditches and by the banks of streams, the species occur in greatest numbers. Grass lands, mosses and fungi, the surfaces of trees and dead animals, bones, chips,pieces of board and excrement, should be searched diligently. Many are thrown ashore in sea-wrack, or occur under the debris of freshets on river banks. Many Carabidae run on sandy shore. Very early in spring, stones can be upturned, ants nests searched, and the waters be sifted for species not met with at other times of the year.

For beating bushes, a large strong ring-net should be made, with a stout bag of cotton cloth fifteen inches deep. This is a very serviceable net for many purposes. Vials of alcohol, a few quills stopped with cork, and close tin boxes for larvae and the fungi, &c., in which they live, should be provided; indeed, the collector should never be without a vial and box. Beetles should be collected largely in alcohol, and the colors do not change if pinned soon after being taken. Coleoptera should be placed *high up* on the pin, (Fig. 21, *Curculio,*) as indeed all insects should. The pin should be stuck through the right elytron so that it shall come out beneath or between the middle and hind pair of legs. Small species should be pinned with No. 19 and 20 pins, which can be afterward mounted on high pins as described for flies. Many coleopterists gum small species, under a tenth of an inch long, upon a small triangular bit of card, placing them crosswise with a cement of inspissated ox-gall, gum arabic and water, or gum mixed with a little sugar. The first mentioned cement is very convenient for mending broken specimens. Specimens thus gummed have some of the best generic characters often concealed, and hence fine pins seem best to mount them upon.

FIG. 21.

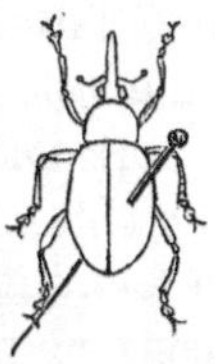

The specimens should be neatly set, in their natural postures. Some individuals should have their wings expanded to show the neuration. Beetles are best arrayed in boxes lined with cork well smoothed and neatly papered, 12 by 9 inches square, and an inch and a half deep. These boxes can be put under cover.

Cicindelidæ. The Tiger-beetle, has a large head, much broader than the prothorax, very long jaws, like curved scissors, and long slender legs. Their colors are green or darker, with purplish or metallic reflection, marked with light dots and stripes. They abound in sunny paths, and sandy shores of rivers, ponds, and of the ocean, flying and running swiftly. Capture them by throwing the bag net quickly over them after they are settled; when abundant remain still in one place, waiting for them to settle near you, thus saving time and trouble. If without a net, throw a handful of sand at one, and thus confuse and catch it in its endeavors to escape. The larvae are hideous in aspect: the head is large, with long jaws, the thoracic rings large and broad, and the 9th ring has a tubercle and hook, by which the grub can climb up its hole, at the entrance of which it lies in wait for weaker insects. These holes are found in sandy banks frequented by the beetles. Either dig the larvae out, or thrust in a straw; which they will seize and often suffer themselves to be drawn out.

This and the four following families are carnivorous, benefiting agriculture from the immense numbers of insects they destroy.

Carabidæ. In this group the head is narrower than the thorax, which is throughout as broad as the abdomen. The powerful jaws are shorter, and not curved as in the Cicindelidae. The body is also flatter and more oblong. They are runners, the under wings being often absent. Their color generally dull. They run in grass, or lurk under stones and sticks, are under bark of trees, and under the debris of freshets, in the greatest numbers in spring. *Lebia* is found in Autumn on trees and tops of composite plants. *Amara* feeds on pith and stems of grasses. Others feed on wheat. They are often attracted by light. *Elaphus*, which is flat, and covered with coarse metallic punctures, runs on the mud flats of rivers, &c.

The larvae are found in much the same situations as the beetle and are oblong, broad, with the terminal ring armed with two horney appendages, and beneath a single tube-like false leg. They are black in color. The larva of *Calosoma* ascends trees to feed on caterpillars.

Fig. 22.

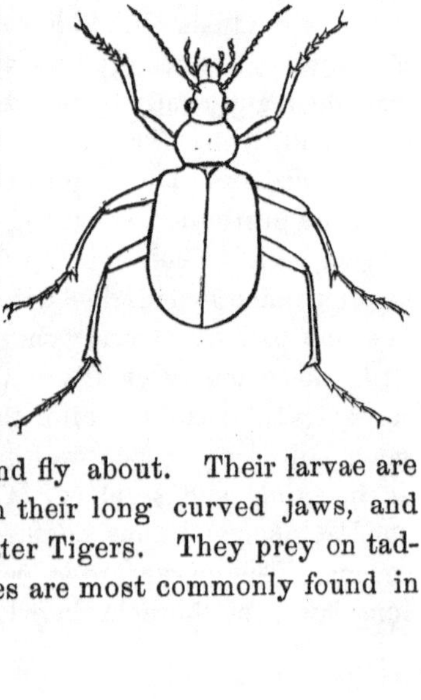

C. scrutator, (Fig. 22,) is our most splendid New England beetle of this family. It has not yet been found in Maine. *C. calidum*, our common golden spotted purple species, digs holes in fields where it lies in wait for its prey.

Dysticidæ, or Diving beetles, are, by their carnivorous habits closely allied to the Carabidae. They are aquatic, flattened elliptical beetles, with their hind legs ciliated, forming a broad surface for swimming. In night time they leave the water and fly about. Their larvae are ferocious looking objects, and from their long curved jaws, and agile and stealthy habits, called Water Tigers. They prey on tadpoles and large insects. The beetles are most commonly found in spring and fall. They can be raised, and their habits observed in Aquaria.

Gyrinidæ, Whirl-gigs, are easily distinguished by their form and habits, being always seen in groups, gyrating and circling about on the surface of pools, and when caught giving out a disagreeable milky fluid. Their short antennae, short mandibles and two pair of ocelli, and bluish black colors, distinguish them from other aquatic beetles. Like the previous family, upon being disturbed, they suddenly dive to the bottom, holding on by their claws to submerged objects. They carry a bubble of air on the tip of their abdomen, and when the supply is exhausted, they rise for more. The larvae resemble a small centipede, with lateral ciliated filaments, serving as organs of respiration.

Hydrophilæ. Carnivorous as larvae, but when beetles, vegetable eaters, and living on refuse and decaying matter; this family unite the habits of the previous mentioned families, with those of the scavenger silphidae, &c. They are aquatic, small, convex oval, or hemispherical beetles. Their antennae are short, and their palpi are long and slender. The allies of the genus *Sphærium,* live in excrements of herbivorous animals.

Silphidæ, Carrion or Sexton beetles, are useful in burying decaying bodies in which they lay their eggs. Smaller species live in fungi, &c. ; other genera live only in caves ; *Catops* inhabits ants' nests. Another genus *Brathinus,* has been found from Lake Superior to Nova Scotia, about grass roots in wet places, and are small shiny insects of graceful form, according to Le Conte.

The group is distinguished by the knobbed antennae. Their larvae are crustaceous, flattened, the sides of the body often serrated, black and of a fœtid smell; or those immersed in the midst of their food have weak limbs and soft bodies. The beetles can be caught on the wing in warm spring days, or taken at light in summer. By placing dead birds and small mammals, &c. in favorable places, they are allured in considerable numbers.

By the *Scydmaenidae* which are minute oval shiny brown insects found under stones near water, in ants nests and under bark, we pass to the *Pselaphidae,* with short elytra, much broader than the prothorax and head, with clavate antennae, and palpi nearly as long, which are found in spring in moss, or swept from herbage or taken while on the wing, we come to the *Staphylinidæ* or Rove beetles, which are long, linear, black, with remarkably short elytra, the abdomen beyond having 7 to 8 visible rings. Though sometimes an inch in length, they are more commonly minute, inhabiting wet places under stones, manure heaps, fungi, moss, under the surface of bark, or leaves of trees. Some burrow in sand, others form galleries under bark; *Stenus* is found running on mud, near water; *Micralymna* is found at low water mark in sea weeds in the larva state. Many species inhabit ants' nests, and should be carefully sought for on dewy mornings, under stones and pieces of wood, which should be taken up and shaken over a white cloth or paper; or the whole nest should be sifted through a rather coarse sieve, when the small beetles will fall through the meshes.

The larvae resemble the beetles, and are difficult to rear.

Histeridæ. These beetles are square or oblong, hard, solid, shiny insects, black, with the prothorax hollowed out to receive the head, which has long prominent jaws. The elytra are usually striated. The antennae are elbowed, club shaped, and the legs are broad and thin. Others are oval and spotted. They are found in excrement and under bark of trees.

Nitidulidæ. Broad oval or elliptical, depressed, the head also received into the excavated prothorax. The three last joints of the antennae are gathered into a broad club. Insects of small size, and found about rubbish, bones, &c. *Ips* has bright colors, often red, is one of the larger genera, and is found under bark and on the sour sap of stumps and trees in the spring. Others are found in fungi and in flowers. The larvae inhabit similar places. They are flattened oblong whitish grubs, the end of the abdomen has four horny conical upturned appendages. The pupae are found loose in rubbish and decaying wood, saw dust, &c.

Of similar form and habits is *Mycetophagus*, and other genera, representing families of small extent.

Dermestidæ. Every entomologist dreads the ravages of *Dermestes* and *Anthrenus* in his cabinet. The ugly bristly insidious larvae which so skilfully hide in the body whose interior it consumes, leaving only the shell ready to fall to pieces at any jar, can be kept out only with great precautions. *Dermestes lardarius* is oblong oval, legs short, black, with the base of the elytra gray buff, covered by two broad lines. It is timid and slow in its movements, when disturbed seeking a shelter, or mimicking death. *Anthrenus musaearum* is round oval, with transverse waved lines. Its larva is thick, with long bristles, which are largest on the end of the body. They eat also the integuments of stuffed specimens, doing great injury. Boxes and drawers should be tight enough to keep them out, or it may be done with camphor or benzine in a sponge or in cotton. Collections which are much infected should be baked.

Byrrhus, which is short, thick convex, is found under stems and on leaves. When disturbed it counterfeits death. Larva long, narrow, oblong. By the small group of Byrrhidae we pass to one of immense extent, and of great importance to agriculturists from the great injury they do as leaf-eaters.

Scarabeidæ, or Lamellicornes, are distinguished by their lamellated antennae, short broad, thick convex form; their legs are flattened, and toothed for the purpose of digging. The tip of the abdomen is generally exposed. The males are often armed with horns on the clypeus. Colors black, dull or shiny, coppery or gaily ornamented. Among them occur tropical insects, such as the Goliath beetles, which are the largest of insects. *Lucanus* has immense jaws; in the males they are like deer's horns. The larva forms a cocoon of the chips it has made in boring into decaying trees. The larvae are thick, cylindrical, soft fleshy grubs, the abdomen incurved, so that the grub lays on its side, the legs being short and weak. They live several years.

Aphodius is a small semicylindrical genus, flying about ordure in spring; of similar habits is *Geotrupes*, a large green or purplish colored genus. *Copris*, called Tumble Dungs, enclose their eggs in pellets of excrement.

Melolontha and allies are leaf eaters, which have long-clawed legs to cling on to leaves, where they are found early in summer. Their larvae eat the roots of grass, and before transforming, form oval earthern cocoons. *Macrodactylus*, the Rose beetle, is found on roses and rhubarb blossoms in gardens.

Lachnosterna, the June bug, does much injury to apple and cherry trees. The males fly in evening in search of the other sex. The large grubs are turned up abundantly in spring, in gardens. Skunks feed upon them, and smaller species are eaten by toads, indeed many rare species of beetles have been found in the stomach of toads and insectivorous birds.

Buprestidæ. Beetles, with elongate, flattened, very solid bodies, often angulated, the antennae slender and serrated, legs short. The head is received into the excavated prothorax. Colors brilliant, often metallic. On being disturbed, the insects draw up their legs and feign death, They creep slowly, flying in the hot sun, and feed on wood, flowers and sap ; being found especially on fir trees. They should be sought for while sunning themselves on trunks of trees, where they lay their eggs.

The larvae are also elongated, the thorax is broad, while the abdomen may be equally broad, or narrow and cylindrical. They are wood borers, and live in this state several years.

Chalcophora virginica is common in May and June. *Dicerca* has the tip of abdomen divided. *Chrysobothris* lives in the apple tree.

Elateridae, or Snapping beetles, are known to many by their power of righting themselves when turned on their backs, by jerking themselves up into the air, since their legs are too short to catch hold of the surface they are upon. They are of a very uniform elongate ellipsoid form, somewhat flattened, the head and prothorax rendered very distinct by a depression of the base of the elytra. Colors are obscure brown, sometimes green with metallic reflections.

They frequent the flowers of Viburnum, of rhubarb in gardens, and are found under bark. The *Eucnemidœ* are rare, being found under bark or on leaves. *Alaus oculatus,* is the larger ocellated species. The larvae are called *wire-worms* from their long cylindrical form. They feed on the roots of grass, grain, &c., often devour turnips, salad, cabbages and pinks, living in the interior of these stems. Moles devour great quantities of them. Other species inhabit rotten stumps. They live several years in this state.

We pass over several smaller groups to the

Lampyridœ, or Glow worm. They resemble the Elaters, but are shorter and broader, and of softer consistence.

The species of *Ellychnia* are found early in spring and fall, on trunks of trees, and they winter under the bark.

The female *glow worm* is apterous, and resembles the larvae ; the end of the abdomen is light colored, and at night this portion sheds a brilliant light at its will. Winged females of other genera emitting a bright light, appear on low grounds in the evenings, at the middle of June. *Drilus* is distinguished by the plumose antennae. The larvae are flatted, the margin of their bodies is serrated, and they are soft and black in color. They are carnivorous and feed on snails, and are found in places frequented by these mollusks, as at the roots of alders and willows, under the bog moss.

Eürypalpus Le Contei is an anomaly, since it lives under stones in rivers and brooks, being oval hemispherical as a larva, the sides of the body greatly extended, resembling some species of crustacea. The beetles are narrow and rather short. The species of *Telephorus* live on leaves of plants, especially the birch. They are carnivorous, often feeding upon each other.

We pass by the *Malachidæ* to the

Cleridæ, which are beetles whose larvae are carnivorous. They are cylindrical, the prothorax narrower than the head. They are fast runners, and run like ants, which they much resemble, over flowers and trees, to feed on the sweets and sap. *Trichodes nuttallii* is blue and red, and found on the flowers of Golden rods and Spiraea. The narrow long pink-colored larvae of *Thanasimus* can be found under the bark of dead pine trees, where it devours the larvae of *Hylurgus* and *Hylobius* ; Clerus and allies are found in bumble bees' nests. In Europe they have been found infesting the nests of mason bees (*Osmia* and *Megachile.*)

Ptinidæ. They also infest herbariums and museums. They are small beetles, of an obscure brown color, somewhat oval, behind truncated, the prothorax slender and receiving the head. The antennae are long and filiform, and in constant motion when the insect walks. Upon being disturbed it feigns death. They are found about out-houses. *Ptinus fur* has done great mischief in eating wheat. *Anobius* is the Death-tick ; the females strike their jaws on the surface of walls, to attract the other sex in the pairing season. The larva are also supposed to make the same noise. When about to change to pupae, they construct silken cocoons. *Bostrichus,* lives in fungi and under bark ; *Cis* in toadstools ; the larvae are fleshy white grubs.

The *Tenebrionidæ,* are apt to be confounded, by beginners, with Carabids, but the prothorax is much narrower than the abdomen, and the head is narrower still. Antennae clavate, feet short, of black or brown colors. The surface is smooth, in *Tenebrio,* or roughly corrugated in *Upis.* They are generally found under stones, logs, and in toad-stools. *T. molitor,* the meal worm, inhabits granaries. Ship bread is eaten by the larvae, which are "about an inch long, of cylindrical and lineal form, very smooth and glossy, of a fulvous color." *Blaps* is found in moist places; other genera, under bark; *Phaleria*, on the sea shore. *Boletophagus*, as the name suggests, lives in fungi.

Passing over several small groups we come to the

Mordellidae, which are wedge-shaped, small, glistening pubescent black beetles, which occur in abundance on the flowers of golden rod and asters, and when disturbed leap like fleas. The larvae of *Mordella* are found in the pith of plants in autumn, and are long, subcylindrical, the sides of the rings furnished with fleshy tubercles.

Meloidae. This and the following family are most interesting, from their parasitic habits, and demand careful study and observation. *Meloe augusticollis*, is an inch long, thorax very small, square; abdomen large and swollen; the elytra are small and oval. The antennae of the male are crooked in the middle. It is of a deep Prussian blue. It feeds on grass in the spring, in the summer it is found in the White Mountains, feeding on Clintonia borealis. The larva is very different from the beetle, and as found parasitic on wild bees, resembles larvae of some Staphylinidae, being oblong, flattened; the three thoracic rings above, of nearly equal size, transversely oblong, the head nearly of the same size, with short antennae; the legs have very long claws, with an intermediate long pad. From the tip of the abdomen proceed two pairs of setae of unequal length. They are found living upon bees between the joints of the head and thorax, their heads immersed in the dense scales of the bee. In Europe this genus has been found parasitic on Cetonia. Our *Cetonia Inda*, and other related beetles should be searched for them. The eggs are laid on the ground, and the active larvae attach themselves soon after hatching, to bees, and to the Syrphus flies, and Muscae.

Cantharis and our *Epicauta*, secrete cantharidine, of use in pharmacy. *E. atrata*, is found in abundance on golden rod, and it is perfectly black, with long elytra. *Rhipiphorus*, is parasitic on the wasp; *Ripidia* on *Blatta americana*, the cockroach.

Stylopidae. The larvae of this most anomalous family are much like that of Melve. They are oval in form. The perfect insects are not a quarter of an inch long. The elytra are pad-like, while the hind wings are greatly developed, expanding broadly, folding when the insect is at rest, along the body. They live but a short time in the perfect state. "They are parasitic in the bodies of species belonging to various genera of aculeate Hymenoptera ; the comparatively large size of these parasites, causes a distension of the abdomen of the Hymenopteron affected, and, on close observation, the heads of the pupa cases may be seen emerging between the segments. The head of the pupa case of the male is convex, that of the female is flat; specimens containing male pupae can be kept confined with proper food, until the parasite is hatched. *Stylops* inhabits bees, of the genus *Andraena.* I have never met with specimens. *Xenon Peckii,* lives in our common wasp *Polistes fuscata.* I have seen stylopized individuals of *Odynerus quadricornis,* and of a large species of Sphex."—*Le Conte.* Stylops has four joints, *Xenos,* six joints to the antennae. There is a species of Xenos, only found, thus far, in Nova Scotia, which must likewise occur in Maine. They are found at different seasons of the year, but mostly in April and May. They have been taken by sweeping grass in August.

The three following families are of great extent, and do great mischief to agriculturists, by the great variety in their modes of attack upon plants.

Curculionidae—(See Fig. 21.) This group is at once recognized, by having the head lengthened into a long snout, near the middle of which are situated the elbowed antennae. Their bodies are hard and round, and often very minute in size. The beetles are very timid, and quickly feign death. The larvae are white, thick, fleshy, legless grubs, with tubercles, instead of limbs, and armed with thick, arched, strong jaws. They feed on nuts, seeds, the pith of plants, leaves or flowers; while some are leaf miners, and others make galls. Before they transform they spin a silky cocoon.

Bruchus pisi is short and oblong, it lays its eggs on the pea, when in flower, and lives in the pea till the following spring.

Anthribus is parasitic in the body of Coccus. *Brenthus* inhabits the solid trunks of oaks. *Apion* inhabits the seeds of clover. *Hylobius pales* is found under the bark of the pine, where *Pissodes strobi* in all its stages occurs. *Rhynchaenus nenuphar* infests the plum. *Calandra granaria,* the grain weevil, is an eighth of an inch long, and consumes the interior of wheat. *Balaninus* forms galls on the willow. *Scolytus, Xyloteres* and *Tomicus* are cylindrical bark borers; "they form galleries in the bark, or sap wood, often causing the disease called fire blight."

Cerambycidae. The Longicorns are insects with long bodies, tapering behind; the elytra broader than the prothorax, the antennae and legs very long, and are large handsome beetles, often gaily ornamented. They fly in hot days about woods and timber. *Orthosoma cylindricum* flies into houses at light in the evening. *Prionus,* and allies, are large, dull colored, flattened beetles, which fly in the evening. The larva is broad and flattened, the head can be drawn in the prothorax farther than usual. It forms cocoons of the chips it makes. *Asemum* flies in hot days, often in great numbers.

Cerambyx, and allies, have the antennae very long, and are highly colored. They are found in trunks of trees, or flying clumsily among the leaves. *Clytus speciosus,* bores in the locust. *Saperda candida,* (Fig. 23.) is the apple tree borer. A species of *Staphylinus* is, in Europe, parasitic upon one of this genus. *Stenocorus putator,* the oak pruner, severs the twigs of that tree, by eating the wood under the bark, which the wind breaks off.

FIG. 23.

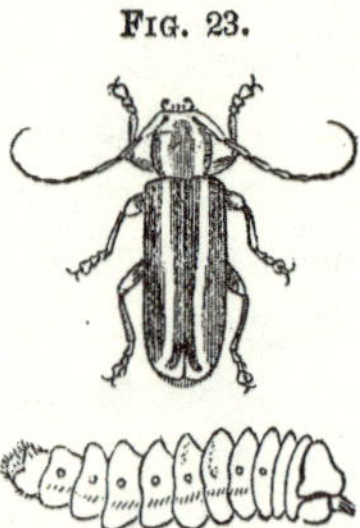

Leptura and the neighboring genera, narrow rapidly at the hinder portion of their bodies, the antennae are rather short, and they occur on flowers, such as Spiraea, &c. *Rhagium lineatum* has a flatted larva which can be found under the bark of pines, in large cells formed of its chips. *Desmoceres palliatus,* the "Purple cloak," is found boring in the pith of elders.

Chrysomelidæ. The insects of this family have hemispherical or oval convex bodies, with small heads sunken in the thorax, and live in all their stages on the leaves of plants. The larvæ have thick bodies, the rings composing it are very convex, and above marked with tubercles and thickened deposits; they are often gaily colored.

Donacia, which approaches the Cerambycidæ in its elongated body and long antennæ, lives as a larva in the stems of aquatic plants; the pupa is found in silken cocoons attached to the roots of the submerged plants. *Lema trilineata,* which closely resembles the squash beetle, devours the leaves of the potato. *Cassida,* or the Tortoise beetle, is round, depressed, and yellow. Its larva is broad and flattened, with lateral ciliated filiaments, and its abdomen is produced into a tail which it holds loaded with its excrement, over its back for purposes of concealment and defence. *Hispa* is a leaf miner, its minute larva making galleries in the leaves of the apple tree, and wild cherry. *Galeruca vittata,* the squash beetle, is yellow with black stripes. The different species of *Haltica* or flea beetles, are little, black colored, most hurtful insects, which destroy young tomatoes, turnips, &c. Several species of *Calligrapha* are found on alders, they are oval and richly ornamented with dots and curved lines:

Chlamys, which is an oblong square beetle, has its convex surface most curiously corrugated; as a larva it lives in a cylindrical case, on the sweet fern.

Coccinellidae (Lady bugs.) They are hemispherical, generally red or yellow, with round or lunate black spots. *Chilocorus* is black with yellow dots. The eggs are laid, often, in a group of plant lice, or Aphides; as soon as hatched the larvae devour them. When about to turn to pupae, they attach themselves by their terminal rings to the leaf they are upon. The beetle is as voracious as the larva. In Europe gardeners take pains to collect and put them on trees infested by lice, which they will soon remove. *Coccinella novemnotata,* (Fig. 24,) is a common species in gardens.

Fig. 24.

Orthoptera.

In studying these insects, the proportion of the head, of the prothorax, of the wings, of the hind legs, and the external genital parts, should especially be taken into account. The ornamentation varies greatly even in the same species, and therefore large numbers of individuals are necessary to ensure a proper knowledge of any species.

The transformation of grasshoppers need careful study. For this purpose their eggs should be sought for, and the development of the embryo in the egg be noted; also the following facts should be ascertained: the date of deposition of the eggs; the manner of laying them; how long before the embryo is hatched; the date of hatching; how many days the pupa lives; also so of the pupa and of the imago, while the intervening changes should be carefully observed. Crows and blackbirds feed on their eggs and larvae, and hens and turkeys feed greedily upon young and old. Ichneumon parasites prey upon them, and also the lower worms, such as Filaria, Gregarina and Gordius and red mites, attack them. Mud wasps provision their nests with their young.

Orthoptera can be easily preserved in strong alcohol, and can afterwards be taken out and pinned and set at leisure. They can be killed with ether or benzine, as coleoptera, without losing their colors, as they would do, after remaining long in alcohol. They should be pinned through a little triangular spot between the bases of the elytra or fore wings, when the wings can be spread to advantage. They are also often pinned through the prothorax, or through the right elytra, as in coleoptera. In pinning these insects for transportation, care should be taken to put in additional pins on each side of the abdomen, and in like manner to steady the hind legs, which are very apt to fall off if too much jarred.

The different sounds produced by Orthoptera should be carefully studied; every species can be distinguished by its peculiar note, and as in different families the musical apparatus varies, so each family has a characteristic chirrup, or shrilling, or harsh, grating, rasping noise.

Forficulidae, Earwigs. Narrow, flattened insects, very unlike other Orthoptera, with short wing covers, like the Staphylinids among beetles; terminal ring armed with a pair of very long incurved forceps-like horny pieces; nocturnal insects, hiding in the daytime between leaves and in flowers, flying about at dark. They feed on the corollas of flowers and on fruit; they will eat bread and meat, &c., and are very troublesome in Europe. Our species has not yet been found in Maine, though inhabiting other parts of New England. An Alpine species lives under stones in Europe.

Blattariae, Cockroaches. Also nocturnal, hiding by day, or as in the wild species, under stones, &c. They are fond of heat. While troublesome from eating garments, &c., they do great service in clearing houses and vessels of bed-bugs, which they prey upon. We have several species in New England which should be carefully sought after. They are found under stones, and are smaller than the house cockroach. They are oval, the head rounded and partially concealed, with long antennae. The fore wings are thickened, the anal stylets short. Color almost invariably a reddish brown. The eggs are laid in large bean-shaped capsules, which are divided into two apartments, each containing a row of separate chambers, about thirty in number, and each enclosing an egg. Many days are required for oviposition. An English writer has stated that in *Blatta* and a species of *Phasma* the larva and pupa state are undergone before leaving the eggs, so that the changes of the hatching belong to the imago state. Future observation must show whether this be generally the case in this suborder. Various Ichneumons feed on the eggs.

Phasmidae, Walking sticks. Our New England *Diaphomera femorata* is four inches long ; linear, wings minute, legs very long and linear, and is found in trees, rose bushes, &c. It is very sluggish and not easily distinguished from the twigs it may be resting upon. The eggs of this group are bean-shaped with scattered dots.

Gryllodea. Crickets are known by their dark colors, depressed oblong form, and long anal stylets, and by their long antennae. The female has an ovipositor nearly as long as her body. They are ground insects and fast runners. The male chirrups to attract the other sex ; the apparatus being a specialization of the membrane and nervures at the base of the wings, so that the rubbing of the wings one upon the other produces a rasping-like noise. The eggs are laid in cases, and the insects come to maturity in the fall. Our common black species is the *Gryllus neglectus.*

Gilbert White says of the English cricket : "When the males meet they will fight fiercely, as I found by some which I put into the crevices of a dry stone wall, where I should have been glad to have made them settle ; for though they seemed distressed by being taken out of their knowledge, yet the first that got possession of the chinks, would seize upon any that were obtruded upon them with a vast row of serrated fangs. With their strong jaws, toothed like the shears of a lobster's claws, they perforate and round their curious regular cells, having no fore-claws to dig, like the mole-cricket. Of such herbs as grow before the mouth of their burrow they eat indiscriminately ; and on a little platform which they make just by, they drop their dung ; and never, in the daytime, stir more than two or three inches from home."

The mole cricket, *Gryllotalpa,* live in wet, swampy soil, by ponds and streams, where they raise ridges, as they make their subterranean galleries in search of insects. Their fore legs are adapted like those of the mole for digging, and are stout and short, much flattened, and armed with solid, tooth-like projections. Their eggs are in a tough sack, containing two to four hundred, it is stated.

"As mole crickets often infest gardens by the sides of canals, they are unwelcome guests to the gardener, raising up ridges in their subterraneous progress, and rendering the walks unsightly. If they take to the kitchen quarters, they occasion great damage among the plants and roots, by destroying whole beds of cabbages, young legumes and flowers. When dug out, they seem very slow and helpless, and make no use of their wings by day; but at night they come abroad and make long excursions, as I have been convinced by finding stragglers, in a morning, in improbable places. In fine weather, about the middle of April, and just at the close of the day, they begin to solace themselves with a low, dull, jarring note, continued for a long time without interruption, and not unlike the chattering of the fern owl or goat-sucker, but more inward.

"About the beginning of May they lay their eggs, as I was once an eye-witness; for a gardener at a house where I was on a visit, happening to be mowing on the sixth of that month, by the side of a canal, his scythe struck too deep, pared off a large piece of tuft, and laid open to view a curious scene of domestic economy. There were many cavern and winding passages leading to a kind of chamber, neatly smoothed and rounded, and about the size of a moderate snuff box. Within this secret nursery were deposited nearly an hundred eggs, of a dirty yellow color, and enveloped in a tough skin; but too lately excluded to contain any rudiments of young, being full of a viscid substance. The eggs lay but shallow, and within the influence of the sun, just under a little heap of fresh moved mould, like that which is raised by ants.

"When mole-crickets fly, they move *cursu undoso*, rising and falling in curves." *White.*

Nothing is known about our New England species, of which we have more than in Enrope.

Œcanthus niveus, is very flat and broad behind, with long legs, and white, colored with yellow; the female is narrower and tinged with green. They live on grape vines, and are easily detected by their loud shrilling. They lay their eggs in the stems of plants, by perforating the stalks with their ovipositor, and they have been found thus perforating the branches of peach trees; they also feed upon the tobacco leaves. It has not yet been observed in Maine.

Locustariae, are large, generally broad-winged grasshoppers, with long, slender legs. The Katydid belongs to this family. It has not yet been found in Maine. But its allies which live in bushes and on trees, such as the large *Phaneroptera augustifolia*, and which make a loud, shrilling noise, are common.

Ceuthophilus maculatus, a wingless species, of a dark brown color, is common under stones; in other parts of the country they are found in caves. A different species from the *maculatus* inhabits the island of Grand Manan.

Acrydii. The common grasshoppers have large heads, rather short and thick antennae, thick, compressed bodies, and the prothorax projects backward conspicuously, aud is often divided by crosswise impressions. The hind legs are stout and thick, adapted to the leaping habits of the insects.

Locusta corallina appears for about two weeks in May in dry fields. *L. sulphurea* and *carolina*, the "quakers," are fall insects; so are the different species of *Chloeältis*, which survive the frosts till late into November. They produce their chirrping noise by rubbing their thighs on the wing covers. Red mites are frequently found sucking the juices beneath the wings. The species of *Tettix* are small, but prodigious leapers, and are characterized by having the prothorax carried out to the end of the abdomen. Toads and frogs devour large numbers of grasshoppers.

Hemiptera.

This suborder has been greatly neglected; these insects are not the favorites of entomologists. In studying the different groups the investigator is aided by the great variation in the general proportions of the body; in the shape and relative size of the head and its appendages. The species are subject to great individual variation, which should caution the student in drawing the limits between them.

Aquatic species should be taken out by the water-net by thrusting it under swiming species, or pushing it among submerged grass or weeds where small species are lurking. Several species of small size are found under logs, &c., in the water. By sweeping grass and herbage as for coleoptera in the last part of the summer, large numbers occur which can only be obtained in this way. Hybernating species are found under leaves in hard wood forests. The large carnivorous kinds are found on bushes frequently with lepidopterous larvae transfixed on their jaws.

The soft bodied species of *Aphis* and allies should be preserved in alcohol. These species should be carefully watched for their parasites, and can be easily kept in slender glass vials through which the insects can be watched. All hemiptera should be pinned through the distinct triangular scutellum in the middle at the base of the wings. The minute hard species of Tettigoniae, Thrips and small Capsidae, may be stuck upon cards as in the coleoptera. When on a collecting tour they can all be thrown into alcohol, and taken out afterwards and pinned and set.

Thripadae. This family has by some authors been considto form a distinct order called *Thysanoptera.* They are minute, narrow and flattened insects, very active in their habits, are found in flowers, especially composite plants, such as the Whiteweed, and when running over the hand cause a severe itching. There are two pairs of long narrow wings without any nervures, which are delicately fringed on the margin, and are laid one above the other over the body. The mouth parts are free, but the mandibles are like two bristles, the maxillae are flat triangular, bearing a pair of palpi. These parts are partially united into a conicle sucker which is folded upon the breast. The prothorax is largely developed ; the legs are short, and the elongated abdomen terminates in several long bristles which are closely united together.

FIG. 25.

Some species are wingless, being found under the bark of trees. They closely resemble their larvae (Fig. 25), which are found in the same situations as the perfect insects, and are distingushed from them by the uniformity of the three thoracic rings, and their similarity to those of the abdomen ; by their softer body, and shorter antennae and legs, and the want of simple eyes. They are often pale yellow, blood red and flesh color. The pupae have "the limbs obscured by a film, and the wings enclosed in a short fixed sheath. The antennae are turned back on the head, and the insect, though it moves about, is much more sluggish than in the other states."

The species are very injurious to flowers, eating holes in the corollas, and sucking the sap from the flowers of wheat, in the bottom of which they hide.

Cicadidae, commonly called "locusts," are large wedge-shaped insects, with a large broad head, prominent eyes, their ocelli on top ; wings transparent with thick veins. The males have a musical apparatus beneath the wings on the hinder ring of the thorax, which acts like a kettle drum, producing the loud, penetrating, shrill sound issuing apparently from trees. *Cicada rimosa,* our smallest species in Maine, begins to be heard a little before the middle of June. The *C. canicularis* is larger and comes later, being an autumnal species. Mr. Verrill has observed this species in

Norway laying its eggs in the stems of Solidago or Golden Rod. It made a longitudinal incision with ragged edges into the pith of the plant, then with its ovipositor forced its eggs some distance down in the pith below the outer opening; there were two rows of eggs succeeding the first single one, each pair diverging outwards, the lower ends of each pair nearly touching each other, and all the pairs were placed very near together. The habits of the seventeen year locust which does not inhabit Northern New England, are well described by Dr. Harris in his Treatise. The young larvae feed on the roots of the oak and apple, clustering upon the roots, and sucking the sap with their beak-like mouth.

Membranicidae. Antennae three-jointed; head broad, with two ocelli. The insects of this family assume the most grotesque forms. They are great leapers. *Ceresa* is broad, wedge shaped, green or brown color, and two species are found in great profusion in bushes in August and September. Different species of *Eutilia*, which are often notched upon the back, are found upon the stems of golden rods and birches, and closely resemble the surface they are upon. They lay masses of white eggs on the plants they frequent. *Clastoptera proteus*, convex above and in front and highly colored, is injurious to the cranberry in Massachusetts. It is a common Maine insect.

Tettigonidae.—Leaf-hoppers. They pass all their lives on the leaves of plants, inserting their beaks into the leaves and sucking the sap, thus causing the leaves to wither and also the twigs, producing what is called "Fire-blight," having much the same effect that the Scolytus produces.

The species of this family are very numerous, and are found hopping on leaves and herbage late in the summer, though a few species are among the earliest spring insects. There are some yellowish species found in moss and grass by the side of pools and puddles in woods just as the snow is going off. The eggs are laid in autumn to be hatched in the spring. A very abundant species on grass, producing what is called "frog spittle," can easily be traced through all its changes by frequently examining the froth which surrounds them. *Tettigonia vitis* is a tenth of an inch long, straw yellow striped with red; it lays its eggs in summer and hides among the dead leaves during the winter. *T. rosae*, a still smaller species, is found on the rose. As a family these insects are characterized generally by their oblong outline, being convex above, the head somewhat triangular or crescent shaped, the prothorax is large and of the same width as the body, and the legs are thickly spined.

Aphidae. Every thing about this extensive group is of the greatest interest, whether it be their structure, mode of growth or habits and relations to other insects. They have soft oval bodies, with two slender tubercles behind, with somewhat square heads and long slender seven-jointed antennae. The beak is often half as long as the body. They are generally colored green, and often have a soft bloom upon the surface. "The brief history of the general conditions of the development of these insects is as follows:—In the early autumn the colonies of plant-lice are composed of both male and female individuals; these pair, the males then die, and the females begin to deposit their eggs, after which they die also. Early in the spring, as soon as the sap begins to flow, these eggs are hatched, and the young lice immediately begin to pump up sap from the tender leaves and shoots, increase rapidly in size, and in a short time come to maturity. In this state it is found that the whole brood, without a single exception, consists solely of females, or rather, and more properly, of individuals which are capable of reproducing their kind. This reproduction takes place by a viviparous generation, there being found in the individuals in question, young lice, which, when capable of entering upon individual life, escape from their progenitors, and form a new and greatly increased colony. This second generation persues the same course as the first, the individuals of which it is composed being, like those of the first, sexless, or at least without any trace of the male sex throughout. These same conditions are then repeated, and so on almost indefinitely, experiments having shown that the power of reproduction under such circumstances may be exercised, according to Bonnet, at least through nine generations, while Duvau obtained thus eleven generations in seven months, his generations being curtailed at this stage not by a failure of the reproductive power, but by the approach of winter; which killed his specimens; and Kyler even observed that a colony of *Aphis Dianthi*, which had been brought into a constantly heated room, continued to propagate for four years in this manner, without the intervention of males, and even in this instance it remains to be proved how much longer these phenomena might have been continued." Dr. Burnett, from whom we quote, considers this anomalous way of increase of individuals as a process of *budding*, and that the whole series, like the leaves of a tree, constitutes but a single generation, which results from the union of the sexes in the previous fall. It has always been supposed that the final autumnal set of individuals were males and females alone. Hear Dr. Burnett again: "The terminal brood has hitherto been considered, as far as I am aware, to be composed exclusively of males and females, or, in other words, of perfect insects of both sexes. I was surprised therefore on examining the internal organs of the non-winged individuals, to find that many of these last were not females proper, but simply the ordinary gemmiperous form. Moreover so great was the similarity of appearance between these two forms—true females and gemmiperous individuals—that they could be distinguished only by an examination of their internal genitalia."

FIG. 26.

Aphides, (Fig. 26,) are found upon every part of plants. Some species which are wingless, are found on the roots of plants, others on the stems or twigs, others roll up leaves, or form gall-like swellings on leaves; the grain aphis sucks the sap of the kernel. Ants are fond of the sweet excretions from the abdominal stylets, and often keep them captives in their nests like herds of cattle. *Syrphus* flies, and *Coccinellae*, keep them within proper limits in nature. Minute species of *Aphidius*, small Ichneumons, kill larger numbers than we imagine. "When an aphis has received an egg from one of these parasites, it quits its companions and fastens itself by its ungues to the under side of a leaf, when it swells into a globular form, its skin stretched out and dried up, and in a short time the perfect parasite escapes by a circular hole, the mouth of which sometimes remains like a trap door."

Eriosoma lanigerum, the American blight, a wooly or cottony covered species, feed on the sap wood of the apple.

Coccidae, or bark lice, are scale-like in form like miniature oyster shells, and live on the bark of trees, or upon the roots. The males alone are winged and pass through the usual changes, while the female only increases in size, preserving its scale like form. "Early in spring the bark lice are found apparently torpid, situated longitudinally in regard to the branch, the head upwards, and sticking by their flattened inferior surface closely to the bark. On attempting to remove them they are generally crushed, and there issues from the body a dark colored fluid. By pricking them with a pin, they can be made to quit their hold, as I have often seen in the common species, *Coccus Hesperidum,* infesting the myrtle. A little later the body is more swelled, and, on carefully raising it with a knife, numerous oblong eggs will be discovered beneath it, and the insect appears dried up and dead, and only its outer skin remains, which forms a convex cover to its future progeny. Under this protecting shield the young are hatched, and, on the approach of warm weather, make their escape at the lower end of the shell, which is either slightly elevated or notched at this part. They then move with considerable activity, and disperse themselves over the young shoots or leaves." *Harris.*

The *cochineal* is prepared from the coccus that lives upon the cactus. In Canada a dye of equal value has been prepared to some extent from a native species of this genus. The minute scales secreting wax that cover certain species in the East Indies, enable the natives to prepare the different varieties of shellac.

The preceding families belong to the order *Hemoptera* of many writers, but it is difficult to draw the line between the two groups of families. As a general thing the following families have the head smaller, the antennae long, and the base of the fore wings thickened; the beak is longer; many of the species are carnivorous. These have by one author been divided into *flower-suckers* and *blood-suckers.* When disturbed they emit a disagreeable odor, and small species are often eaten with fruit, producing a particularly offensive and lasting taste. Various genera, such as *Velia, Gerris* and the *bed-bug,* often have no wings when merely perfect insects but pads instead, as all hemipters have when in the pupa state; but as the functions of reproduction are carried on, they have by some writers been called different species from the fully winged individuals.

Notonectidae, or water-boatmen, are like *Tettigoniae,* but their legs are ciliated and formed for swimming. The different species of *Corixa* are common in every pool. Their motions are rapid, diving suddenly to the bottom and holding on to submerged objects when disturbed. They fly well, but walk with difficulty.

Nepidae. This group comprises, among others, two singular genera. *Belostoma,* containing the largest species in the suborder, often measuring three inches in length. They may be seen in winter swimming beneath the ice of ponds. *Ranatra* is long linear, a water walking-stick. The head is small, the forelegs enlarged and adapted for seizing insects, as they creep about the roots of aquatic plants.

Hydrometridae. The genus *Gerris* which represents this family in Maine, is long, narrowing alike towards both ends, being shaped like a wherry, and with their long legs they run over the surface of ponds and streams, moving backwards and forwards with great facility. They are among the earliest spring insects.

The following families are terrestial, living for the most part on plants :

Reduviidae. Insects with rather long, somewhat flattened bodies ; the beak is much curved ; the head is narrowed behind ; the eyes are very prominent, and the prothorax is much raised in the middle, with a thin, often serrated ridge. The European *Reduvius personatus* feeds on bed-bugs, its larva and pupa concealed in a case of dust, the better to approach their prey. *Ploiaria* is very narrow, with very long legs ; it is common in gardens, and is found as late as the middle of November. *Nabis ferus* is stouter, and very common in gardens.

Pentatomidae. This is a large family of insects, of bright colors, and often of large size. The head is received into the large, broad, short prothorax, and the scutellum or the triangular piece at the base of the wings is large and distinct ; they are generally oval in form. They are found in shrubs, sucking the leaves, or often seizing some caterpillar with their hooks. De Geer describes the eggs as being generally of an oval form, attached to leaves at one end by a glutinous secretion, the other being furnished with a cap, which the larva busts off when it hatches out. The larvae are rounded oval.

Coreidae. These insects are narrower than the preceding group; they are flat above, and beneath convex. They run and fly well, their habits being generally very active. They are the most gaily colored, perhaps, of hemiptera. The larvae differ very little from the perfect insects. They are found on plants, or at their roots. *Phytocoris lineolaris* is is our most abundant and injurious insect. It appears early in spring. *Coreus tristis*, the squash-bug, (Fig. 27,) collects in numbers around the stems of squash vines next to the roots.

Fig. 27.

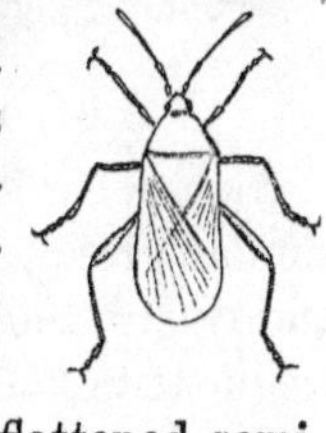

Tingis hyalina represents another family of broad, flattened semi-transparent hemipters. The *hyalina* is very abundant on the willow early in summer.

Capsus is the type of another family, which consist of small species, with soft, rather narrow bodies, and long beaks and legs. They are very active, flying readily. They are found in flowers, and on fruit, such as raspberries.

Cimicidae. The bed-bug, (*Cimex lectularius*,) has a small, somewhat triangular head, orbicular thorax, and large, round flattened abdomen. It is generally wingless, having only two small wing-pads instead. The eggs are oval, white; the young escape by pushing off a lid at one end of the shell. They are white, transparent, differing from the perfect insect, in having a broad, triangular head, and short and thick antennae. Indeed, this is the general form of *lice*, to which the larva of Cimex has the closest affinity. Some Cimices are parasites, infesting pigeons, swallows, &c., in this way also showing their near location to lice. The cockroach is the natural enemy of the bed-bug, and destroys large numbers. Houses have been cleared of them after being thoroughly fumigated with brimstone.

Pediculi, Lice. These degraded, wingless forms of Hemiptera, still preserve the mouth parts in the form of a sucker, but it is fleshy and retractile. The triangular head has two simple eyes. The body is rather long, the abdomen oval. They are generally white, and of minute size. The metamorphosis is very incomplete —that is, there are but slight differences between the larva and the imago. The species of *Pediculus* are blood-suckers, and parasitic upon Man and some of the Mammalia; different species being found upon different regions of the body. Different varieties are found living upon the bodies of different races of men.

Mallophaga, bird-lice, live on hair of mammalia and feathers of birds. In this group there are distinct jaws. Nearly every bird and mammal has its parasite, so that the number of species is actually very large.

Neuroptera.

As a suborder these insects are the most aquatic of any other similar group, and it is swampy low grounds, the banks of pools and rivers, the thick dense damp forests, that the collector must frequent to find them. The large Dragon-flies when taken by the net must be killed by brushing with alcohol or benzine carried in a vial, and then the wings can be folded together and the insects be placed in bags, or pieces of paper, as directed for putting up Lepidoptera. The smaller, more slender and delicate species should be pinned directly in the collecting box, &c. Many species are caught at light in the night time, such as *Polystoechotes nebulosus* and the *Phryganeidae;* and a bright light placed in damp situations by streams, &c., will attract large numbers. Like moths the smaller species are attracted a great distance by light. Other species of this family so numerous in New England, are found in great numbers floating in the lakes and ponds of the wild lands of Maine that are rare elsewhere. For the proper study of the genera of these insects, and often of the species, they should be collected in alcohol, so as to be studied in a flexible state.

The aquatic larvae and pupae can easily be reared in aquaria in jars and tumblers, taking care that the weaker species are separated from those more powerful and bloodthirsty. The little entomostraca or water-fleas serve as food for the smaller species. With very little care many species can be raised in this way, and so little is known of their transformation that figures and descriptions would be of great value. The interesting and varied habits of the different families can also easily be noted. They can be called summer insects, since few are found late in the fall or early in the spring. Hemerobius and several species of Phryganeids are found ere the snow has gone in the spring,—a few species of the latter family are found in November.

Termitidae. White ants, so called, from their resemblance to ants, and the snowy whiteness of their wings, and the pale colored female, like the true ants, are social, living in communities; while the majority are wingless males, often called neuters. In the winged individuals the wings are much larger than the body, being folded, when at rest, one upon the other. The wingless individuals have an enormous head with scissor-like mandibles. The American white ant, *Termes frontalis,* has been found in Massachusetts ruining the roots and stems of the grape vine. The insect is careful to conceal its work by leaving the outer crust intact. It feeds on dead wood, eating the inside of the sill of the house next to the grape vine.

Psocidae. These little insects when winged, as most usual, and flying about in August, have a remarkable resemblance to Aphides. The body is soft and short; the head is broader than the thorax;

the wings are broad, the second pair much smaller than the first, both having raised nervures ; the prothorax is very short.

Atropos divinatorius is the little wingless louse-like insect always running over the leaves of books, and about dusty places, and they feed on cabinet specimens, sometimes doing considerable injury. These little soft insects should be gummed on pieces of cards, or put into alcohol; while the winged species can be pinned with small pins.

Phryganeidae. (Caddice-flies, Case-worms.) The imago has a rounded body, with moderately broad, parallel veined wings, which are folded on the sides of the body, and the head is provided with long antennae and palpi. Both larvae and pupae are active. The smaller species are often hardly distinguishable from many small moths. The female lay their eggs in gelatinous masses on aquatic plants, above or beneath the surface of the water. The larvae are found abundantly in the bottom of ponds, in cylindrical cases of grass or stems of reeds, or bits of sticks, sand, minute shells, &c. They assume different forms, sometimes a long, conical shape, or imitating snail shells. The larva lines the interior with silk, and by bristles on the side of the body and a pair of anal hooks keeps its body adhering to the sides of the case while it drags it over the bottom. They eat large quantities of minute water fleas (entomostraca) and small insects, while many are herbivorous, the larger ones eating whole leaves that have been submerged, while the smaller ones leave the veins entire. When about to change to pupae, the larva closes up the mouth of its case with a net-work like a grate for the passage of the water for respiration. When about to leave the pupa state they crawl up stems of plants, or the smaller species use their light cases as rafts to rest upon as their wings are drying.

Fig. 28.

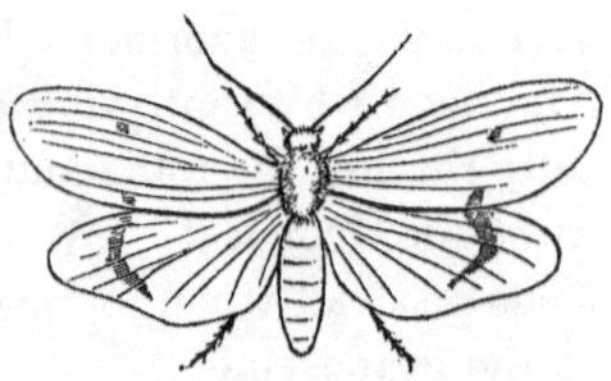

Neuronia semifasciata, (Fig. 28,) is our largest species, and is taken away from damp places; but the smaller species are only taken on leaves of bushes and herbage by streams and ponds. They run swiftly, but fly with some difficulty. The species are numerous. They should be pinned as moths, and their wings set carefully.

Perlidae. Long, flat neuroptera, whose hind wings are largest, the abdomen with two terminal long filiform appendages. The females of Perla are shorter and have much smaller wings than the males. The pupae are active, with prominent wing-pads, they are found in rivers under stones, while the imagines fly on the bank, or are found resting on leaves, always in damp low situations. *Pteronarcys* is distinguished from other genera by its large size, and possession of several pairs of outer tufts of filaments serving as organs for respiration.

Myrmeleon, the *Ant-lion* is the type of another family, very carnivorous in their habits. They resemble the Libellulidae very much except in having long antennae. The larvae, on the contrary, bear a close resemblance to that of Chrysopa figured below. It makes a pitfall in sand in which it hides, only showing its large jaws open to seize any insect that may fall into them. These insects have not been found in Maine.

Hemerobiidae. Aphis-lions, Lace-winged flies. *Chrysopa,* here figured, has a slender body, delicate, gauze-like wings, and is generally green, with golden eyes. When disturbed it throws out a fetid smell. They are very abundant in summer wherever plant lice are found, laying their eggs placed on long pedicils on leaves. The larvae (Fig. 30) feed ravenously on the lice, and when other food is wanting, on each other. They turn to pupa late in summer and pass the winter in that state. Gardeners in Europe search for these Aphis-lions and put a pair or two on trees overrun with lice which they soon depopulate. *Hemerobius* proper, has broad pale rings, and is of smaller size than Chrysopa.

Fig. 29.

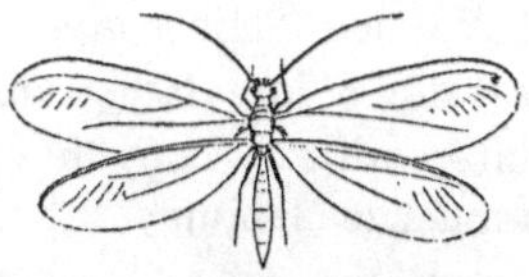

Fig. 30.

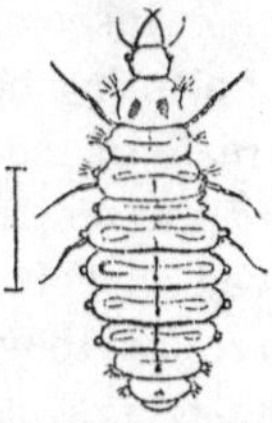

Sialidae. This group comprises aquatic, sluggish insects of moderate or of immense size. They have large heads with large jaws, square thoracic rings; and the abdomen in Corydalis cornuta has long anal filaments. This genus expands five or six inches, and the head is armed with immense horns, besides the long antennae, while the long wings are folded horizontally. In *Sialis americana,* an insect not an inch long and found resting on leaves of trees in their perfect state, the wings are deflexed on the sides, as in Chrysopa.

Panorpa represents another family, which have the head long and narrow, wings narrow and banded, and the tail armed with a forceps-like apparatus. It is common in woods and feeds upon other insects.

Libellulidae. Dragon-flies. Devil's-darning-needle. Musquitoe-hawks. *Demoiselles* in France. The head is large and globular, eyes immense, encircling the head; thorax square, wings large net-veined, equal; abdomen long linear, cylindrical. They are continually flying over pools, hawking for smaller insects in hot summer days, flying often till dusk. Though dreaded by most persons, they are perfectly harmless, though giving a sharp bite with their powerful jaws when held in the hand. They are difficult to kill, and should be brushed with alcohol or benzine, or killed by ether. The *Agrionidae* are small slender species of graceful form, and blue, green or bronze or red colored, flying away and alighting upon rushes in the water, and are easy to catch; they must be pinned carefully, and are very brittle when dry. The large species are hard to catch; patience and swiftness in the use of the net will soon render the beginner dexterous. These insects have also their *subimago* state. They should be described in life, as the colors fade rapidly after death. The larvae (Fig. 31) are interesting. They have large jaws, marked by an immense labium, otherwise the mouth parts are much like grasshoppers, &c. The larva of Agrion is slender and long, with thin caudal lanceolate plates. They all walk over the bottom in search of other insects, and propel themselves more rapidly by ejecting behind them, with considerable force, a stream of the water that has been used for respiration.

Fig. 31.

Ephemeridae or May flies, as their name implies, are very short lived insects. They have weak slender bodies, obsolete mouth-parts as they take no food in the perfect state, minute antennae, the wings are very unequal in size, and the abdomen has two or three long appendages. The May flies soon after leaving the pupa case with their wings of full size, cast off a thin pellicle. This moulting is attended by a change of color and of increase of length of the tail-like appendages, and this period is called the *subimago* state. They fly towards evening in large numbers. The larvae while resembling the imagines, have long antennae, mandibles for chewing, lateral ciliated filaments along the sides of the body for breathing organs, and three caudal filaments. They live, it is stated, two or three years. They either live in burrows, under stones, or among grass and weeds, when they may be taken with the water net in great abundance, and are beautiful objects for aquaria. The perfect insects should be preserved in alcohol for study, as they shrivel up when pinned. They should be described when alive if possible.

Thysanura.—Spring-tails. These interesting, minute, wingless forms, which seem to afford a passage into the Myriapods by the uniform size of their ings, which form a continuous series, from their head to the extremity, without showing the usual divisions into three divisions of the body, seem to be but a degraded form of neuroptera by their resemblance to the larvae of Perla and Ephemera; for like them they have long antennae, distinct jaws and maxillae, and also caudal setae or bristles on the terminal ring of the body. Their limbs also strongly resemble those of Perla. Moreover they undergo no metamorphosis, the larva gradually assuming the adult form by successive changes of their skin. The species are found abundantly in moist, dark places, under sticks, stones, among fallen leaves, or under bark of trees, while some occur in great profusion about manure heaps and hot beds in early spring.

Podura. This genus is rather broad, the body is hairy with a few scales, antennae short and few jointed; the head is separate from the thorax, and the abdomen is provided with setae converted into a forked tail bent beneath the body, used for leaping to a great distance. They are found in gardens, hot beds, or leaping on the surface of the water in quiet pools.

FIG. 32.

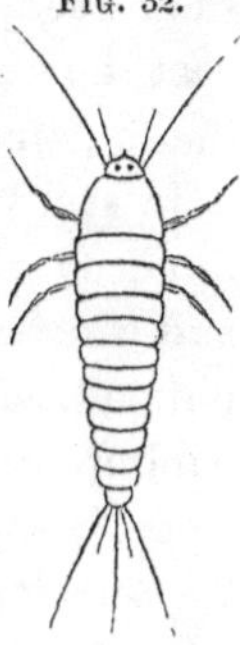

Lepisma, (Fig. 32,) is long, and covered with minute silvery scales; the antennae are rather short, and the abdomen has three long bristles. The species run rapidly and are found in old books, in woolen cloths which they eat, in mould and under bark, &c.

ARACHNIDA.

Spiders have no antennae. Their legs present seven distinct joints, and the tarsi are two jointed. At the base of the mandibles is a vesicle filled with poison, which can be poured into the tips of the jaw, and thus poison the insects bitten by the spider. This bite, except in rare instances, is harmless to man. "Scorpions are viviparous. With the egg-laying spiders, the egg, under the changes of development, slowly loses its previous form, and almost assumes that of a spider, indicating all the external parts of the enclosed animal. At length the shell bursts on the thorax, and the spider, first with the head, and afterwards with the thorax, comes to view ; then follows the abdomen, to which, however, the egg-membrane, like a scale, continues attached for a time ; then comes the feelers and feet. The young spider, through whose integument the granules of the yolk may be clearly distingished, is not yet in a state to weave a web and catch its prey ; for the spinning organs are still concealed beneath the common integument. After the lapse of a week, or, in some species, a longer time, during which the spider takes no food, it casts its skin for the first time, and is, as it were, born for the second time. The young spiders now quit, on some mild day in May or June, the web in which the mother had hidden her eggs ; they allow themselves to fall on the ground by a thread, and begin at once to weave their nets, or in some other way, according to the instinct of their kind, to watch for small insects corresponding to their age and powers.

"Most arachnids feed on other animals, which they either swallow alive, or whose blood and fluids they suck. Usually after their escape from the egg, they undergo no metamorphosis. They cast, however, their skin more than once, and are commonly after the fourth or fifth moult, in a state for pairing." *Van der Hœven.*

In studying spiders, of which we have in New England over two hundred species, the number and relative situation of the eyes, and the relative length of the different pairs of legs should be noticed. Their web and the manner of constructing them ; their habitats, whether spreading their webs upon or in the ground, or in trees, or on herbage, or whether the species is aquatic, or whether the species is erratic, and pursue their prey without building webs to entrap them, should be observed. So, also, how they deposit their eggs, and the form and appearance of the silken nidus, and whether the female bears her eggs about her, and how this is done, whether holding on to the egg-sac by her fore or hind legs, should all be carefully noticed. Care must be taken not to mistake the young for full-grown, mature species, and describe them as such. Spiders can reared in boxes as insects. The only way to preserve them is to throw them into alcohol ; when pinned, they shrivel up and lose their colors, which keep well in spirits.

The colors of spiders vary much at different seasons of the year, especially during the frosts of autumn, when the changes produced are greatest. All spiders are directly beneficial to agriculture by their carnivorous habits, as they all prey upon insects, and do no harm to vegetation. Their instincts are wonderful, and their habits and organization worthy of more study than has yet been paid them. We have no species poisonous to man, except when the state of health renders the constitution open to receive injury from their bite, just as musquitoes and black flies often cause serious harm to some persons.

The Arachnids are divided into two groups of families: First,

PULMONARIA,

which have pulmonary sacs for respiration, and six to twelve ocelli. This group includes two families, one consisting of the true *Spiders*, the other of the *Scorpions*.

Araneidae. Spiders. Palpi simple pediform; mandibulae armed with a moveable and perforated claw, emitting a poisonous liquid. The genera have been divided by Walckenaer:

1. Into those that incessantly run or leap about the vicinity of their abode to chase and catch their prey. *Mygale* hides in holes in the ground or among stones. The largest spiders are found in this genus. *Filistata* forms white silk tubes in walls and crevices of rocks. *Dysdera* is found in silken tubes under ground. *Segestria* makes silken tubes under the bark of trees. *Lycosa* is found under stones, in holes, &c., bearing their cocoons attached to their anus, and carrying their young on their back. The Tarantula of Italy belongs to this genus. *L. fatifera* lives in holes nearly a foot deep. These holes seem to be dug by the spider, and to be increased gradually, as its size may require; the opening has a ring of filaments woven by the spider to prevent the filling up of the cavity by rain. In *Dolomedes* the female of one species constructs a web not unlike that of *Tegenoria*. They wander near streams or ponds, often hiding under the surface of the water, or rambling on trees. *Sphasus* makes no web, except when the female makes her cocoon. The species wander in quest of prey about the trunks of small trees, or upright trees, and when at rest, spread their feet like many species of Thomisus. I have reason to think that the young are carried on the back of the mother as in Lycosa. (*Hentz*, in whose words most of these remarks are given.) *Attus* leaps prodigiously after its prey. Some species closely resemble ants.

2. Into those species which wander abroad and are incessantly

spying out for prey. No fixed residence except at the period of oviposition. They also walk and run sideways or backwards; occasionally throwing out threads to entrap their prey. *Thomisus* wanders after its prey on flowers, rails, trees, &c.

3. Into those that prowl about the neighborhood of their nests, or near the threads which they throw out to catch their prey. *Clubiona* forms silken tubes in leaves which they twist, or under the bark of trees. Most species fly about in the air, by means of a long thread, at the end of which they suspend themselves, and which is borne by the wind, sometimes raising them to a great height. *Herpyllus* makes no web or tube for its dwelling, but wanders for its prey, and runs with great velocity. *H. atec* is a small black species found under stones in highways; *H. ecclesiasticus* is blackish with a white band on the head-thorax, a band on the abdomen, beginning at base and reaching the middle, and a spot near the apex white. This one attains to a great size, and is found in houses, under stones, planks, the bark of decaying trees, &c.

4. Into those which spin large webs to entrap their prey, lying in wait in the middle or at the side. *Agelena* makes in the fields a web which is spread horizontally, and at the upper part of which is a tube for the retreat of the spider. *Theridium* makes a web formed of threads crossed irregularly in every direction. Most species of this genus are the common prey of the several species of *Sphex*, called *sand-daubers*. *Pholcus* inhabits the ceiling of houses. *Tegenaria* makes in houses, cellars and other dark places the common webs, which are spread horizontally, and have a tube, usually concealed in a hole or crevice, for the reception of the spider. This is the common house spider, the web of which is narcotic and has been administered internally in some cases of fever with success. *Epeira* is the common large grey species with a full round abdomen which makes its large circular web in corners, &c.

5. Into those which swim in water, and then spread their filaments to entrap their prey. *Argyroneta* lives in fresh water. "One species spins a bell-shaped, water-proof web that is filled with air, and open below; this it attaches to water plants by threads." We have a species perhaps of this genus that collects on the leaves of water plants, and when disturbed plunges to the bottom, carrying with it a bubble of water. We have one species of spider which makes a noise somewhat resembling the purring of a cat; during the production of the sound the body makes a tremulous motion it is said.

The second group of families, which is called

TRACHEARIA,

embraces those arachnids which breathe by means of tracheae, or air tubes and do not have more than four ocelli.

Pseudo-scorpionidae. This family includes *Chelifer*, a small scorpion-like animal, which has a large, broad, flattened abdomen, distinctly ringed; and the palpi are much enlarged, bearing a claw at the extremity much like that of a lobster. A species is very common in books and dusty boxes, drawers, &c.

Phalangidae. Harvest-men, Daddy long-legs. The common long, slender legged, round, oval-bodied spiders, so abundant everywhere out of doors in corners and damp places, and often called by the names above given, are known to every one. The legs come off easily, and when separated from the body for some time show considerable irritability.

Acarina. Mites have the head-thorax joined in a mass with the abdomen, and not divided apparently into rings. They are all of small size, some very minute. *Trombidium* has two horny mandibles, which are clawed at the end, included in the labium, which in the mites surrounds the mouth parts, thus forming a tube-like organ. This genus includes the little square velvet red mite, seen generally in the spring in flower beds, or in moss, &c. Another similar kind of red mite is common about decaying matter under stones and sea weed between tide marks on the sea shore. They are mostly parasitic, such as the itch mite. *Ixodes*, the tick, lives in woods and attaches itself to animals. Many species (*Gamasus*) are found on insects, especially beetles. The species of *Hydrachna* live on water-bugs, &c. In coming to maturity it passes through forms which have been described as distinct genera by authors. They should be preserved in small vials of alcohol, or mounted for the microscope.

MYRIAPODA.

All the species, of which we have but a few in New England, live hidden under stones and sticks, leaves, &c., The larvae when hatched have generally nine rings which afterwards increase in some cases to eight times that number.

The families are divided into two suborders, of which the first, the *Chilopoda*, comprises those myriapods which have the body flattened, with a limited number of rings, each of which has a single pair of legs articulated to the sides, of which the last pair is largest and extended backwards. The antennae are long and with numerous joints.

Lithobiidae. Lithobius, (Fig. 33,) called in this country Ear-wig, is our most common genus, and is found every where, under sticks and about manure heaps, where they feed upon insects and earth worms, and are in turn devoured by the red back salamander. The head is large orbicular, antennae forty-jointed, long and filiform, and there are sixteen rings in all. They are fast runners.

Fig. 33.

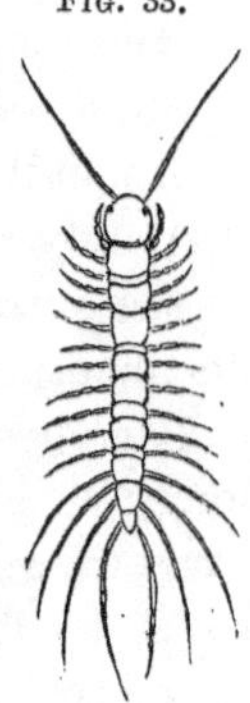

Scolopendridae. *Scolopendra,* the Centipede, has twenty rings besides the two that form the head; antennae 17-20 jointed. A rather slender species about three inches in length, is found in Maine, under dead leaves.

Geophilidae. Geophilus is greatly elongated and slender, with many rings, from thirty to two hundred. A small, slender species, is common under leaves, and debris of freshets, where so many varieties can be found.

Those Myriapods included in the second suborder, *Chilognatha,* have a greater number of rings, each of which bears two pairs of legs, and few jointed short antennae. In Polydesmus the body is still flattened and the legs articulated upon the sides of the body. A species occurring in considerable abundance with the myriapods is about an inch long and of a pale brown color.

Julidae. (Thousand-legs.) *Julus* is found commonly under sticks, &c. It is long, cylindrical, hard, with numerous feet, short and weak, attached to the under surface of the body nearly in the middle of the abdomen. The antennae are short and filiform. They crawl rather slowly, and at rest curve the body into a ring. They live on vegetable substances, or eat dead earth worms or snails. "In the spring the female deposits her eggs in masses of sixty or seventy, in a hole excavated for the purpose under the ground; after three weeks or more the young make their appearance, but still continue to adhere for some days by a string to the shell, which has burst longitudinally without motion, and surrounded with a proper membrane; at that period they have no legs at all; as soon as they have got three pairs of feet, they separate themselves from the shell; they have now a great resemblance to the larvae of some Coleoptera; soon the number of rings and feet begins to be [periodically] increased in that part of the body which is seated in front of the penultimate ring." *Van der Hoeven.*

Entomological Journal. Every collector should keep a daily diary of his captures and observations, noting down every fact and hint that falls under his notice. In this book, commenced as soon as the season opens in early spring, can be placed on record the earliest appearance, the time of greatest abundance, and the disappearance, of every insect in any of its stages. Also the descriptions of larvae and observations upon their habits, with sketches of them ; though drawings had better be kept upon separate pieces of paper for easier reference. The insects when captured and unnamed, should be numbered and refer to corresponding numbers in the note book. At the close of the season one will be surprised to see how much material of the kind has accumulated. He can then make a *calendar of appearances* of perfect insects and larvae, so as to have the work of the next season portioned out to him ; he will thus know when and where to look for any particular insect or caterpillar.

Cabinet. After the insects have been thoroughly dried they may be transferred to a chest of drawers of a convenient size, say eighteen by twenty inches and two and one-half inches deep, corked upon the bottom and glazed above, and thus rendered as nearly air tight as possible to keep out Dermestes, mites and moths. The insects should be arranged neatly in rows, labelled or numbered with small pieces of paper attached to the pin. Enough individuals should be selected to illustrate the sexes and variations of the species. Boxes three inches or more deep and twelve by eighteen inches square, rabbetted around the edges rather broadly, are very convenient. Cork in sheets can be had of R. Beeching & Co., Commercial St., Boston. It can be cut in strips or the whole surface covered and fastened down with glue, or better still with tacks. The pith of elder, corn stalk, or felt or palm wood, are substitutes for cork. For transporting specimens for exchanges, they should be pinned securely in boxes, lined with compact cork, and the boxes wrapped in cotton, covered thickly with paper, and then placed in a larger box.

For guarding with some success against the attacks of insects, the drawers should be provided with camphor or little bottles containing spirits of turpentine or benzine, to be kept always full.

Specimens can be relaxed by exposing them to steam or hot water. Lepidoptera can be softened and their wings expanded, after having been laid on moist sand for a few days, or confined in a vessel of warm water on the surface of which they can be floated on bits of cork.

The *strongest alcohol* is necessary for preserving insects; and when a bottle has been filled, the old alcohol should be poured out and kept for other collections, and its place filled by fresh alcohol.

When the collector has no box with him his captures can be wrapped in papers or stuck on his hat, or in the lining within. Lepidoptera can be very easily laid in papers a little longer than broad, which should be so folded that the opposite corners can be laid one upon the other, leaving a margin on the under side which can be folded upon the upper side, thus making a triangular paper case, in which the insect soon dries. In this way many specimens can be easily transported.

Entomological Works.

The best introduction to American Entomology is the new edition of Dr. Harris's Treatise on Insects. It not only classifies and describes many of our New England insects, illustrating them with colored engravings and wood cuts in great profusion, but is of special value to farmers, from the great amount of information about the habits of noxious insects. Dr. Fitch's Reports on the Noxious and Beneficial Insects of New York, with some illustrations, and accounts of the habits of many insects not especially noticed in the former work, is a very necessary book to have. Kirby and Spence's Introduction to Entomology, and Westwood's Introduction to the Modern Classification of Insects, are still more general works, almost indispensable to the beginner.

Very many of our American insects have been collected by Europeans, and described by their entomologists in the transactions and proceedings of learned societies, which are to be found only in our large libraries. There are also many large and expensive general works, including those of Linnæus, Fabricius, Count De Geer, Palisot de Beauvois, Drury, Bosc and Coquebert, which include many North American species.

St. Fargeau, Newman and Haliday, in the Entomological Magazine; Smith in the British Museum Catalogues of Hymenoptera, and M. De Saussure in his Monographs of the Vespidae, have described many of our hymenopters. Hübner, Cramer, Madam Merian, and more lately Herrich-Schæffer, Doubleday and Westwood, have published large illustrated works, containing many of our Lepidoptera. Guéneé has published five illustrated volumes where hundreds of our moths are first described. Likewise, for the Diptera, the special works of Desvoidy, Macquart, Meigen, Wiedemann, Zetterstett and Loew, are necessary to identify North American flies.

For Coleoptera, which have been largely described abroad, the standard authors are still more numerous. The names of Aubé, Bonelli, Erichson, Dalman, Dejean, Illiger, Klug, Knoch, Eschscholtz, Forster, Germar, Gravenhorst, Guérin, Hope, Lacordaire, Newman, Paykull and Schönherr, can only be mentioned. Burmeister in his Hand-book of Entomology has described many of our beetles, Orthoptera, Neuroptera and some Hemiptera. Stoll, Herrich-Schaeffer, Hahn and Haliday have also described more Hemiptera. Serville, in his Natural History of Orthoptera, mentions many American grasshoppers. There is also the general work of *Rambur* published like those of St. Fargeau, Macquart, Guéneé and Serville, in the Suite à Buffon in Paris, with those of De Selys Longcamp on Libellulidae. Pictet has written on the Perlidae and Ephemeridae, while several papers of Hagen treat of the Neuroptera. The British Museum is publishing catalogues of the various suborders containing great numbers of American insects.

Of those works treating of American insects exclusively, the rare and costly work of Smith and Abbot on the Rarer Lepidopterous Insects of Georgia, delineates the metamorphosis of many southern butterflies and moths. More lately Boisduval and Le Conte issued an Iconography of North American butterflies, giving drawings of the metamorphosis of many species. This important work leaves the Hesperidae unfinished. In 1817–18, Thomas Say published his American Entomology, which includes insects of all the suborders, in three finely illustrated 8vo. volumes, accompanied with a glossary. This, with Say's miscellaneous papers, which chiefly appeared in the Journal of the Philadelphia Academy of Natural Sciences, have been re-printed under the care of Dr. Le Conte. Through the Transactions of the American Philosophical Society, the Journal and Proceedings of the Philadelphia Academy of Natural Sciences, the Annals of the New York

Lyceum of Natural History, the Proceedings and Journal of the Boston Society of Natural History, and the Proceedings of the Philadelphia Entomological Society which has lately been established, are scattered memoirs and tracts by Melsheimer, Ziegler, Hentz, Harris, Haldeman, and the two Le Contes, which are mostly upon Coleoptera. Dr. Randall published a paper describing many new beetles from Maine in the Boston Journal. Dr. Clemens has published in the Philadelphia Journal a synopsis of North American Sphingidae; and in the Proceedings of the same Society descriptions and notes of the habits of the small moths. Mr. Scudder has printed in the Boston Journal "Materials for a monograph of North American Orthoptera;" and Messrs. Uhler and Walsh have writen upon the Neuroptera of the United States. Mr. Norton has described in the Proceedings of the Boston Society descriptions of new Hymenoptera. Baron Osten-Sacken has printed in the Phil. Proceedings an elaborate paper on the *Limnobiae*, a group of Tipulidae, and also his researches on Gall-flies and their products.

The insects of British America have been treated of in Kirby's Fauna Boreali-Americana. This well illustrated quarto volume is of special value, since it describes so many insects which are found in Maine. In the New York State Natural History Reports, is a quarto volume, with many plates illustrating the injurious and beneficial insects of that State, by Dr. Emmons. Mention should also be made of the writings of Mr. Townsend Glover on the Cotton and Orange insects of the Southern States, which appeared in several volumes of the Patent Office Reports, and of several papers by Le Conte, in the Reports of the Pacific R. R. Exploration, and Stansbury's Report on the Salt Lake.

There is still needed a general work to combine these scattered materials, and the results of further investigations. The Smithsonian Institution is in a great measure supplying this deficiency, and promoting a zeal for these studies that is being manifested throughout the country. Catalogues of the Lepidoptera, and also a compilation of all the descriptions of the Lepidoptera of North America as far as the Bombyces, by Dr. Morris; of the Diptera by Baron Osten-Sacken, with a treatise on the *Cecidomyœ* and their galls; together with Monographs of several Dipterous families by an eminent European Dipterist, M. Loew; and of the Coleoptera by Dr. Melsheimer, revised by Le Conte and

Haldeman, and also a work entitled the Classification of the Coleoptera of North America, (Part I,) by Dr. Le Conte, together with a synopsis of the Neuroptera of North America, by H. Hagen, an accomplished Neuropterist of Belgium, have been issued under the auspices of that Institution. Similar works on Hymenoptera by M. De Saussure of Geneva; and on the Hemiptera by Mr. Uhler of Baltimore, are in course of preparation for the Smithsonian Collections. A list of described North American Hymenoptera by Mr. Cresson, is now appearing in the Proceedings of the Entomological Society of Philadelphia. H. C. Wood, Jr., has written in the Philadelphia Journal, an account of the Chilopodous Myriapods of the United States. The Spiders of the Southern States more especially, have been described and beautifully illustrated by Prof. Hentz, in the Boston Journal of Natural History.

Addendum.—Add the following to the catalogue of Amphibians of Maine, upon page 142. The specimen upon which our authority for this addition is based is in the Cabinet of Amherst College, and was obtained by C. B. Adams:

Amblystoma Jeffersoniana. Rare.

PART II.

REPORTS UPON THE GEOLOGY OF MAINE.

GEOLOGY OF MAINE.

BY C. H. HITCHCOCK.

PRELIMINARY REMARKS.

The Second Annual Report of the Geology of Maine will differ considerably in character from the *General Report* published last year. Certain elementary remarks were there introduced, and the geology of the whole State was treated of in a strictly systematic form. Consequently it is not necessary now to present elementary principles for the proper understanding of the details, nor would their repetition be judicious. It is impossible to present anything like a *system* of the geology of the State for similar reasons. We shall not be able either to draw upon other sources than our own explorations for material, as we did last year; hence the present report cannot be so lengthy as the previous one, and we fear that the multiplicity of scientific details will not be interesting to many. Our only apology is to be found in necessity.

We have found it quite difficult to settle upon a satisfactory mode of arranging our materials. Reports are in hand from every quarter of the State, and the nature of the rocks treated of is very diverse; still no exhaustive classification is possible. The best method we can devise is the following, which we shall attempt to follow strictly; although we are reminded by its heterogeneous combination, of the distorted, elongated and flattened pebles composing a singular conglomerate rock in the State, presently to be described.

A. Geology of the more southern and settled portions of the State. This will include the results of an unusually protracted examination of the rocks on the west side of Penobscot bay, particularly about Rockland, Thomaston and Camden, illustrated by a geological map. We shall endeavor to give in this sketch some idea of what a Final Report upon the Geology of the State should be; that is, of the particularity with which descriptions of every formation should be given. Annexed to this sketch will be a notice of the rocks of Vinalhaven, by George L. Goodale. Then will follow a detailed account of the measurement of a geological section by us the past summer between Mount Desert and the Canada line, by the way of Bangor and the Forks of the Kennebec river. This is really one of the most valuable results of our summer's work.

Under this head Dr. Holmes will give some account of the geology of Oxford, Franklin and Kennebec counties; being the results of many years observation. Possibly we may add a few words to it.

B. Reports upon the geology of the Schoodic waters, near the eastern boundary. These include the eastern Schoodic lakes, the western Schoodic lakes, and a portion of the St. Croix river, and consequently pertain chiefly to an unsettled region. Perhaps we may make a few remarks upon the geology of a portion of New Brunswick in this connection.

C. Reports upon the more northern and unsettled portions of the State. These are upon the region of the great lakes in northern Oxford and Franklin counties; upon the vicinity of Moosehead lake; upon the country watered by the upper portions of the west branch of the Penobscot and St John rivers; upon the Alleguash lakes; upon some portions of northern Aroostook county, particularly the results of Dr. Holmes' explorations among the Lower Helderberg limestones in search of the marble layer.

D. Surface geology in general; or a summary of our observations upon the geology of the Alluvial Period, in addition to what was said respecting it last year. We shall be able also to present something respecting the microscopic animals and plants of the infusorial deposits found sparingly beneath some of our ponds.

E. Descriptions of new species of Devonian and Silurian fossil plants and animals, by Dr. Dawson and Mr. Billings.

F. Mineralogical Notes.

G. Economical Geology.

After having travelled over the State the second time, we were surprised and gratified to learn how few errors of statement and generalization are to be found in our preliminary report. Where one is obliged to infer from the observation of others so much as was done in that, there is great danger of misstatement. Wherever errors of any consequence occur, we feel bound to correct them; for we are striving to attain a knowledge of the true distribution and relation of the rocks, and the exhibition of an error is often as important as the discovery of a new fact. Errors in the premise lead to false conclusions, therefore great care must be taken to present the facts without obscurity, prejudice or ignorance. Hence, in accordance with these remarks, we will proceed to point out the most important corections that are needed in the Preliminary Geological Report.

Our later explorations go to show that a large portion of those rocks called Talcose schist in the last report, belong rather to the Mica schist; for example, the great belt upon the river St. John. Likewise a considerable portion of the immense Clay Slate region in the central portions of the State proves to be micaceous. These rocks are very obscure, and we shall say more about them further on.

The carefully-measured regions in Machiasport, described on pages 186–8, 235 and 246, have been re-published in the Proceedings of the Portland Society of Natural History, with some additions, more particularly references to the catalogue of the collection where the specimens illustrating each stratum may be found.

Upon page 224, line 25, 1785 should be inserted in the place of 1857, as the freshet there mentioned belonged to the past instead of the present century.

We find, too, that we gave unintentionally a false impression respecting the character of the rocks composing the Quebec Group. The inference from the language on page 227 is, that the fossils near Quebec were obtained from sandstones. This is not true. They were derived from limestones chiefly—as we have been privileged to know the past season by actual inspection of the localities in company with Professor T. S. Hunt, who very kindly pointed them out to us. These limestones contain some of the forms of life which are found in the typical sandstones further west, although Mr. Billings suggests that the Quebec fauna presents some resemblance to the fauna of the calciferous sandrock.

A re-examination of the section from Charlotte to Presque Isle, figured upon page 381, shows us that the amount of clay slate there represented is nearly twice what it ought to be, and that none of it has the south-easterly dip there represented. The error arose from a too hasty examination at the first.

Dr. Holmes informs me that he was unable to find a large bed of limestone described by me as occurring near the First Seboois Lake, in No. 7, R. 7, upon page 413. Nor were gentlemen familiar with that vicinity able to point out the locality. We can only refer to our authority for the statement: viz., to *Jackson's Second Annual Report on the Geology of the Public Lands, pages* 29, 30, where he says, "This bed is no less than 90 feet wide, and large masses project above the soil. It is situated near the mouth of the first Seboois Lake, upon township No. 7, 7th range," etc.

Last year we prepared a Geological Map of Maine upon an uncolored copy of Chace's map, thus exhibiting at a glance all that was known respecting the geology of the whole State, and presented it, with the Report, to the Legislature. We shall do the same thing this year, and shall be able to color many spaces that were left blank before. It will be upon exhibition in some portion of the State House, while the Legislature are in session.

We shall also exhibit at the State House an enlarged colored copy of the section we measured through the widest portion of the State, viz., between Eden and the Canada Line. It will be suspended very near a set of the specimens which were collected along the route. It is designed both to show the character of the rocks as they appear at the surface along a given line, and to represent the supposed relation to one another which the respective formations have beneath the soil.

It is possible that persons in different parts of the State have been expecting to see us in their towns during one of the past two seasons, and not having seen us, may have thought we have not appreciated the importance of their rocks. We can only say, if such be the fact, that we have been as expeditious as possible in our field-work, and that our apparent neglect has arisen from the impossibility of being in two places at the same time. If any persons are very anxious to have the rocks of their neighborhood examined in the future, we would suggest to them the desirableness of informing us of their wishes, so that our plans of action may be made to coincide with their wishes.

WALDO
MONTVILLE
MORRILL
BELFAST CITY
BELFAST
Poors Mills
N. Searsmont
Belmont Cor.
BELMONT
Browns Corner
LIBERTY
St. Georges P.
Montville
Woodmans Mills
Searsmont
SEARSMONT
NORTHPORT
Saturday Cove
Knights P.
Newbert P.
APPLETON
Smith
LINCOLNVILLE
Duck Trap
Washington Pond
Andrews P.
Centre
Spruce Hd.
North Union
The Beach
HOPE
Hope Cor.
Sennebec Pond
UNION
Union
Lincolnville Canaan
Hobbs P.
S. Hope
E. Union
Camden
Camden Har.
CAMDEN
Negro I. Lt.
Barrets Pt.
Round P.
Crawfords P.
Seven Tree P.
Simontons Cor.
Rockport
Ingrahams Cor.
White Oak C.
Beach Hill
Rockville
Clam Cove
Rockport Har.
Monument
WARREN
Warren
ROCKLAND
Blackintons Corners
Jamesons Pt.
Crocketts Pt.
Owls Head B.
ROCKLAND
Ingrahams Pt.
Owls Head Light
Monroe I.
THOMASTON
Thomaston
SOUTH THOMASTON
S. Thomaston
Sheep I.
Broad Bay
FRIENDSHIP
CUSHING
Cushing
GEORGE RIVER
ST. GEORGE
St. George
Ash Pt.
Ash I.
Thorndike Pt.
Dix I.
Seal Harbor
Seal Har. Light House
Muscle Ridge
Friendship
Long I.
Cranberry
High I.
Northern I.
Tenants Har.
Southern I. Light H.
Moose Pt.
Mosquito I.
Hoopers I.
Mark I.
The Brothers
St. Georges Islands
Davis I.
St. George Har.
Allen I.
Burnt I.
SCALE OF MILES.
1 2 3 4 5
EXPLANATION of the MARKINGS.
Granite and Syenite
Gneiss and Feldspathic Mica schist
Saccharoid Azoic Limestone
Taconic Quartz Rock
Taconic (Eolian) Limestone.
Taconic Argillo-Micaceous schist
A. MEISEL LITH. BOSTON

A. GEOLOGY OF THE MORE SOUTHERN AND SETTLED PORTIONS OF THE STATE.

According to the plan proposed on a previous page, the description of the Geology of the more southern and settled portions of the State will embrace three districts; first, the country adjacent to Rockland and Camden; second, the section from Eden to the Canada Line, in Sandy Bay; and third, a large region west of the Kennebec River. The Geology of the first district is interesting because the relations of the (so-called) Taconic limestones, quartzites and schists are discussed, both in a scientific and practical manner. The second subdivision is occupied entirely with the description of a Geological Section, the longest and perhaps the most important one that could be measured anywhere in the State.

I. Region of Rockland, Camden and vicinity.

Perhaps it will be well at the outset to give the geographical limits of this region. The accompanying Geological Map specifies them very definitely. Beginning in St. George and Cushing, the portion explored widens in looking northwards, including Rockland and Warren; grows still wider where it includes Appleton, Hope and Lincolnville, when it begins to diminish, and comes to a point in Waldo. In brief language, it is a strip of country from five to twenty miles in width, on the west side of Penobscot Bay, between Belfast and St. George. The geology of Vinalhaven will also be described briefly, although not delineated upon the map.

This region was explored by C. H. Hitchcock and G. L. Goodale during the first twenty days in July last. We were greatly assisted a portion of the time by J. O. Robinson, Esq., of Thomaston, who often travelled with us to point out localities, and to communicate facts respecting the quarries of limestone. A great deal of our time was spent among the limestone deposits of Thomaston, Rockland and Camden, since the most difficult stratigraphical problems to be solved were connected with them. I may say truly that I never before found a region where it was more difficult to ascertain the true position of the strata than here.

I have worked for a long time among the same rocks in Western Vermont, and though I did not profess to have arrived at definite conclusions there respecting the time, age and position of the strata, I found it much easier to be satisfied with my explorations. Evidences of disturbance are much more prominent among the limestones of Penobscot Bay, and they are far more abundant. We had the maps of the Vermont rocks given us by our predecessors to correct, and the writings of eminent geologists to place distinctly before us the salient points to be examined. But here the field is a new one. None who have gone before us in Thomaston ever attempted a section of the strata, or described the rocks with any precision. We had, however, a poor section of the Camden rocks, which was of a little value to us.

For two reasons we undertook the exploration of this region; first, we wished to explore some limited portion of the State more carefully than ordinarily, that the authorities and citizens might know by the inspection of its results, how we wish to examine and describe the rocks of every portion of the State in a Final Report; second, we wished to "take the bull by its horns," as it were—that is, to grapple with one of the most difficult and at the same time interesting geological problems to be solved in Maine. To a large proportion of scientific men the description of the Taconic rocks will be the most interesting part of the Report.

Like the discussion of certain important theological doctrines, the discussion of the age of the Taconic rocks of North America bids fair to be continued from one generation to another. Those who first debated the question are now well advanced in life, while the discussion is mainly carried on by those who were either too young to appreciate it in 1840, or have but recently taken up the subject. The rocks of similar apppearance, and nearly the same age in England, the Cambrian strata, have excited the attention of their geologists for thirty years; and the history of the discussion shows that the question of their age and position has apparently been decided several times; but they are not yet all of them agreed as to their nomenclature. We anticipate a longer period of discussion upon this subject in America than in England. We cannot see that the debate has yet more than fairly commenced among us.

It is unnecessary for us to go into a detailed history of the Taconic controversy, for it has already been given, so far as is needful for our present purposes, in the Preliminary Report, (page 225, etc.) The Taconic rocks recognized in Maine on Penobscot Bay, are the Quartz Rock, Eolian (Stockbridge) Limestone, and the pecular Argillo-micaceous and Argillo-talcose schist, erroneously denominated "Magnesian Slate." None of them, in this region, are fossiliferous in the slightest degree, so far as is now known.

The rocks in this district, according to our conclusions, are the following: 1, Granite and Syenite; 2, Gneiss and Mica Schist, older than the Taconic group; 3, Saccharoid Azoic Limestone; 4, Taconic Quartz Rock; 5, Eolian Limestone; 6, Taconic Schists, mostly Argillo-micaceous; 7, Alluvium. This is the arrangement adopted in the map, only that Alluvium is not there represented. We will now proceed to describe these rocks in order.

1. *Granite and Syenite.*

Five patches of granite and syenite are represented upon the map. The first occupies the greater portion of St. George, with a part of South Thomaston. The second is really a part of the same mass, although separated from it by the St. George river. It is in the central part of Cushing. The third patch is in Warren, forming Congress mountain. The fourth is a small development of granite constituting the backbone of a ridge of hills in the north-west corner of Hope. The fifth patch lies chiefly in Camden, and runs under Megunticook mountain, the highest eminence in this district.

At Ash Point the granite crops out at the ocean's edge, while the greater part of the point away from the shore line is covered by drift. The island off the point is also granite. Upon Spaulding's Point, to the south-west, the granite changes its mica for hornblende, and hence becomes syenite. The rocks on these two points would make an enduring building material. The constituent crystals are often quite large.

At the village of South Thomaston one may see the boundary line between the poorly characterized mica schist and the granite. This line runs north-westerly for two or three miles towards Thomaston village, and then continues to the St. George river in a south-westerly direction, crossing over to the middle of Cushing. The western border of this granitic mass is composed of hornblende rock and granite. Passing to the middle of the granitic expanse, we find the rock sometimes porphyritic—i. e., containing large crystals of feldspar—but more commonly an ordinary granite. At the "Landing" in South Thomaston the rock is extensively quarried, and its quality can safely be recommended. Perpendicular seams or *joints* running N. 80° E. are quite numerous here, and remind one of strata very much. Jointed seams of this character which have been superinduced after the formation of the rock, as here, are commonly as true and evenly parallel as if elaborately chiseled out and smoothed by the hand of skillful lapidaries. The greater part of St. George is underlaid with granite, as the map and some future remarks will show.

The islands off the coast of St. George are chiefly of granite, and yield much fine stone for the market. Dix island, Seal Harbor islands, White Head islands, Clarke's island and High island, are certainly granitic. We are not informed as to the rock upon the islands adjacent to the *Muscle Ridge*, but suppose them to be composed of granite.

In Cushing the granitic rocks are chiefly confined to a narrow belt crossing the town a little below its geographical centre. Going south from Maple Juice Cove we come to syenite at a saw mill not far from the Cushing church, and this soon is replaced by granite, the syenite being upon its western border. The granitic belt is nearly two miles wide and extends into Friendship, where its extent is unknown to us. But the mica schist south of the granitic belt, and the gneiss in the north part of Cushing, abound with immense veins of granite, almost of sufficient importance, to be delineated upon the map. One in particular, of considerable width, extends from the village of Cushing to South pond in Warren, and is beautifully porphyritic. Other large veins of granite occur in the gneiss west and north of South pond in Warren.

At the corners in the road beyond the south-east end of South pond, is a large ledge of granite, and in the whole town of Warren are many similar ones, not represented upon the map, still of considerable dimensions. The second patch of granite, composing Congress hill and vicinity, in Warren, is probably a patch of this general description, only that is much larger. According to the views presented in the Preliminary Report, (page 204,) concerning the origin of granite, we should expect frequently in a gneiss region to find large veins of granite. Congress mountain must be about 900 feet high, and is in the north part of Warren. The granite is not beautiful, and the constituents are coarse. My impression is that this granite extends a greater distance than is represented, or into the valley of St. George river towards the village of Warren.

Another limited patch is exhibited in the north-west part of Hope, and its stratigraphical relations appear in Fig. 35. We noticed it upon the road crossing Muddy mountain. It is a coarse variety, generally possessing inferior economical qualities. It appears to be an immense bed in gneissic strata.

The last development of granite to be noticed lies mostly in Camden, forming the base of Megunticook mountain. Part of it occurs on the west side of Lincolnville or Canaan pond, part lies on the promontory running into the pond, but the greatest portion is about the mountain in Camden. It is a beautifully white variety, extensively quarried from the large fragments that have fallen down the steep side of the mountain. A portion of it is represented in a section in Fig. 35, where it seems to form a synclinal axis with the granite of Muddy mountain. The precise extent of this granite into Megunticook mountain is unknown to us, as we had not time to ascend it. It is a fact worthy of remembrance that the granite of Knox county is disposed to arrange itself into mountains and mountain ranges. The scenery about Megunticook mountain is quite wild and interesting.

The gneiss, (mica schist often,) of Lincolnville and Northport, contains many beds of granite. One of them near the Witherby House in Northport, is a beautiful variety of tabular granite, and is comparatively quite large. As we explored these towns very little, larger masses of granite may exist there, which escaped our notice.

2. *Gneiss and Mica Schist.*

The greater portion of the rocks of the region now being described are obscure gneiss and mica schist. So much of the schist prevails that it is difficult to avoid calling the whole deposit by that name. The micaceous rocks which are evidently connected with the Taconic series are excluded from this designation. It is doubtless the case that in our anxiety not to detract from the limits of the Taconic series, we have included some of the older mica schists among them. It cannot be doubted that the strata described under this head are older than the Taconic rocks, whatever age may be assigned to the latter.

The general character of these older schistose rocks may be briefly summed up by describing the three most common varieties. These are, first, an obscure gneiss, generally of an extremely uncouth and ugly appearance, rarely appearing in beautiful ledges, but more commonly protruding a short distance above the surface with a weather worn and dilapidated aspect; second, mica schist, generally showing scattered crystals of feldspar upon its edges, though not readily seen when the specimens are viewed with the flat surface up: this rock is often beautifully foliated and handsome in appearance; third, beds of granite, sometimes tabular, sometimes finely grained, rarely porphyritic, but commonly an easily decomposing, very plain, rusty looking variety, fit only to disintegrate and form new soil for vegetation.

Four deposits of mica schist and gneiss in this region deserve our notice. The first is gneiss, in the south part of St. George. The second is mica schist, in the south part of Cushing. The third is the greatest of all, extending from Cushing to the corner of Belfast, occupying the principal parts of Cushing, Warren, Union, Hope, Camden, Lincolnville and Northport. The fourth is sparingly represented upon the map in Waldo and Morrill, but belongs to a gneissic area much larger than the one just described. If we should speak of the geology of the whole country between Kennebec and Penobscot rivers, we should call the last the great deposit, of which the third was only a spur. The third and fourth deposits are probably connected together by a synclinal axis, upon which the Taconic schists running south-west from Belfast are superimposed.

Having seen it stated that the whole of St. George was composed of granite, we were quite surprised last July to find the whole of the south part of the town composed of gneiss, although the strata were often very much contorted and the rock uninteresting. In leaving the granite of St. George, one first sees the feldspathic mica schist near Tennant's Harbor, where it dips 85° N. 80° W., or nearly perpendicular. At Mosquito Harbor the gneissic type is predominant, and of an unusually clean appearance. It is full of what we regard as pebbles altered and distorted by metamorphic action and pressure, as discussed in our Preliminary Report, page 178. The work, however, is carried to an extreme here. The strata dip 50° N. 70° W.

Outcrops of gneiss are common all the way to the Light House at Herring Gut. Here are several trap dikes, running N. 70° E., being vertical. The strata of gneiss are wonderfully convoluted, and very fine specimens to illustrate them in Cabinets might be obtained here. A few layers of hornblende schist of a handsome variety are interstratified with the gneiss. The strata have an average dip of 40° south-easterly, and thus we have the evidence of an anticlinal axis running through this patch of gneiss in the south part of St. George. Curious veins of syenite abound in the trap near the Light House, the syenite appearing conglomerated. It is very rare elsewhere to find syenitic veins in trap rock. The trap dikes are only a few feet wide, and the syenitic veins as many inches.

In the south part of Cushing we regard the mica schist as predominating, insomuch that upon the large map of the Sate it is not embraced in the same formation with the gneiss of this region. The texture of the rock is very fine and the strata very thin, as if coinciding with planes of cleavage, and the rock altered from clay slate. Veins of granite are remarkably abundant, and many of them are very tortuous. Most of the granite in the veins is very coarse-grained. Fig. 34 represents a portion of a tortuous granite vein from this vicinity, the block being twenty-six inches long, and the vein varies from half to three-quarters of an inch in width. The straight lines represent the strata of schist, and the crooked ones the vein. The strata appear not to have been at all affected

Fig. 34.

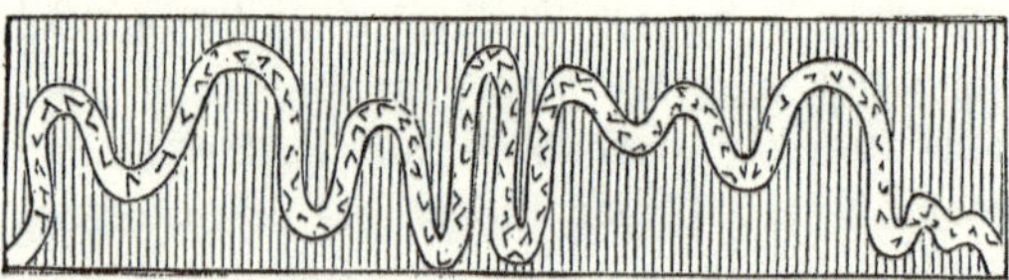

Granite veins in mica schist.

by the protrusion of the granite. It is difficult to conceive how such a crooked fissure could have been formed at the outset; and then to imagine how the crevice was filled so compactly and apparently quietly. The whole vein as measured is thirty feet long, and it divides into branches, tapering finally to a point. It is wonderful how numerous these contorted veins are at the extreme south point of the promontory.

Much of the mica schist in Cushing is rapidly decomposing. The dip of the strata varies from 75° to 80° south-east. At the end of the promontory we noticed an interesting band of conglomerate twenty feet wide. The cement is mica schist, while most of the pebbles are elongated and flattened, and entirely siliceous. They illustrate the different phases of conglomerate metamorphism more finely than the example in Washington county adduced last year. Gay's Island, which is a large island connected with the main land during low tide, is also composed of mica schist. Whether the St. George's Islands or any in Muscongus bay have a similar basis, we have no means of ascertaining. To learn the geological character of these and all the islands scattered along the coast of Maine will require a considerable time with a sail boat.

Adjacent to the granite in St. George and South Thomaston is a narrow belt of obscure mica schist, which really belongs to the deposits of this age, although the paucity of our observations there and the concealment of the ledges by drift makes our knowledge of them very limited. It is at least a mile in width where we crossed it after leaving the quartz rock on West Keag river. Its relations to the adjacent rocks are given in Fig. 37.

We cannot find much that will interest in the details of the distribution of the gneissic rocks of the third area specified, or that including the north part of Cushing, Warren, etc., to Belfast. The rocks in Cushing are mostly distinct gneiss but of uninteresting appearance. Where the rocks of this group first show themselves on the road to Cushing from Thomaston, they consist of gneiss and hornblende rock with a little granite. The strata are very much plicated. A synclinal axis runs along near the eastern shore as far south as the rock extends. In the north-west part of Cushing and the north part of Friendship, the dip is uniform, and the gneissic character of the rock prevalent, interspersed with numerous granite veins.

In the south part of Warren, Mr. Peter Wallace's house stands upon the eastern border of the gneiss. About South pond the dip is variable, and the gneiss alternates with granite, often very beautiful. The uniformity of the gneissic rocks in the other parts of Warren is relieved by the occurrence of occasional beds of saccaroid azoic limestone.

In Camden there is greater variety. Between Simonton's Corners and the Harbor the rock is a handsome variety of the mica schist, sometimes suitable for paving stones. On Roach hill and westwardly the gneissic type predominates. A section from Ingraham's Corner to Union shows both mica schist and gneiss with several folds in the strata. This road passes by a pond, on the opposite side of which interesting white rocks crop out on the sides of steep mountains. These are probably granite, though they resemble beds of limestone showing their edges in true sedimentary style. In the north-west part of Camden the mica schist occasionally appears with the gneiss—also a single ledge of hornblende schist with a small northerly dip. A range of high mountains in the west part of the town are gneissic, running north-easterly from Congress mountain in Warren. The names of several peaks are Spruce mountain, Ragged mountain and Bald mountain. The last two are appropriately named. Immense disturbances must have accompanied the elevation of these mountains, for at the base of Bald mountain, east of Hosmer pond, the strata run at right angles with their course at the summit of the ridge; and neither of the two courses are the common ones in this region; for the one is east and west, and the other north and south, while the common strike is a north-east and south-west course.

In Union we find more regularity in the position, although folds are more common. Cobb's hill, in the village, is an anticlinal knob of highly ferruginous mica schist. Near the north line of Union, on the east side of Sennebec Pond, the rock is largely quartz—particularly in the vicinity of the iron ore. Near the quartz the strata are mostly of mica schist. In other parts of the town we saw only the gneissic variety.

This formation, as it runs into Appleton, is mostly mica schist. In Hope, the gneiss predominates. Near Hobbs' pond there is pyritiferous mica schist, and mica schist with granite is developed adjacent to Bartlett's limestone.

East of Megunticook and Battie Mountains in Camden, a sort of spur of an imperfect gneiss may be seen. Following up the coast to the "Beach" in Lincolnville, a siliceous slate is found, which we suppose to constitute a portion of the gneiss formation. In the south part of Northport feldspathic mica schist appears; and to the north some talcose rocks are interstratified with the gneiss. In the principal portions of Lincolnville and Northport gneiss abounds.

The last belt of gneiss represented on the map is in Waldo. All that we saw was of an inferior quality. Some large veins of granite appear in it. Ten miles west of Belfast there is a coarse mica schist containing garnet, tourmaline, hornblende, and large masses of feldspar.

For the convenience of those who are interested in tracing out the stratigraphical relations of this deposit of gneiss, we give here in a table all the observations of the position of the strata which we have. To ascertain the strike, it is necessary to calculate it from the dip; remembering that the course of the strata is always at right angles to the dip:

Locality.	Dip.	Remarks.
Cushing, north-east part,	80° S. 60° E.,	Strata plicated.
Cushing, Maple Juice Cove,	75° S. 75° W.,	Anticlinal with previous [observation.
Cushing, west border,	South-easterly.	
Friendship, east part,	Strike N. 10° E.	
Warren, south-east of South pond,	65° S. 68° E.,	Near Peter Wallace's.
Warren, south of South pond,	50°–65° S. 78° E.	
Warren, west of South pond,	85° N. 78° W.,	Interstratified with granite.
Warren, west part,	75° S. 12 E., and 85° S. 12° E.	
Warren, do.,	40° S. 75° E., and 65° S. 12° E.	
Warren, near the village,	65° S. 60° E.,	Much contorted.
Warren, east from the village,	55° S. 70° E.,	At a limestone bed.
Warren, at saw mill on Oyster river,	75° S. 30° E.,	Near Thomaston line.
E. Union,	S. 70° E.	
E. Union,	Both east'ly & west'ly,	An anticlinal.
Union,	80° NW.,	At Bullen's limestone bed.
Union common,	50° S. 70° E.,	At Batchelder's " "
Union common,	An anticlinal,	On Cobb's Hill.
Union,	High to S. 60° E.,	At Miller's quarry.
Union, north part,	75° westerly,	At Iron Ore Bed.
Camden, Ingraham's corner,	Perpendicular,	Strike E. and W.
Camden, do., one mile west,	55° S. 20° E.,	On top of mountain ridge.
Camden, Roach hill,	48° S. 70° E.	
Camden, south-west part,	75° S. 20° E.	
Camden, Simonton's corner,	60° N. 30° E.,	Mica schist.
Camden, north of do.,	N. 30° W.	
Camden, west part,	Westerly.	
Camden, north-east part,	43° easterly, 70° S. 20° E., and 30° N. 70° W.,	East of Mount Battie.
Camden, Hosmer pond, north side,	45° S. 30° E.	
Camden, south of Hosmer pond,	– –	Strike E. and W.
Camden, top of ridge,	45° E.,	West of Hosmer pond.
Camden, head of Megunticook river,	25° N.,	Hornblendic rock.
Camden, north-east part,	27° S. 60° E.,	Anticlinal with previous [observation.
Hope, Hobbs' pond,	40° S. 80° E.	
Hope, west part,	South-easterly.	
Appleton, south-east part,	High to S. E.	

LOCALITY.	DIP.	REMARKS.
Appleton, north-east part,	80° S E., and 60° S. E.	
Lincolnville pond,	50° N. 20° W.,	In Camden.
Lincolnville, Beach,	Dip easterly.	
Northport, south part,	70° S. 36° E.	
Northport,	75° N.	
Northport, north part,	80° S. E.	

The following results flow from this table: 1. The most common dip in the whole gneiss district is to the south-east. 2. A short anticlinal axis occurs in the north-east part of Cushing, and of course a synclinal of equal length on the west side of the anticlinal. 3. There is a very important synclinal between Meguntícook mountain and the Taconic schists in Appleton; and the granite of the mountain itself appears to be in the ridge of an anticlinal. 4. There are two anticlinals and two synclinals between Ingraham's corner in Camden and Union common, as seen along the road passing through E. Union. 5. The variations of the dip in the west part of Camden among the high mountains, result from displacement, rather than the usual plicating forces. 6. The other exceptions to the usual dip are either local matters, or their relations have not yet been traced out.

FIG. 35.

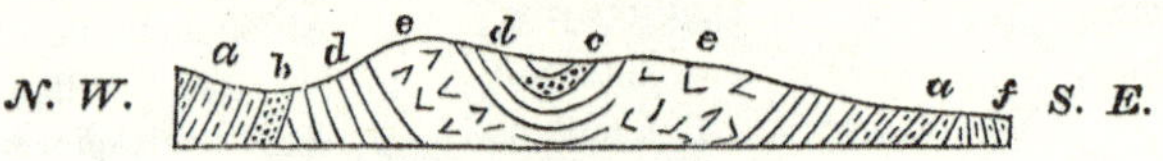

Section from Appleton to Camden.

aa. Argillo-mica schist, (Taconic.)
b. Azoic limestone, Smith's Mills.
c. Azoic limestone, Hope Corners.
dd. Gneiss.
ee. Granite.
f. Camden Harbor.
N. W. North-west.
S. E. South-east.

Fig. 35 illustrates the relative position of the gneiss strata with the adjacent rocks on a section from Camden harbor to the north part of Appleton, exemplifying the third conclusion derived from the Table of Dips. At the northwest end of the section are the Taconic schists, *a*, with a high north-westerly dip of 65°, often leaning slightly to the northwest. Passing into the valley, we find near Smith's mills a very wide bed of Azoic limestone, *b*, supposed by us to be imbedded in gneiss. But it seems to dip north-westerly, averaging 75°, and hence under the schists, just as if it belonged to the Taconic series. It is also of great longitudinal extent. The gneiss, *d*, dips south-easterly at first 80° and then 60°. Passing to the top of the hill or Muddy mountain, in Hope, we find granite, *e*, at its summit. The gneiss appears again in Hope. East of the Corners appears another bed of Azoic limestone, *c*, dipping N. 10° W., so that we have a synclinal axis. We have not seen the westerly outcrop of this limestone, as represented in the Figure, but cannot doubt its existence, though it may be concealed by drift deposits. East of the limestone more gneiss and granite, *e*, appear, apparently the same layers repeated that were just passed over. They continue into Camden, until we come to the Taconic rocks again, *a*, which appear to be disposed in an anticlinal form. These Taconic rocks here often perplexed us by their similarity to the older schists.

Under *Economical Geology* we will describe a valuable vein of iron ore in the gneiss in the north part of Union.

3. *Saccharoid Azoic Limestone.*

We draw a distinction this year, which was barely hinted at last year, between the Saccharoid Azoic Limestones and the Taconic Limestones. As far as lithological appearances go, it would be impossible often to distinguish between the two, but when the question of geological age is taken into consideration, the distinction becomes plain. In the first group we include all the beds of limestone which occur in the older gneiss and mica schist. The Taconic (Eolian) Limestones are connected with quartz rock and argillo-micaceous schists, and are generally of greater width and length.

The Azoic Limestones are found in Warren, Union, Hope, Camden and Lincolnville. We shall describe only those of which we have positive knowledge. The beds in many of these towns are so large, and the stone so firm, that blocks of marble have frequently been taken from them. Some of them are now being quarried for the kiln, but at others these operations have ceased to be carried on profitably. We had time to examine only two beds of limestone in Warren, although others more valuable exist in the town. One is about a mile north-east from the village, and the rock is inclined 55° S. 70° E. It is white, of good quality, of great amount, and quarried. The other bed noticed was not as good, and is in the north part of the town near Congress mountain. The other beds in town have been thus described: "On the western side of the St. George river on Mr. A. Starrett's estate, there are two beds of limestone, the largest of which is twenty feet wide, dipping 78° south-easterly. They are dolomitic, and contain crystals of galena and zinc blende. On the eastern side of this river upon Mr. A. Starrett's land, there is a similar bed of limestone, 99 feet wide, and 150 feet long, as exposed. Another bed belonging to John Starrett, is 122 feet wide, and has been uncovered for 220 feet. It is inclined 55° S. E. On Benjamin Starrett's land there are two beds of limestone, one of which is 30 feet wide, dipping about 55° S. E. On a high bank of St. George river a cliff of limestone is exposed, 26 feet high, and traceable laterally half a mile. It is half a mile north-west from A. Starrett's bed." All these beds of limestone are white, and often highly crystalline. One would think that a better quality of lime could be manufactured from such pure material, than from the darker-colored and coarser-looking limestones of Thomaston, etc., which are so celebrated in the market. Experience, however, is the only reliable test of the best quality of rock for the manufacture of quick-lime.

Quite a number of excellent beds of limestone are found in Union. Large beds are found upon Harding's point at the south-west side of Crawford's pond, in the south part of the town, one of which is nearly a hundred feet wide and over three hundred feet long. Near Union Common, upon Capt. N. Batchelder's estate, is a large bed thirty feet wide and nine hundred and twenty-four feet long. It has a bluish tint, and is very saccharoidal. In the south-east part of the town, near Mr. Miller's, are two large beds of limestone, both being in the same range. It has been dug into in several places. The dip common to all the outcrops is about 65° S. 60° E. A large granite vein crosses the limestone, the first instance of the kind we have ever seen. The limestone is a fine-grained light gray rock, and the width of the bed varies from ten to thirty feet. It must be twenty rods long. Near it are scattered smaller beds, often of no value. Upon the Bullen farm, to the north-west, is a bed of limestone fifty feet wide, quite firm, and might be wrought to a limited extent as a marble. It has been traced over a thousand feet. It dips 80° N. W., and has been quarried in three places. Another bed appears on the north side of Hope lake. This is in the town of Hope. Upon Ephraim Bartlett's farm, near Hope Corners, is a very large bed of limestone, dipping from 12°–20° N. 10° W. Granite and mica schist are the wall rocks. It has been quarried at two different places.

Near Ingraham's Corners in Camden, there is a large bed of limestone, if not two, side by side, which closely resemble the Taconic limestone, and indeed it is barely separated from the west end of the Camden belt of Eolian limestone. It is not impossible that the two belts once joined each other, and have been separated by the forces which elevated Bald mountain and Ragged mountain, although their strikes are at right angles to each other. The bed is in mica schist, about three-quarters of a mile south of the Corners, and dips southerly about 75°. It is quarried very extensively, and burnt.

At least four large quarries of limestone are wrought in the south part of Lincolnville, viz: Young's, Coleman's, Heal's and Ball's. We suppose them all to belong to the same belt of rock. They dip 38° S. E. They are inexhaustible. We heard of limestone also at Brown's Neck, and on Jonathan Moody's land, but had not time to visit the localities.

A very large bed of limestone runs from Appleton into Searsmont. It is on the east side of the St. George river, and is quarried at several openings at Smith's Mills and in Searsmont. It is several miles long, and the belt is upwards of a quarter of a mile in width. Its dip is 75° N. 50° W., or apparently beneath the Taconic schist upon the hill west. We found no rock between it and the first Taconic outcrop on the hill, and do not know but that it may possibly prove to be of Taconic age.

All the beds of Azoic limestone now described, either have been, are now or are capable of being quarried extensively for the manufacture of quicklime. The manufactured lime must be equal if not superior, as a general thing, to that made in Rockland. Being at a distance from the sea-shore it cannot be furnished to the market so cheaply, and hence the inland quarries are not worked with so much vigor as those in Thomaston and Camden.

A slight examination of the map shows us that there are three if not four belts of the azoic limestone, yet the inclinations are such as to suggest whether they may not be repetitions of the same belt. One range is in the St. George valley in Hope, Appleton and Searsmont. Another runs from Ingraham's corner to Warren. The third extends from south of Crawford's pond to north of Hope Lake. If there is a fourth it embraces the Hope and Lincolnville beds. It is curious that the dip of the different beds in all these lines is different. That in the valley of the St. George river dips in opposite directions away from each other, making an anticlinal; so do those on the Crawford pond range; while the Lincolnville and Hope beds dip towards each other, and the Ingraham corner and Warren beds dip in the same direction essentially. Fig. 35 shows that the Hope and Appleton beds are distinct from each other. The Crawford pond and St. George river ranges may make a great synclinal axis with each other; and if the former is connected at all with the Ingraham corner range the connection is not obvious. The study of the connection between different beds of limestone may not only indicate the number of folds in the older strata, but may also lead to the discovery of more limestone now concealed from view by the soil. We hope that these glimpses of connection between the various limestone beds of this vicinity may lead some resident to work out the relations more fully.

4. *Quartz Rock*, (*Taconic.*)

Three belts of this Quartz Rock are represented upon our map. One is between the two belts of Eolian limestone in Rockland and Thomaston; another extends from South Thomaston, generally along the shore of Penobscot bay, with an occasional submergence beneath the salt water, to Rockport; and the third is an isolated outlier in Camden, forming Mount Battie. In the west part of Thomaston is a considerable thickness of quartz rock not represented.

The best characterized locality of quartz rock is near the south end of the first belt, near West Keag river in South Thomaston. The hill south and west from the dolomite quarry, (Marsh quarry,) is the place where it crops out so finely. The rock is a sandstone, very much like the white purely silicious Potsdam sandstone about Lake Champlain. It has numerous minute crystals of magnetite scattered through it. It dips 75° N. 60° W., and its relations to the adjacent strata are given in Fig. 37. It appears to overlie some of the older mica schist, and to underlie the Taconic schist and dolomitic limestone. This locality is the most promising place we have seen in this rock for the occurrence of fossils.

In the very south-east corner of Thomaston, we see more of this quartz rock, dipping 55° N. 40° W., and in the south part of the city limits of Rockland the dip is 60° N. 70° W., and 40° S. 70° E. The latter example is the furthest south. Much mica is associated with the quartz here, and all along through Rockland. The city is situated upon this rock, and we find indications of an anticlinal axis there from our notes—one observation of the dip being 60° N. 60° W., and the other 60° southeasterly. It is exhibited in Fig. 38. The quartz rock beyond Rockland passes beneath the bay till it rises again at Jameson's point. It is supposed to occupy the whole of this projection of land, and to be cut off again at Clam cove in Camden. Figs. 39 and 40 show its relation to the schists. North of Clam cove the quartz layers dip 70° N. W. Where this rock appears still further north, to its utmost limit, it is a hyaline or glassy variety, precisely like the typical "granular quartz" of Emmons. A great boss of it a mile south of Rockport dips apparently 30° S. E. It may be traced along the whole shore to the inmost corner of Rockport harbor, and would make at almost any portion of its course, an excellent material for the manufacture of glass.

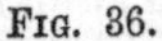
Fig. 36.

Explanation of the Figure.

The scale of this map is the same as that of the Map of Waldo County. The space covered by crosses represents gneiss; the perpendicular or N. & S. lines represent quartz rock; the horizontal lines represent Eolian limestone; the checked surface represents silicious slates and slaty quartz; the lines inclining to the right represent a micaceous quartzite or schist; the lines inclining to the left are Taconic schists. The oval space in the centre represents Lily pond.

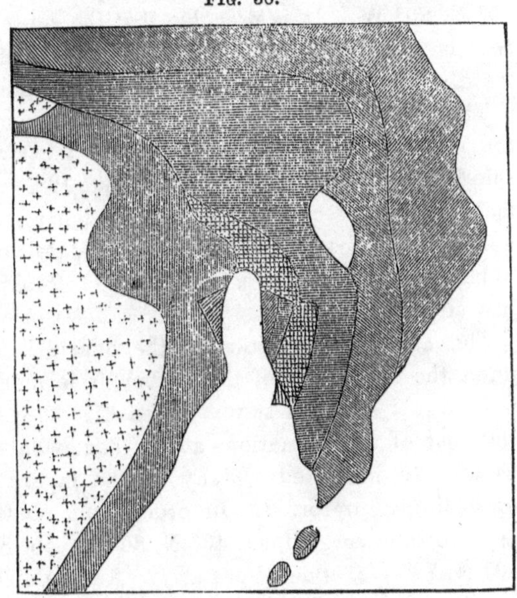

Geology of Rockport and vicinity.

About Rockport the details of its distribution are more interesting. In general the shape of the formation may be seen in Fig. 36. Just south of the village its width quadruples; and in less than two miles to the west it has thinned out entirely. It is an important fact, as learned from this map, that the formation makes a bend of a right angle, running westerly instead of northerly. Another interesting fact is illustrated by a small patch of quartz rock on the east side of the harbor. It is triangular in shape, and not more than half a dozen rods wide, but it has been elevated so as to stand side by side with a newer rock—a mass of strata that has been elevated vertically. Consequently there is what is called a *fault* between the two rocks, and we can see here the smoothed vertical surfaces of both rocks where they rubbed against each other during this vertical motion. Another fault is also shown upon both sides of the harbor in connection with the quartz, where the argillo-micaceous schist is displaced. Here too it is the quartz that appears to have been elevated.

In the first case the course of the fault is N. 10° E.; the quartz rock dips 30° S. E., while the adjacent wrinkled slaty rock dips 20° N. 80° E. In the second example the dip of the two rocks is equally variable, the former dipping 60° S. 70° E., and the latter 50° N. 20° W. It is probable that the cause of these dislocations is to be found in the bending of the course of the formation westwardly. Perhaps its original course was more to the northward.

The characters of this rock show finely south of Goose river, on the road to Rockland. Adjacent to the river the quartz is very thin bedded, and dips northerly about 12°. A few rods south the massive hyaline quartz dips 40° northerly. This rock appears in numerous bosses among the houses and gardens in this part of the village. The position a mile west of Rockport is similar to that just noticed.

The more slaty portion of the quartz is above the rest; and upon the north side of the harbor, it is more largely developed, often a real silicious slate. Upon Fig. 36 it is distinguished from the rest of the formation, and is somewhat disturbed by faults, as already indicated. Many of the layers are calcareous, and exceedingly contorted. In order from south-east to north-west are the following dips: 40° N. 80° E., 55° S. E., 20° N. 10° E., 40° N. 70° W., and 78° easterly. As the distance passed over is less than half a mile, these various dips indicate disturbances. Between Lily pond and a small church this rock occurs, and has interstratified with it a few layers of conglomerate, some of whose constituent pebbles have been flattened and elongated in the direction of the strike. They dip 45° N. 70° E. That part of this rock which is situated nearest the river, in the central portions of the village, merges into a greenish schist, having a small easterly dip; while the more northern portions, as on the road to Camden, are dark silicious slates and black slaty quartz rock, dipping 45° N. 40° E. West of the bridge in Rockport the dip of this slaty quartz is northerly.

Two islands off Beauchamp's point are represented upon our map in Fig. 36, the lowest one, Lowell's rock, being entirely composed of quartz rock; and the other, Indian island, having a micaceous schist with the quartz.

Concerning the belt of quartz rock in Rockland which lies between two belts of limestone but little need be said. A reference to the map will show its situation, and Fig. 39 will show its stratigraphical relations. It appears to constitute an anticlinal axis, underlying both bands of limestone. This rock constitutes the foundation of the highest hill north-west from Rockland, or where the Fourth Regiment M. V. M. was encamped last summer.

Mount Battie is an immense pile of consolidated distorted pebbles, and belongs to this formation, since the rock is pure silica, and seems to be of the same age. Last year we said something about this mountain under the erroneous name of Megunticook mountain. The latter name belongs to the granitic pile behind Battie, of still grander proportions. Mount Battie is exceedingly precipitous—it being impossible to ascend it on the east and south sides. The strata are very obscure, although at almost every step the pebbles composing them are obvious. Upon the south and east sides of the mountain the strata appear to dip southerly and easterly at a small angle, and they are thus represented in the section, Fig. 41. Professor Emmons estimates the dip at 25° S. E.

Those who are interested in the exhibition of distorted pebbles in conglomerate may find examples upon Mount Battie. We would not represent this case as a *typical* one; *i. e.*, one where the phenomena are in the greatest perfection. Still the proximity of this locality to frequented routes of travel makes it a good one to visit. The finest examples of these distortions occur in the less frequented parts of the State, as in Rangely. These pebbles on

Mt. Battie are all small, and are not always elongated or flattened. But the localities of the altered ones are so common that we think no one could fail to discover them. Good examples were seen on the top of the mountain, both at the south and northern ends. Two sets of joints cross the strata—one running N. 75° W., and the other N. 10° E.—which sometimes cut the pebbles in two.

Our last year's estimate of the thickness of the strata of this mountain at 500 feet we will not change. The height of the mountain is about 1,000 feet above the bay at its base. One can hardly resist the conviction that Mounts Battie and Megunticook have been elevated to their present height by a "convulsion of nature," rather than the quiet and gradual way in which ranges of mountains are normally elevated. The section in Fig. 41 confirms this view; for what is more likely than that the quartz rock in Rockport, *bb*, forms a great synclinal axis with *l*, the conglomerate of Mt. Battie? The fault then must be found, if anywhere, between *k* and *l*.

In Fig. 38 a narrow band of quartz rock is exhibited west of the Meadow's quarries in Thomaston. Whether this is to be considered a part of the formation we have been describing, or as a member of the Taconic schists, we know not. We had not time to trace it through its whole extent, still we think it quite limited. It dips 50° N. 50° W., or away from the limestone. Still it may be repeated by an anticlinal, which we suppose to exist, although the rocks cannot be seen, before the limestone is reached, and thus underlie it in the normal way. Some of these quartz layers are calcareous, and resemble dark siliceous slate. Others are bright colored; and some show the constituent grains of silica distinctly. The mica schist to the west has a still higher dip in the same direction. The belt is a dozen rods wide.

5. *Eolian Limestone.*

The name suggested by Professor Emmons for the principal belt of Taconic Limestone, was *Stockbridge Limestone,* from the town of Stockbridge, in western Massachusetts, where are large quarries of marble. For geological reasons my father suggested* a change

*See Final Report upon the Geology of Vermont, Vol. 1, page 395.

in the name of the group to *Eolian Limestone*, and we adopt it here. The name is derived from Mount Eolus in Dorset, Vt. Emmons applied the name Stockbridge Limestone to the limestones in Thomaston and Camden twenty years ago. There are three large belts of Eolian limestone in this region of country to be described--and, indeed, we know of no others in the whole State. The largest extends from the State's Prison in Thomaston nearly to Chickawakie pond in Camden. Another is in Rockland. A third is in Camden. Besides these, in Thomaston and South Thomaston are several large and small beds of limestone and dolomite in the schists.

We present here four sections, (Figs. 37, 38, 39 and 40,) crossing all the limestone belts and beds in Thomaston and Rockland. They will give the relations of the different rocks to one another much better than detailed descriptions.

These sections are very important, as they show the relations of all the beds of limestone in Rockland, Thomaston and South Thomaston, to all the rocks in the vicinity. They are all drawn parallel to one another,

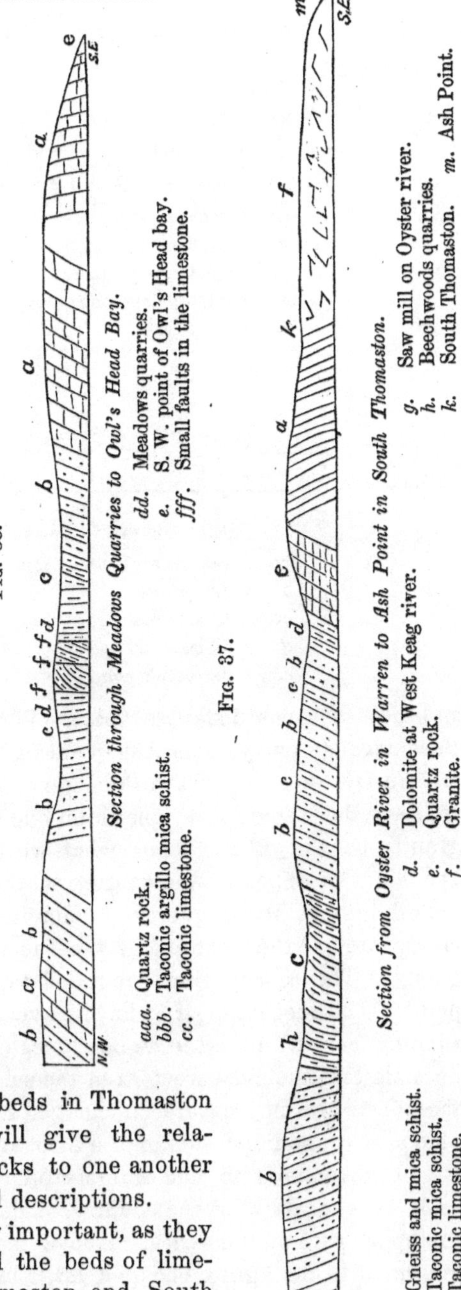

FIG. 38.

Section through Meadows Quarries to Owl's Head Bay.

aaa. Quartz rock.
bbb. Taconic argillo mica schist.
cc. Taconic limestone.
dd. Meadows quarries.
e. S. W. point of Owl's Head bay.
fff. Small faults in the limestone.

FIG. 37.

Section from Oyster River in Warren to Ash Point in South Thomaston.

aa. Gneiss and mica schist.
bbb. Taconic mica schist.
ccc. Taconic limestone.
d. Dolomite at West Keag river.
e. Quartz rock.
f. Granite.
g. Saw mill on Oyster river.
h. Beechwoods quarries.
k. South Thomaston.
m. Ash Point.

FIG. 39.

Section through Ingraham's and Rankin's Quarries.

aa. Taconic argillo-micaceous schist.
c. Ingraham's limestone quarry.
d. Limestone—north end of the Meadows quarries.
e. Rankin's limestone quarry.
f. Quartz rock.

FIG. 40.

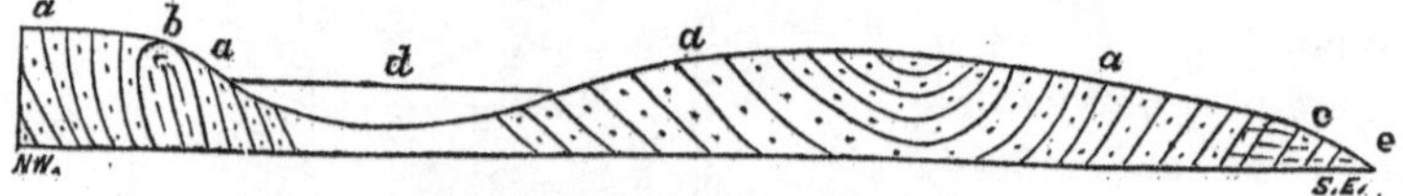

Section across Chickawakie Pond.

aaaa. Taconic argillo-micaceous schist.
b. Limestone.
c. Quartz rock.
d. Chickawakie pond.
e. Jameson's point.

and not at great distances apart. Fig. 37 extends through the Beechwoods quarry, from Ash point in South Thomaston to a saw mill on Oyster river, just in the edge of Warren. At the two ends we have the Azoic rocks, gneiss on one side, and mica schist with granite on the other. Then what we suppose to be the oldest Taconic rock appears—the quartz rock, near the West Keag river, underlying the Marsh quarry of dolomite. This rock may possibly be repeated at the west end of the Taconic schists adjacent to the gneiss. The rocks being concealed we could not determine this point. The rocks appear to have a synclinal structure, the quartz running beneath both the Eolian limestone and the schists. It is possible that the most western of the schists are synchronous with the quartz; or may have been changed into the latter by metamorphic action upon one side of the axis and not upon the other. It is not uncommon to find interstratified with the Taconic quartz rock, in other parts of the country, schists very similar to those of the upper part of the series. Above the quartz rock, and to the left hand in the figure, occurs a large bed of dolomite; then succeeds the Taconic schists with at least two beds of limestone, owned by Charles E. Butler. At length we cross the largest belt of limestone in this region. At first the dip is quite large, then it becomes smaller, and finally becomes large again. The most western exposure on the section is at the Beechwoods quarry; where the dip is smaller, the exposure is at the Fulling Mill quarry. Succeeding the limestone are Taconic schists dipping south-easterly, and towards the limestone.

Fig. 38 presents a section of the rocks from the extreme south-west angle of Owl's Head Harbor in South Thomaston through the Meadow's quarries to the north-west part of the town of Thomaston. Only Taconic rocks appear upon this section, as well as in Figs. 39 and 40. At the south-east end there is an anticlinal axis of quartz rock. Next is a thin band of Taconic schist over the quartz and underlying the Eolian Limestone. The dip of the limestone is here more irregular but seems to form a synclinal axis. Taconic schists and a little quartz rock succeed on the west; probably in an anticlinal form. The westerly part of the dip is certain, the eastern conjectural.

Fig. 39 presents a section of the rocks from the bay in the north part of Rockland city to the town line between Rockland and Thomaston immediately west of the west branch of Mill river. Quartz rock at the east end of this section forms a synclinal. Next appears the eastern belt of limestone, particularly in the latitude of Rankin's quarry. It dips east. Next succeeds an anticlinal of quartz rock, followed by the second band of Eolian limestone, at the north end of the Meadows quarries. West of this to the end of the section the schists prevail, interrupted at Ingraham's quarry by an *outburst*—this word seems to express the physical appearance of the rock—of limestone. Hence we have two anticlinal and one synclinal axis at this end of the section.

Fig. 40 gives a section of the rocks from Jameson's point across Chickawakie pond in Camden. We have at the east end quartz rock dipping north-westerly, then a synclinal of Taconic schists before coming to the pond south. West of the pond occurs an inverted anticlinal of limestone, almost concealed, wrapped in the Taconic schists.

Having now described the sections in general, we will proceed to give all the particulars needful respecting the character and position of the Eolian limestone at all its exposures.

The most southern exposure of the largest belt is at the State's

Prison quarry in Thomaston with the strata inclined 80° S. E. If the rock extended further, the valley of the St. George river would be the place to find it. Northeasterly from the prison the limestone extends into Miller's and the Beechwoods quarries. At Miller's the rock is whiter than is common in this vicinity, dipping from 65°–70° N. 40° W. Further on the dip is ten degrees greater. It now begins to curve to the east. At the south end of the Beechwoods it runs N. 80° E.; southeast from here it runs N. 75° E., and finally at the Fulling Mill quarry, not far from the east end of Thomaston street it dips 40° N. Then it bends back again to its former course, and passes into the Meadows quarries, unless there is a break, which is improbable.

At the Fulling Mill quarry the rock is unusually dark. At the mill itself it is yet darker and makes a good marble, slabs of which have been sawed out in times past, some of which remain here showing their excellent character. Below the Fulling Mill quarry on Mill river is an old quarry of white marble of excellent quality, dipping 20° N. 30° W. This is in Mr. Jacob's pasture. Above the marble is a bed of schist thirty feet wide in the limestone, with a greater dip, viz., 60°. In returning to the village from the marble quarry one passes by several exposures of limestone. At a quarry where a siliceous limestone was being blasted for paving and building material the dip is 60° N. W., and again 50° N. W. The nearest exposure of the limestone to the mass of schist at the east end of Main street, is near Elm street, and the strata are inclined 25° N. 30° W. These details are of practical interest, aiding in the development of the rock suitable for the manufacture of quicklime.

Both a north-westerly and a south-easterly dip may be found in the different openings of the Meadows quarries. Along the line of the section in Fig. 38, we found the following positions. At the south-east side of the limestone, the dip is 75° N. W. At the calcite locality the dip is 80° S. E. Between this opening and the one on the west side of the north and south road where a variety of white talc occurs, the dip is north-westerly, but variable, from 20°–80°. Several slips are noticeable, and we have attempted to give an idea of the irregular dip and the slips in our section. On the west side of the road the strata dip 75° S. E., becoming steeper according to the depth in the quarry. But no particular dip is constant through all these openings. The total length of this large bed of limestone is six and one-half miles. Its northern limit is near Chickawakie pond.

The section in Fig. 39 shows us a limestone anticlinal west of this principal range at Ingraham's quarry. A similar mass is exhibited on the hill west of Chickawakie pond, Holmes' quarry, in Fig. 40. The two masses must have been elevated by the same agency, and are in the same line of strike, three miles apart. Ingraham's quarry is on the east slope of Marsh's mountain, and west of Mill river. The strata are almost perpendicular, but show an anticlinal structure. Between two openings in the quarry there is an interesting bending of the strata, too small to be represented upon the Figure. On this hummock the strata dip south-easterly about twenty-five degrees; while beneath them a few feet the layers are nearly perpendicular. We suppose this is not an example of mere bending, but of slipping for a short distance also. At first we were inclined to believe that the bending was the result of the drift force; but the dip forbids this supposition. The drift agency would have bent it in the opposite direction, had it operated upon the ledge. Of course this hummock is rendered of inferior quality for the kiln. An interesting black mineral occurs in a portion of this limestone very abundantly, giving it a different character. And a few rods east of the quarry may be seen a narrow band of dolomite, just as there is east of the Beechwoods quarry. This limestone, as well as that at Holmes' quarry, differs in no respect from the common Eolian limestone of this region.

At Holmes' quarry in Camden, the amount of limestone is smaller. It is situated on the side of the hill directly west of Chickawakie pond. No one would suspect its existence from the adjacent rocks—and we only stumbled upon it accidentally. It could hardly have shown itself at the surface before it was quarried; but now that it has been opened, it is found to increase in width in descending. The limestone is perpendicular. A few strata of schist are folded around the calcareous mass, and dip sixty degrees away from it on both sides. But on the west side the dip very soon changes to correspond with the easterly dip on the other side. No other mass of limestone appears upon either side of Holmes' quarry on the line of strike.

We suppose the curious position of these two quarries of limestone must be explained in this manner. The schists naturally overlie the limestone, and had no plicating forces ever operated in Maine, these beds would have forever remained concealed. The force in this region seems to have been exerted more powerfully than in some other parts of the State. Now when this force began to act—supposing it to operate slowly and gradually—these two kinds of rock were pushed up so as to form a very pretty anticlinal curve of considerable width. As this agency continued to crowd the sides of the anticlinal from the south-east, the dips became greater and greater, until they become perpendicular, standing side by side. The pressure at Ingraham's quarry seems to have elevated the strata to this pitch; while at Holmes' quarry the force operated still further and inverted the fold, so that the strata of schist all dip in the same direction, while the limestone has not been inverted. Were it not for these masses of limestone there would be no indication remaining of this anticlinal fold. How many more such inverted folds may exist in this region we have no means of ascertaining in the absence of such protuberant masses as this limestone.

An important practical inference results from this theory. If these two projections of limestone indicate an anticlinal ridge, then all along the line connecting these two quarries we may expect to find more of the limestone. These quarries have been opened on the very crest of the fold; and there must be some approximation towards horizontality in the top of the ridge, and consequently if not at the surface certainly at a short distance below it, the limestone may be found.

Several beds of Taconic limestone east of the principal Eolian belt should be noticed. One of considerable importance is near the east end of the principal street of Thomaston, to the south, and was formerly worked as a quarry. The rock is whiter than the common limestone to the north. It dips 25° N. 20° W. Its limits are not known; but it may ultimately be found to be a spur from the grand deposit.

Two beds are found upon Charles E. Butler's farm, upon the north and south of his house, in the south-east corner of Thomaston. The northern one is located on a bluff facing the Meadows quarries; an immense tract of lowland occupying the country between. An old quarry here was in a blue limestone similar to the common rock of the other quarries; while underlying it is a white dolomite dipping 50° N. W. and resting upon the Taconic micaceous schists and quartzite. The other bed is of the blue variety and is of less importance. It is several feet thick, enclosed in schists.

Brenton Butler owns a quarry of dolomitic limestone near a large marshy bog on West Keag river in South Thomaston, hence it has been called the Marsh quarry. It was not being worked when we visited it in July. Dr. Jackson pronounced this bed to be two hundred and fifty feet wide, and its outcrop forty or fifty rods long. It is white, granular and compact, producing a lime containing much magnesia which does not injure the quality to any important extent. The dip is about 60° N. W. At the quarry we observed some peculiarity in the stratification, perhaps of the nature of a slide and consequent dislocation of the strata, but had not time to investigate it. Another body of limestone is said to exist north-east of the dolomite, which we did not examine.

Another extensive deposit of Eolian limestone over two miles in length, and upon which important quarries are located, occurs in Rockland. The most southern exposure of this belt is at Rankins' quarry. The rock is the same as that in the main belt, as is all the rock of this belt. It is traversed by an immense number of small veins of calcite running about north and south. The limestone dips 70° S. 75°E. At Blackington Corners a number of quarries have been wrought in the limestone, so that very large holes have been excavated. Different observations here show that there is an anticlinal axis in this belt, although the inclinations in both directions are very steep. There is a very large boulder in one of these quarries underlying the road to Rockland. It is thirty-five feet in length and rather ovate shaped. It is a fact to be remembered that the southern end of this belt of limestone as far as it is worked, corresponds very nearly in latitude with the northern end of the Meadows quarries in the principal belt. The limestone has not been traced a great distance north of Blackington Corners. What if a transverse fault should be indicated here!

We had intended to prepare detailed, historical and statistical notices of all the quarries in Rockland and Thomaston, but owing to unforeseen circumstances have been prevented from obtaining the desired information. Such as we have will be presented under *Economical Geology.* The following are the names of the principal quarries and their locations. In Thomaston, on the principal belt are the State's Prison quarry, in the yard of the State's Prison, near the west end of the main street; Miller's quarry, north-east from the State's Prison; the Beechwoods quarries, which are still further north-east; the Fulling Mill or Mill River quarries, just north of the east end of the street; a large number of small openings between and in the vicinity of the last two quarries; and the southern part of the Meadows quarries, which lie partly in Rockland to the east of Mill river. These are worked the most. In Rockland there is Ingraham's quarry on Marsh's mountain, with Rankins' and the Blackington Corners quarries on the eastern belt. In South Thomaston is the Marsh quarry. The history of all these quarries would be extremely interesting.

The limestone common to this group is of a bluish or bluish gray cast generally, coarse in its appearance, but nearly pure carbonate of lime, having about two per cent. of impurities. Several different qualities are known among the quarrymen, such as the "birds-eye" and the "lump.".

We regret very much not having prepared in season a wood cut to illustrate the geographical distribution of those bands of the limestone which are the most highly esteemed for the manufacture of lime; since we think the study of them would show all interested parties in what direction to look for further discoveries of the most valuable portions. Such a chart was prepared for private study with the assistance of J. O. Robinson, Esq., of Thomaston, and from it we glean an important suggestion. From the State's Prison quarry a belt of the most valuable layers extends in a straight line to the Beechwoods quarries, and then turns easterly to the Fulling Mill. Thus far the quarries and openings are numerous. But from this point to the south end of the Meadows quarries, nearly a mile in distance, we could not ascertain that a single opening had ever been made. To be sure the ledges are concealed. But let any one consider the relations of these two lines of quarries, and we think he will begin to see the probability of the existence of other beds of limestone over this space of equal value with those now being wrought. A reference to the geological map will show that the belt of limestone, at this gap in the line of quarries, is wider than anywhere else in its whole extent.

A glance at this chart shows us also that there are really two lines of quarries in the Meadows; while often between them the rock is of an inferior quality. This may be due to folding, probably a synclinal. It is of course implied in these remarks that it is only a portion of these Eolian limestones which can produce the best quality of quick lime; and also that the space occupied on the map by the marking for this rock includes the *formation*, not the lime belt merely. The relations of the better to the poorer qualities, however, has not yet been made out. We trust that the results of our explorations now set forth, will be a contribution to so desirable a result.

Camden Belt. The large geological map, and also Fig. 36, show the distribution of another important belt of Eolian limestone in Camden, whose lithological and geological character ally it to those just described in Rockland and Thomaston. The belt commences in Rockport harbor, passes northerly beyond Lily pond, and then turns west, terminating finally near Hosmer pond in the west part of the town.

As we first see this limestone on the coast, it is whitish and rather friable, but of good quality. Its connection with an overlying slaty rock is very plain. Kilns of it have been burnt here. As all the rocks upon this promontory have been much disturbed, so the dip of this limestone is variable. The following are the dips observed in order from right to left: 75° S. E.; 80° S. 80° E.; 65° N. 20° W.; 80° N.; 45° N. 10° E. This variation is rather from dislocation than folds. A part of this belt seems to have been torn from it by a fault, and is displayed in the village of Rockport and on the shore to the west of the principal belt, (See Fig. 36.) This indicates the existence of a lateral fault, which we have represented imperfectly.

Passing up to the carriage road south of Lily pond, we find the limestone dipping 60° N. 30° E. It is exposed in large coarse-looking hummocks on both sides of the road. The quarries now worked in Rockport, are chiefly upon the south-west side of Lily pond, in a high bluff. The rock is really a conglomerate of limestone pebbles; yet the stratification is extremely obscure. The dip appears to be about 50° N. E. It is chiefly the lower part of the limestone that is conglomeratic. It is possible that these supposed pebbles are concretionary. We now come to the curve in the formation, and the numerous observations all show the gradual change of the strike. The formation is twice as wide here as in any other part of its course. To the north of the pond are several quarries, and also to the north-west, whose quality is precisely similar to the Thomaston stone. The angle of inclination varies from 35° ro 80°, and essentially in the same direction. With the names of the different quarries we are unacquainted.

More exposures of limestone appear on the roads west of Rockport, some of which are quarried. We find the conglomerate west of the village dipping 60° N. 10° E.; and north-west of the village, on the west road to Camden, there are quarries. The layers appear to be vertical, with the strike of N. 65° W. Some slaty layers adjacent dip 60° N. 70° E. At the quarries still further west, near Simonton's Corners, the dip is 70° N. 10° E. Beyond these, which are very extensively worked, we measured strata dipping 75° N. 60° E., and N. 75° E. The rock has been traced to within a short distance of Hosmer pond. As the strata are thrown about in large masses near Hosmer pond, we conclude that the limestone terminates here, because of a dislocation, rather than on account of the gradual thinning out of the strata.

We introduce here a table of the dips of the Eolian limestone in all the localities where they have been observed, whether already noticed or not:

The Large Belt in Thomaston and Rockland.

Locality.	Dip.	Remarks.
State's Prison quarry,	80° S. E.	
Miller's quarry,	65°–70° N. 50° E.	
North of do.,	75° N. 40° W.	
Beechwoods quarry,	N. 30° W.,	South-west end.
Beechwoods quarry,	75° north-westerly.	
Beechwoods quarry,	N. 1[illegible]° W.,	A few rods east.
Fulling Mill quarry,	62° N. 10° W.	
North of do.,	20° N. 20° E.,	At the old dam.
Below Fulling Mill quarry,	20° N. 30° W.,	Marble quarry.
Near the marble,	60° north-westerly,	Pavement quarry.
Near Thomaston village,	50° N. W.,	Near do.
Near Elm street, Thomaston,	25° N. 30° W.,	On Mr. Jacob's land.
Meadows quarries, Talc locality,	75° S. E.,	And grows steeper.
do. east of do.,	20°–80° N. westerly,	In a section.
do. Calcite locality,	80°–82° S. E.	
do. S. E. part,	75° N. W.	
Day's ledge,	74° S. 35° E.	
	Other Belts.	
C. E. Butler's two ledges,	50° N. W.	
Brenton Butler's, S. Thomaston,	About 60° N. W.,	The Marsh quarry.
S. E. from Thomaston village,	25° N. 20° W.,	An old quarry.
Ingraham's quarry,	82° N. W., and 20°–70° S. E.	

Rankin's quarry,	70° S. 75° E.	
Blackington Corners' quarries,	N. Westerly 65°, and 55°–60° S. 70° E.	
	Camden Belt.	
On Vauchamp Point,	70° S. E.,	Near "Last House."
do. further west,	80° S. 80° E.	
Half a mile west of last,	65° N. 20° W.	
Near the last,	80° N., & 45° N. 70° E.,	Slaty.
South of Lily pond,	60° N. 30° E.,	On road to Camden harbor.
West shore of Lily pond,	50° N. E.,	At quarries.
Between Lily pond and the road west,	35° N. E.,	Conglomerate limestone.
Quarry north of Lily pond,	55° N. E.	
Quarry west of the last,	70° easterly.	
Smart's quarry,	70° to 80° N.	
Near Rockport,	45° N. 40° E.,	At the junction with silicious slate.
do. west,	60° N. 10° E.,	Conglomerate.
North-west of Rockport,	Str. N. 65° W., 90°.	
do. further on,	60° N. 70° E.,	Slaty layers.
Simonton's Corners' quarries,	70° N. 10° E.	
West of do.,	75° N. 60°–75° E.	

A highly important section is illustrated in Fig. 41. It extends from Mount Battie southerly beyond Rockport, a distance of four miles, but it crosses every member of the Taconic series in Camden. At the bottom we find the older mica schist, in Rockport. Above it in nearly regular succession are the successive members of the Taconic system. They are in the ascending order, Hyaline quartz, dark colored quartz rock, dark colored silicious slate, quartz rock and conglomerate, limestone conglomerate, thick-bedded limestone, (the last two members are Eolian,) micaceous quartz rock, wrinkled clay slate, argillo-micaceous schist. The four first members are properly parts of the formation quartz rock, and the last three are parts of the Taconic schists. Estimating the thickness trigonometrically, we should say the quartz rock is *at least* 1,084 feet thick, the Eolian limestone 630 feet, and the schists 1,690 feet thick. The conglomerate upon Mount Battie is supposed to be the equivalent of the quartz, and is 500 feet thick, which is very nearly that of the lowest member of the formation in Rockport. Hence the total thickness of the series in Camden is at least 3,404 feet. We have taken care not to over-estimate.

This section illustrates the fact that none of these rocks are repeated by axes, save some unimportant flexures in the upper schists. As elsewhere explained, a fault undoubtedly exists between *k* and *l*; and the upheaval of the mountain may have been the reason why the schists are more elevated and disturbed than the underlying members. This may be styled a *classic* section for the Taconic rocks on Penobscot bay, since it is the best one for showing the natural order that can be found in this region, and it will be prized as such.

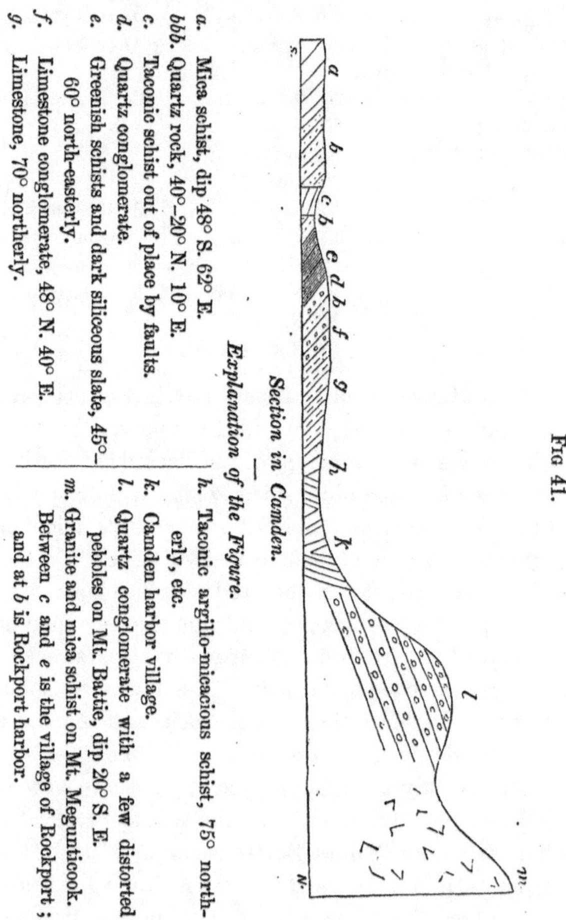

FIG 41.

Section in Camden.

Explanation of the Figure.

a. Mica schist, dip 48° S. 62° E.
bbb. Quartz rock, 40°–20° N. 10° E.
c. Taconic schist out of place by faults.
d. Quartz conglomerate.
e. Greenish schists and dark siliceous slate, 45°–60° north-easterly.
f. Limestone conglomerate, 48° N. 40° E.
g. Limestone, 70° northerly.
h. Taconic argillo-micacious schist, 75° northerly, etc.
k. Camden harbor village.
l. Quartz conglomerate with a few distorted pebbles on Mt. Battie, dip 20° S. E.
m. Granite and mica schist on Mt. Meguntiocook.

Between *c* and *e* is the village of Rockport; and at *b* is Rockport harbor.

One end of the Camden belt of Eolian limestone now appears to terminate in the bay. We have no doubt it continues for considerable distance under water; and suspect that it is directly connected beneath the schists with the Thomaston limestones; perhaps very much in the same way that we thought a connection existed between the limestones exposed at Holmes' and Ingraham's quarries.

Trap Dikes. In the principal quarries of limestone in Rockland and Thomaston interesting dikes of trap occur. To the workmen they are a nuisance, but to the geologist are full of interest, as indications of ancient molten masses like lava, in this vicinity. When the rocks about here were being tilted up at various angles, folds were formed and many large masses were broken apart. Molten matter from beneath then crowded up and filled many of the crevices. When this lava had cooled, probably quite slowly, these dikes were found to be the result. These dikes we have seen in the Beechwoods, several of the Meadows, near Rockport, and in the Blackington Corners quarries. They are seen now as walls upon the sides, in the middle or crossing the quarries, being composed of a dark, tough, very fine grained rock. Occasionally a row of trap nodules seems to take the place of a dike.

Fig. 42.

Trap Dike in the Beechwoods Quarry.

An interesting example of a trap dike at the Beechwoods quarry is represented in Fig. 42. It is in the sketch seen to form a wall, which curves, and at the right hand side passes round a pillar of limestone facing the observer. It then crosses to the other side of the quarry, and continues on in a course parallel to its first appearance. That part seen in the sketch resting against the framework of limestone, is composed of numerous small columns, piled up with great regularity, as if done by hand, just as many sticks of wood are laid up with the sawed ends in front. At another quarry the limestone has been excavated from both sides of the trap, leaving the dike as a high wall in the centre of the excavation.

A few other dikes were noticed in this region in other rocks. Near Owl's Head Light House we saw a dike of quartz eight feet wide, not perpendicular like those in the Eolian limestone, but inclined 70° S. 20° E. Pieces of the wall rock are found in it, as if torn off and mixed in the fluid mass at the time of its injection. Several quite large dikes of trap traverse the gneiss at Herring Gut Light House in St. George. They run N. 70° E., the same as the strike of the quartz dike. A small dike of white quartz, running N. W. and S. E., cuts across the Taconic schists on Negro Island in Camden harbor. Several dikes (or almost beds) of trap occur in the Taconic schists and quartz rock in Belfast harbor.

6. *Taconic Schists.*

We now come to the consideration of the last solid formation of this region west of the Penobscot bay, the schists belonging to the Taconic system. In Emmons' scheme these were called "Magnesian slate," an inappropriate term, since both the term slate is improper, and they contain no magnesia. In Vermont we styled them Talcoid schists. But in Maine the Talcoid character is very obscure, and in its stead we find the mineral mica abundant. We do not suppose that it is essential to theory, that these schists should every where be talcose in their appearance, because they are so in Massachusetts and Vermont. At all events in Maine so much mica is present frequently that they cannot be distinguished from the older mica schists, and on account of this difficulty it will not be strange if we have improperly distinguished between them in our geological maps, particularly in South Thomaston.

Three general deposits of these schists occur on the west side of Penobscot bay; first, those associated with the Eolian limestones of Thomaston and Rockland; second, a deposit in Camden; and third, a still larger deposit of interstratified schists and quartz rock running south-westerly from Belfast. To expedite the description of these deposits we will give a table of all the observal dips:

Rockland Deposit of Schists.

LOCALITY.	DIP.	REMARKS.
Owl's Head,	75° N. 10°–40° E.,	Quartz rock.
Owl's Head promontory,	75° S. 40° E.,	At west end.
Thomaston,	84° N. 40° W.,	East end of village.
Thomaston, toll bridge to Cushing,	60° N. 20° W.,	East side.
do.	12°–30° N. 20° W.,	A local curve.
do.	75° N. 40° W.,	West side.
Thomaston, south of St. George riv.,	53° N. 40° W.,	Half mile south.
do.	90° N. 40° W.,	Still further south.
Thomaston village,	70° N. 40° W.,	At Mr. Jacob's garden.
Thomaston,	60° N. 30° W.,	Near Fulling Mill quarry.
Thomaston,	75° S. 20° E.,	West of Beachwoods quarry.
Thomaston,	75° N. W.,	East of Meadow's quarries.
Thomaston,	50° N. W.,	At Chas. E. Butler's.
Thomaston, west part,	65° N. 50° W.,	Near bed of quartz.
Thomaston, on Rockland line,	Anticlinal axis,	North of last observation.
Marsh's Mountain,	Perpendicular.	
North of Blackington's Corners,	45°–50° S. 60° E.,	Quite argillaceous.
Rockland, N. E. part,	79° N. W.	
Camden, near Clam Cove,	40° S. E.,	At fork in road.
Camden, Clam Cove,	70° N. 20° W.	
Camden, north of Clam Cove,	48° N. 70° E.	
Chickawakie Pond, S. W. part,	50°–60° S. 70° E.,	See Fig. 40.
Chickawakie Pond, N. W. part,	48° N. 30° W.	
Chickawakie Pond, north of,	30° northerly.	
do. do. a mile,	50° northerly.	

Camden Deposit.

LOCALITY.	DIP.	REMARKS.
Simonton's Corner, north of,	High dip to N. 30° W.	
Simonton's Corner,	60° N. 30° E.,	West of kilns.
Between Camden and Rockport,	50° N. 20° W.,	On the west road.
do. at last house in Camden,	75° N. 12° W.,	On the middle road.
Rockport, south of Goose river,	50° N. 70° E.,	Brought up by fault.
Rockport, on north shore of harbor,	30° N. 20° E.,	" " "
Indian Island lighthouse,	30° northerly.	
At junction with Eolian Limestone on Beaucamp's Point,	50° S. 80° E.	
do. 20 rods south-easterly,	40° S. 70° E.	
Deadman's Point,	35°–70° N. 60° W.	
do. further east,	40° S. 60° E.,	Anticlinal.
Above Deadman's Point,	55° S. 60° E.	
Lily Pond, east side,	50° N. 30° E.	
Lily Pond, north of,	50° N. 40° W.,	Near quarries.
One-eighth mile north-east of last,	40°–70° S. 80° E.,	Argillo-micaceous.
Hill south of the harbor,	40° N. 60° E.,	On east road to Rockport.
Negro Island,	45° N. 70° E.	

Mouth of Megunticook river,	60° N. 70°–80° E.	
At Methodist Church, Camden,	70° northerly.	
Camden Village, foot of Mt. Battie,	75° S. 70° E.	
Camden, west base of Mt. Battie,	8°–10° south-westerly.	
	Belfast Deposit.	
Belfast, Steamboat Wharf,	63° N. 40° W.	
do.	28°N.20°W.& 45°N.W.	
Belfast city, S. W. part,	75° N. 30° W.	
Belfast, mouth of Little river,	60° N. 20° W.	
Belfast, west of city,	High to S. E.,	Quartz rock.
Belfast, south-west line,	High to N. W.	
Waldo, east edge of,	70° N. 10° W.,	Quartz rock.
Waldo, east part,	45° S. W.	
Waldo, middle,	75° southerly.	
Belmont Corners,	60° S. E.,	One and a half miles north.
Appleton, west of Smith's Mills,	75° north-westerly.	
Appleton, near McLain's Mills,	65° N. 30° W.,	On top of hill.
Stockton, N. E. part,	80° S. 60° E., and 90°.	
Stockton, north line,	80° N. 60° W.	Anticlinal.

Figs. 37, 38, 39 and 40, show the relations of the Taconic schists to the adjacent formations. If these sections illustrate the natural order of the formations, then we must say at once, that a part of these schists are older, and a part more recent, than the Eolian limestone. We rely upon such cases as the exhibition of the strata in Holmes' and Ingraham's quarries, (Figs. 39 and 40,) to prove the schists in some cases the more recent of the two. And by the supposition of inverted folds, we may regard all the schists as the newest. For example; in Fig. 37, suppose the dolomite at the Marsh quarry to be the equivalent of the other belt of limestone at the Fulling Mill and Beechwoods quarries. Then the intervening schist is newer than both kinds of limestone, and the western mass of limestone contains two or three inverted folds. Similar suppositions can reduce the other sections to the same order. It will be interesting, in this connection, to compare the theoretical order of these formations with the actual existing order in Camden, as shown in Fig. 41. There the schists are all above the limestone, except a few that have been disturbed by a fault.

A little study of the preceding table of dips will show us a few axes besides those figured. An interesting one is situated west of Marsh's mountain, both in Rockland and Thomaston. It extends for a considerable distance—a mile or so we have followed it; and the crest of the ridge has been worn away while the sides were not removed, so that there is now a valley along the anticlinal ridge. This is what is called an *anticlinal valley.* Between Chickawakie pond and Clam cove there is a synclinal axis. It is remarkable that a northerly dip should be the prevailing one north of this pond, extending even into the mica schist of the older formation. At the north-west side of the pond is a north-westerly dip which is connected with the inverted anticlinal at Holmes' quarry further south. There is no axis in the Camden deposit; although there is some diversity of dip, it is very difficult to find anything more than local folds, of which the one in Camden village is a sample. The variations generally are all contained in a single quadrant. The small south-westerly dip on the west side of Mt. Battie is probably an isolated remnant of the original position of the belt. The Camden schists are all in one belt, then, dipping northerly and north-easterly, but making a great bend like the Eolian limestone and quartz rock beneath.

In the Belfast belt we get a few axes. At Appleton none were observed, but in the vicinity of Belfast we have two: a synclinal west of Belfast city, which we have traced ten miles into Belmont; and an anticlinal still further west in Waldo. An anticlinal exists in the north-east part of Stockton in similar rocks, and most probably the very same belt.

The schists on Owl's Head promontory are largely quartzose. Indeed, at the end of the point the rock is entirely quartz. The western portion contains a great deal of mica and presents a woebegone appearance. The quartz contains tubercular veins of a whiter and purer hyaline quartz. A few veins of syenite are also present. The greater portion of the schists in Thomaston and Rockland are very handsome readily cleaving layers of a pretty well characterized mica schist, generally too much penetrated by jointed planes to allow very large plates to be exhumed from the ledges. At the east end of Thomaston village, on the west bank of Mill river, and lying exposed very prominently in the street, is a large ledge of a very argillaceous schist, much like the characteristic Taconic schists in Massachusetts and Vermont. It is seen again in the bed of the river lower down. The rock in Mr. Jacob's garden is similar, and it crops out again at the Toll Bridge to Cushing. On the west side of the bridge the rock is largely quartzose, and may be in the same layer with the ledge of quartz on the road to the Beachwoods quarry, seen after leaving the main street.

This layer seems to dip under the limestone. A bed of schist thirty feet wide appears near the Fulling Mill quarry in the limestone. If the schists between the West Keag river quartz and South Thomaston are Taconic, they will be ranked equivalent to the quartzose rocks of Owl's Head.

Two varieties of the Taconic schists are designated in the map of the Camden rocks in Fig. 36, the lower layer being very micaceous, while the upper layer, or that furthest to the right, is chiefly argillaceous. In the table of dips a multitude of observations are given. We can see the micaceous rock exposed all the way from the end of Beauchamp point northward and westward as far as the limestone extends, being adjacent to it. On Deadman's point an anticlinal is found in it. Here many layers of quartz are interstratified with it. Near the junction with the limestone on the shore, the layers of both rocks are very much disturbed. The other argillaceous belt follows around the shore from the interior of Camden harbor nearly to Deadman's point, (the point shown on the map very near the extreme southern point of the large promontory in Camden.) In some places it is impossible to determine the stratification, so homogeneous is the rock and so obscure the layers. This is the peculiar rock spoken of by Emmons as a "wrinkled magnesian slate." Any one who looks at it cursorily, will see at once the appropriateness of the term wrinkled. A good place to see this schist in perfection, is on the hill in the east part of Camden village. Being an enduring rock it retains very well the diluvial markings. This belt we suppose to overlie the other variety, and to be the uppermost member of the group.

We hesitated somewhat how to represent the Belfast belt of Taconic rock. Emmons expressed some doubt whether these were truly Taconic, but we think there can be very little question that they belong to the same age as the Camden group. Our difficulty was another; whether it was best to distinguish between the quartz rock and the *argillaceous schists*—if the expression be proper. At the steamboat wharf and to the east, many layers of quartz are interstratified with the more common slates and schists. To the east the schistose rocks predominate ; while to the west of Belfast the layers are mostly purely silicious, even to the very border of the formation. Probably upon future maps we shall be able to draw a clear distinction between these two varieties of rock. An interesting fold appears in the rocks near the wharf in Belfast.

The general dip is north-westerly, and the curve is crowded between the highly inclined strata, and is rendered a sharp curve in consequence of the great pressure to which it was formerly subjected. We mention it particularly as it is exposed in a favorable locality for inspection by those who are interested in viewing how the forces of Nature have operated. We intended to have measured very carefully the dip, thickness and precise mineral character of every layer that is exposed along the shore in Belfast, but time prevented. It is the most favorable place for such measurements we have seen among any of the Taconic schists. The careful measurements should extend to the older gneiss in Waldo.

7. *Alluvium.*

Quite a number of observations were made among the alluvial rocks and markings in this region; but as we have thought it best to group together under *Surface Geology*, our descriptions of alluvial phenomena from every portion of the State, we will therefore refer to that heading for the description of the alluvium of the country about Penobscot bay.

Mr. George L. Goodale spent a few days, at our request, in Vinalhaven, and we now present his description of the rocks there seen. We suppose these rocks, with those of North Haven and Islesborough, to be entirely Taconic, and to belong to the schists, or the upper portion of the series.

To C. H. Hitchcock, *A. M.:*

Dear Sir:—In accordance with your wishes, I visited the Fox islands in Penobscot bay during the early days of July. Owing to limited time I was unable to make the examination as thorough as I could have desired. But through my friend Dr. Delaski, who has studied carefully the geology of this neighborhood during his residence of several years at Carver's Harbor, I was enabled to pursue my investigations to better advantage than would have been possible in the case of a perfect stranger. Dr. Delaski has paid particular attention to the records of the drift phenomena at Vinalhaven, and he has kindly promised to furnish a detailed account of his views upon this highly interesting, and, as yet, obscure subject.

Although his opinions may differ materially from those entertained by the Geologist of the Survey, I am sure they will be received with pleasure as coming from one whose observations are acute and accurate.

The Fox islands, in the wide bay of the river Penobscot, comprise two large and several small islands. The largest of this collection of many islands is called Vinalhaven from the name of one of the earliest inhabitants. The largest island is surrounded by an almost perfect breakwater of many smaller islands, so that there are two excellent harbors and many inferior ones upon the south and south-west. Although the islands composing this chain are exposed on their front to the full force of the swell from the Atlantic, they yet exhibit few marks of the power of tide-water and waves. This is due to the firm character of the rocks of which they are made up. It is a peculiarly fine-grained syenite of good color and containing little or no oxidizable iron. To this "granite," as it is commonly known in the bay, the little village of Carver's Harbor owes its present prosperity. The granite is easily worked into a tabular form even where the natural rift is considered most unsatisfactory, and these merchantable blocks are quarried within a short distance of the wharves.

The specimens of granite from this island compare very favorably with granite from any other part of Maine. There are three large quarries on the island and some minor ones, all of which were busily engaged in furnishing rock for government uses. The north island, or "North Haven," as it is generally called, is separated from the island just noticed, by a thoroughfare varying in width from two miles to very much less. The village is situated on the north side of the thoroughfare. I have deemed this brief topographical description of both islands of much importance in understanding their prominent geological features.

Leaving the drift, the surface geology, in the able hands of Dr. Delaski, let us notice briefly the rocks as we pass from Carver's Harbor in a northerly direction. We find nothing save syenite till we reach the school-house on Calderwood's neck. Here is a peculiar schist, apparently micaceous in its character, with a dip of 85° and a strike N. N. E. Near this rock and further from the granite, occurs a silicious slate of great density and containing a small per cent. of lime. Owing to the white weathering of this slate, it was thought by many of the inhabitants formerly that it might, by calcination, become a fair lime for rough work. Some of this slate contains more lime than the Kittery slates, so called, and would be more likely to deceive those unpracticed in the matter of selecting limerocks. But I have little hesitation in expressing my opinion that lime of a poor quality might be manufactured from certain rocks occurring in this silicious range, and one is the more confident in regard to this when it is remembered that the great lime beds of Rockland lies much less than twenty miles west of the islands. North of the point last mentioned we come upon a singular conglomerate belonging perhaps to the Megunticook series. In its lithological characters it is not much unlike the conglomerate in Roxbury, Mass., and occurs in much the same manner. This rock varies very much as we push farther north, becoming, in North Haven more like an indurated limestone containing occasional pebbles of a fine-grained, slaty character. This conglomerate contains considerable lime, which is not without its effect upon the fertility of the soil. While the south of Vinalhaven, underlaid by syenite, is comparatively infertile, the island of North Haven contains many excellent farms. The peculiarity of the conglomerate and the amount of calcareous matter present in it, forms a subject of much interest when considered in connection with the lime bearing rocks of Thomaston and Rockland. My duty in the hasty examination of these highly interesting islands, was that of a collector. The specimens placed in your hands, Mr. Hitchcock, will give a correct idea, I trust, of the lithological character of the islands. In my opinion the north of Vinalhaven, and the North island itself, are underlaid by rocks of the same age as those in the Taconic basin of Knox county.

I am, sir, your obedient servant,

G. L. GOODALE.

Portland, Nov. 1, 1862.

II. Description of a Geological Section from Eden to the Canada Line.

By a *Geological Section*, we mean an exhibition of the stratigraphical relations of the rocks to one another over a given line. This line should always cross the strata as nearly at right angles with them as possible. It is just as if one could see the edges of the strata deep down in the earth, where an immense trench had been dug along the line of the section. Such cuts are frequently exhibited to us in the strata of gravel and sand exposed in bluffs by the crowding action of rivers. The geologist reasons out the position of the solid strata beneath the soil by an attentive observation and comparison of the dips where they are exposed above the soil in ledges. He supposes that if strata dip 40° north-west into the ground in some prominent ledge, that the rock will continue to dip so two hundred feet below the surface, unless there be some reason to the contrary. And if for the distance of one mile, he finds four ledges of rock with equal intervals between them, and all possessing the same inclination, then with reason he infers that the concealed layers between the ledges have the same dip. But if he finds two of these ledges to dip north-west, and the other two to dip south-east towards the former, then he concludes that the strata meet under ground and form a *synclinal axis* or a sort of basin. Of if they dip away from each other, then he concludes they must meet in the center, much like the roof of a house, only that the top is generally rounded instead of angular, and much of the crest has been worn away. Such an axis is called an *anticlinal axis*. By such processes of reasoning is every observation accompanied, only the geologist does not consider it necessary to repeat them with every measurement. He takes it for granted that the reader will always supply this deficiency in his own mind, if need be.

When the geologist has travelled over a section many miles in length, it is highly probable that he will have to record several axes, either anticlinal or synclinal. And it may be that he must notice the occurrence of unstratified rocks along with the stratified. These cannot be represented as occurring in layers, but the nature of their junction with the stratified ones must be carefully noted. Again, the geologist may be at a loss to know certainly how certain formations are disposed beneath the surface, especially upon high land, where the number of ledges are very scanty. Hence he need not fill out the space allotted to him by the altitude, but simply locate in their proper places the respective dips.

It is necessary often to employ two different scales in protracting sections from notes—one for the horizontal and the other for the perpendicular distances—otherwise the hills will not be conspicuous. In the section about to be described, our vertical section is one inch for every one thousand five hundred feet of elevation, and of course for the smaller altitudes proportional parts of an inch. The horizontal section is nearly an inch for every twelve miles. Hence the outline of the surface must be distorted, but this distortion will not generally be obvious to the observer.

The section extends from Mount Desert Island to the Canada

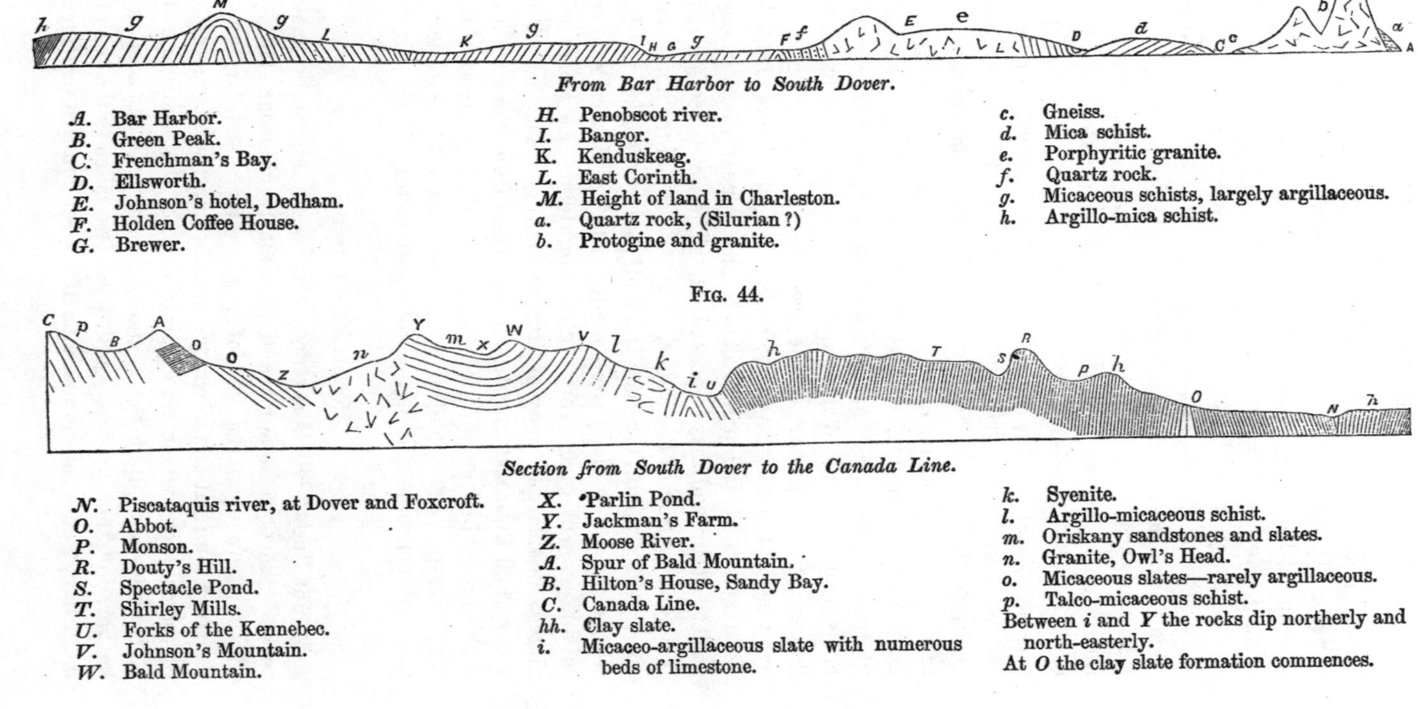

FIG. 43.

From Bar Harbor to South Dover.

A. Bar Harbor.
B. Green Peak.
C. Frenchman's Bay.
D. Ellsworth.
E. Johnson's hotel, Dedham.
F. Holden Coffee House.
G. Brewer.
H. Penobscot river.
I. Bangor.
K. Kenduskeag.
L. East Corinth.
M. Height of land in Charleston.
a. Quartz rock, (Silurian ?)
b. Protogine and granite.
c. Gneiss.
d. Mica schist.
e. Porphyritic granite.
f. Quartz rock.
g. Micaceous schists, largely argillaceous.
h. Argillo-mica schist.

FIG. 44.

Section from South Dover to the Canada Line.

N. Piscataquis river, at Dover and Foxcroft.
O. Abbot.
P. Monson.
R. Douty's Hill.
S. Spectacle Pond.
T. Shirley Mills.
U. Forks of the Kennebec.
V. Johnson's Mountain.
W. Bald Mountain.
X. Parlin Pond.
Y. Jackman's Farm.
Z. Moose River.
A. Spur of Bald Mountain.
B. Hilton's House, Sandy Bay.
C. Canada Line.
hh. Clay slate.
i. Micaceo-argillaceous slate with numerous beds of limestone.
k. Syenite.
l. Argillo-micaceous schist.
m. Oriskany sandstones and slates.
n. Granite, Owl's Head.
o. Micaceous slates—rarely argillaceous.
p. Talco-micaceous schist.

Between *i* and *Y* the rocks dip northerly and north-easterly.

At *O* the clay slate formation commences.

Line, crossing the strata nearly at right angles over its whole course. It passes in order through the towns of Eden, Trenton, Ellsworth, Dedham, Holden, Brewer, Bangor, Glenburn, Kenduskeag, Corinth, Charleston, Dover, Foxcroft, Guilford, Abbot, Monson, Shirley, East Moxie, Forks plantation, No. 1 R. 5, No. 2 R. 6, Parlin Pond, Jackman, Dennis, Holden and Sandy Bay. This line is nearly one hundred and eighty miles long by the map. This route was selected in preference to any other, because it was the longest one in the State which could be travelled over with a carriage, and it passed through a fossiliferous region of great importance; and it will give a good general idea of the relative position of the most important formations in the State. In future it is desirable that a large number of sections be measured parallel to this. This should be done, since the strata commonly possess a north-easterly course, and an exploration of the geology of the State by transverse lines parallel to one another, about fifteen miles apart, will afford the quickest and cheapest mode of learning its geological structure. The preceding figures will give a bird's-eye view of the position of the rocks along the route of this section. Both stratified and unstratified rocks appear upon it, and every variety of dip and axis is presented. The lower section (Fig. 44,) is a continuation of the upper, (Fig. 43,) while the numerous references will explain the names of all the rocks and the localities where the observations were made; and the description, which is to follow, will give every detail minutely. As previously intimated, an enlarged copy of this section, colored, and with specimens of all the rocks mentioned by name from the very localities, will be on exhibition at the State House during the session of the Legislature. As a matter of course the representation of the larger dimensions will be the most satisfactory.

Mount Desert Island.

Mount Desert Island is chiefly composed of granitic unstratified rocks. An occasional mass of an obscure siliceous slate appears, but nowhere upon the line of the section: and for the sake of showing the relations of a more than ordinarily interesting deposit of quartz rock, which may possibly be of the same age with the slates elsewhere upon the island, we commence the section at Bar harbor, and then proceed south-west two miles to the top of Green mountain, where the regular line of the section is intersected. We were agreeably disappointed in discovering at Bar harbor in Eden so interesting a deposit, although its dimensions in Eden are very small. The salt-water of the ocean seems very often to preserve the primitive character of many rocks within the reach of its influence, while that part of it which is exposed only to the action of atmospheric agents is woefully weathered and obscure. This fact is well illustrated in Machiasport, where it is utterly impossible to do anything with the weathered siliceous slates upon the crest of the promontory, while within the reach of tide-water the character of the rock and the dip of the strata are distinctly preserved, as well as the drift striae. So it is in a less degree in Eden. At Bar harbor, opposite Bar island, for the distance of one and a half miles, an interesting quartz rock appears on the shores, dipping at a very moderate angle and exposed in very interesting bluffs. It appears to belong to the same formation as the Rock of Flint island, described last year. We suggested a Lower Silurian age for that upon no very reliable grounds—more of fancy than real argument, because it reminded us so much of the Potsdam sandstone in its external appearance. This rock is less flinty and more sedimentary and micaceous in its character. Ripple-marks are found upon the layers, and also the curious cylindrical stems so common at Flint island. The layers dip 12° N. 20° W., quite the reverse of what they appear upon this section. But it was necessary to represent the dip in the section as south-easterly in order to show the true relation of the strata to the underlying granite—dipping away from rather than underlying the granite. The rock upon five islands lying between Bar harbor and Gouldsborough, viz: Bar island, Long Porcupine island, Ironbound island, and two others, is probably the same, and it may extend into Gouldsborough. Bar harbor is only twenty-five miles from Flint island in a straight line, and it would not be strange if these rocks are connected together under the water. We notice that quite a large promontory in Gouldsborough has the name of Grindstone point, which causes us to conjecture that a sandstone rock exists there to give the name to the point.

Going back from the shore this quartz rock is very much acted upon by the weather, and would not be recognized commonly as the same rock, and the marks of stratification are overshadowed by the planes of cleavage, so that the former are rarely noticed. Another patch of this siliceous rock not on the line of the section, skirts the shore of the north part of the island between Hull's Cove and Saulsbury Cove. A dike of trap containing veins of limestone appears at Hull's Cove. Probably this rock does not extend inland further than a line connecting the two coves. The dip of the strata could not be satisfactorily determined.

The first salient point in the section is Green mountain, 1,533 feet high by our Barometer. The rock at the summit is protogine, having a talcose mineral substituted for the mica of granite. The color is red. No doubt a great portion of the so called granite of the whole island is protogine. Large plates of the protogine at the summit dip 60° N. W., while near Eagle pond they are perpendicular. A few dikes of trap are found on the mountain also. The view from Green mountain is delightful. No other peak of the same height can be found on the Atlantic coast of the United States, from Lubec to the Rio Grande, nor from any eminence on the coast, can so fine a view be obtained. We were never charmed by any view so much as by this. The boundless ocean upon the one side contrasted with high mountains upon the other, and along the shore line numerous islands appearing like gems set in liquid pearl, form the most prominent features in the scene. White sails dotted over the water glide slowly along. We know not what view can be finer than this, where the two grandest objects in nature, high mountains and a limitless ocean occupy the horizon. The emotions awakened by this scene cannot be expressed, but the scenery remains in the memory, the never dying reminescence of a beautiful portrait placed in nature's gallery by the Master Painter. The name of Eden is truly appropriate to this beautiful place.

Following the line of the section we found quite a high mountain of granite in the north part of Eden. At Thomas and Hadley's Mill, on the north shore at the mouth of a stream, there is a band of granite. The spur of the land running out to the Toll-gate is composed of a variety of gneiss, appearing rather talcose, yet abounding in feldspar. The first dip is 35° S. 20° E., or beneath the granite rocks. At the first island on the bridge road the dip is 30° S. E. At the Toll House is a narrow slaty ledge, perhaps the first of the succeeding formation, dipping 45° S. 12° E. Upon Uncle Israel's point, a mile east of the Toll House, the rock is very distinct gneiss, dipping 37° southerly, and is intersected by four trap dikes, two of which are only one foot wide, while the third, a mass of columns, is three feet wide, and the fourth, a crystalline trap, is eighteen feet wide.

Some of these dikes resemble indurated slate, but their course is north and south, or at right angles to the course of the strata in the vicinity. West of Captain Thompson's house is an old quarry of syenite, which is very beautiful but extremely tough ; so much so as to prevent its use. A company once operated here, and our impression is that $33,000 were wasted on the enterprise.

Mica Schist Formation.

For two or three miles in Trenton the rocks are concealed on the direct route of the section, but it is not so at Oak point. The dip there is small, the layers of mica schist dipping 20° S. 40° E., almost the whole length of the west shore of Trenton above the point. Probably the gneiss does not extend into Trenton on the line of the seaboard, so that we are at once introduced into a formation of mica schist, often talcose in its appearance, particularly in certain layers. Near S. Crippen's house in Trenton is the first ledge of schist seen by the road side, which dips 30° N. W., making an anticlinal axis with the schists to the south, which must be in place beneath the soil. Following the road running nearly due west to the very corner of the town on Union River Bay, we find the dip changes there, also, from the southeasterly dip at Oak point, so that this cross road is constructed very nearly along an anticlinal line. The ledge near Crippen's may be seen in the road south of his house, but to the best advantage a few rods east of the road on the west bank of Jordan river, where some of the layers are calcareous. The locality was shown to us with the inquiry whether the lime was not sufficiently abundant to warrant the erection of a kiln. The rock is not adapted to be burnt for lime, nor have we seen any bed of limestone in the whole of this terrain of mica schist.

Following the road to the north we see an occasional ledge with the northwesterly dip. Near D. Higgins', at the triple branching of the road the rock is argillaceous and thin bedded. Between here and the south part of Ellsworth the ledges are obscured by drift. One mile from the village the ledges are numerous, dipping 43° N. W. This varies very little from the last observation on the section, hence we conclude that all the concealed layers between Ellsworth and the triple branching of the road in Trenton have about the same inclination. This is confirmed by observations upon both flanks of the section: for in the south part of Ellsworth, for three miles along Union river, the strata dip northerly from 30° to 50°, and upon the east side in Hancock, the strata dip 15°–20° northerly. At the shore of Union river south of the Post Office in the village these mica schists dip 45° N. W.

We now come to a change in the dip. In the north part of the village on a hill the strata dip 70° S. 15° E., so that we have here a synclinal axis, whose line runs along the course of Main street in Ellsworth village. It is because the strata have been pushed up almost upon their edges on the north side of this axis, that the hill in the north part of the village exists, although further north it has been largely worn away by the waters of Union river. Essentially this dip prevails in the rock as far as its northwestern limit, near C. Jarvis' house, a trifle more than three miles from Main street, or just about half the distance from the Trenton Toll Gate. It is worthy of notice that there are just as many feet thickness of the strata on one side of this axis as on the other, since where the dip is half as great the distance is double. A reference to the section in Fig. 43, will show this fact most clearly.

We have calculated the thickness of the strata upon both sides of this axis, and will give the process. We shall be careful to have the result too small rather than too large. Taking the west side first—we have before us the problem, given the dip of the strata for a known distance to find the thickness. It is done by the solution of a simple case in Trigonometry. Given the hypothenuse and one of the angles of a right angled triangle to find the base, which is the thickness of the strata. Now the dip is 70°. Deduct from this ten degrees for the possible original inclination of the layers from the first deposition, and five degrees more for possible errors, (for we find ourselves always inclined to state the dip too high,) and we have 55° as the true angle for calculation, and three and a quarter miles for the hypothenuse. The result is 13,000 feet, in round numbers, for the thickness of the north side of this axis. Now the average of the dips upon the south side, taking into account the very small dip at Oak point, which must be supplied for a small part of the distance, and then deducting the fifteen degrees as before, must be 25° as the true angle for calculation. Taking six and a half miles for the hypothenuse, we find the base to be 13,600 feet in length in round numbers, or a little more than two and a half miles of vertical thickness of strata. The results agree within 600 feet, which is a small error considering the roughness of the estimate.

Last year we spoke of the great interest attached to this nearly isolated basin of mica schist, on pages 160, 161. This interest is increased by running a section across it, and ascertaining its synclinal form. Our observations of this year will enable us to give its limits with more precision on the second maps, and we need not here give all the details of its limits and position. The basin is probably of Cambrian age, judging from lithological characters alone.

Porphyritic Granite.

The next formation is porphyritic granite, which is very extensive, being *e* of the section. There is very little to say about it. Every ledge showed a distinct porphyritic character, and it is wonderful that such an immense amount of this variety of granite could exist. The band is eleven miles wide here, and extends to an unknown distance northeasterly and southwesterly. This region is very mountainous, many of the peaks being fully as high as those in Eden. The mountains are scattered peaks, not one continuous range, although the whole granitic mass may be considered as belonging to one general range. Another feature of this region is the presence of an immense number of huge boulders of porphyritic granite. This is a characteristic of almost every granite formation in the State. The boulders being mostly of the same material as the underlying rock, cannot have been transported a great distance. The soil over most of this belt is rather inferior. The route of the section is along the Telegraph road from Ellsworth to Bangor, passing through the Falls village, North Ellsworth and Dedham, east of Fitz's pond. Between Orland and Ellsworth this granite belt is twelve miles wide, and is not all porphyritic. The dissemination of large crystals of potash feldspar (orthoclase) through this rock, makes it a beautiful variety to look upon, and we noticed many places by the way where good quarries might be located.

Quartz Rock.

Very near the boundary line between Dedham and Holden, the granite disappears and quartz rock succeeds, dipping 70° S. 70° E., as it were, beneath the granite. Some planes of a jointed structure dipped 70° S. 20° W. in this vicinity. This rock merges into siliceous slate. At Graves' Coffee House in the east part of Holden, considerable mica is present in the quartz rock, with a dip of 88° S. 60° E. Beyond this hotel the rock is very much contorted and a local variation in the dip is 75° N. 60° W. But the real north-westerly dip is apparent near A. B. Farrington's house, two miles west. Here we have an argillo-micaceous rock dipping 50° N. 40° W., and a few miles further the ledges are entirely quartz rock as far as the middle of Holden, dipping 65° N. 40° W. Cleavage planes are also present in great abundance, dipping 80° S. E. We regard all the rocks mentioned thus far, away from the granite, as essentially one formation of quartz rock, and forming an anticlinal axis. The rock in the centre of the axis is somewhat micaceous, and more nearly resembles the rocks west of Holden village. If this is the true order of things, then we have found a quartz rock underlying the great mass of schists between Holden and Dover. Hence if future researches shall reveal occasional bands of quartz rock among these schists, especially if they have an anticlinal form, we shall have a safe criterion to inform us respecting the number of foldings in the whole area. Upon the section *f* and *f* show the position of the two sides of the quartz anticlinal, while the first *g* shows the more micaceous axis. We do not suppose this axis can be of precisely the same age with the micaceous rocks to the westward, because it underlies them in association with quartz.

We suspect that this quartz rock is the continuation of the quartz rock of the Taconic series in Belfast, described in a previous part of the report. That was associated with schists just like this, and we find on a comparison of various disconnected observations made between the two places, that a quartz rock, more or less obscure, can be traced with its associate schists all the way from Belfast to Holden. This is a discovery of some importance, as will be seen hereafter.

Argillo-mica Schist.

We next come to the largest and widest-spread of any formation in the State,—to a rock that would receive different names from different geologists. It would be called clay slate, talcose schist or mica schist, according as the observer happened to inspect different portions of it. Last year we ranked it all as clay state, specifying many localities where talcose and micaceous varieties abounded. But this year, after a further examination of this rock, we shall call it *Argillo-micaceous schist,* coloring it on the map as mica schist. Inspection of all the varieties discloses the presence of minute scales of mica. They are found even in the roofing slate of Brownville, which is associated with the schists, and by their presence throw light upon the mineral structure of the whole series, showing it to be micaceous rather than talcose. What we described as one formation of clay slate last year, we now divide into two, the clay-slate proper and the Argillo-micaceous schist.

The manner in which the boundaries of this sub-division were suggested to us, is quite interesting and valuable, as indicating the direction to be taken in studying these rocks in future. We possessed a series of observations of the position of the strata, crossing the whole argillaceous belt in several places, and mostly radiating from Bangor. Upon comparing these sections with one another, we found them to agree essentially at the same distances from Bangor, or from the south-east side of the formation. The material from which we drew is mostly contained in our last year's report. We there described a section from Bangor to Patten; another branching off at Mattawamkeag up the East Branch of the Penobscot; another from Bangor to Brownville. This year we explored one from Bangor to Moosehead Lake; and also another from Shirley to Brighton. At Bangor the dip is north-westerly, but at a few miles distance on every route it changed to south-easterly, thus making a synclinal axis. This synclinal line, then, we found to run, (so far as our meagre observations allowed us to judge,) from the mouth of Sunkhaze stream in Milford westerly through the north parts of Oldtown and Pushaw Lake, thence curving south-westerly it passes west of Kenduskeag village, and probably to Carmel, and N. E. Dixmont. Upon the east side of this line the dip is north-westerly, on the line of our principal section, as far as Holden center; and upon the west side the dip is south-easterly as far as the north part of Charleston, thus making an enormous basin, twenty-nine miles wide, whose thickness must be seven miles on the lowest estimate. The anticlinal line west of the first synclinal was first observed near Passadumkeag village, and can be traced westerly through Edinburgh, Lagrange and Bradford, till we find it rising into a range of mountains, which continue through Charleston, Garland and Dexter. This is a very distinct axis, as it is coincident with a mountainous range for so great a distance. It runs towards the mass of granite in Enfield, which most probably was forced up along the anticlinal line, as the rock would naturally be weakest there. Very likely the anticlinal described last year in Weston, (page 384,) is the continuation of this anticlinal line.

The next basin is very narrow, and the rock is more argillaceous than in the previous basin. The synclinal line runs along the valley of the Piscataquis river, even as far up as Parkman, and then it must run on the west side of Penobscot river a great distance. We think that its position is indicated near the Five Islands in Winn, by the change in the dip. Of course these lines must extend further in both directions than we have indicated, but we point out the lines only so far as we have knowledge of them.

Next we come to another change in the dip, with clay slates prevailing on one side and argillo-mica schists upon the other. Hence we do not regard it an anticlinal, but a change in the dip incident to different formations, the slates overlying the schists, perhaps unconformably. This line, which upon our large map we have for the present established as the boundary line between the two formations, is first recognized in the north-east, in No. 1, R. 5, in Aroostook County, on the Aroostook road. It can be traced through Molunkus, the south-east corner of Medway, (formerly called Nickatou,) thence in a straight line to Medford, when it takes somewhat of a westerly course through Milo, Sebec, Foxcroft, Guilford and Abbot. Here it resumes the south-westerly direction, and we have traced it through Kingsbury, Brighton and Bingham, to the Kennebec river. The axes on the various radiating sections correspond with one another no further than to this boundary, but the rock on the north-west side of this line is almost entirely clay slate, and is the only belt in Maine from which roofing slate is now obtained. The variations in dip in this clay slate formation we conceive to be due to various causes more or less local, and not to be treated of here. Scarcely anything has been discovered during the Survey which has given us greater pleasure than these axial lines. It is a very important onward step in the progress of our knowledge of Maine rocks, and a faint shadow of what would be developed by a series of comprehensive parallel sections.

Returning to the details upon our principal section, we find the dip to vary somewhat over the first half of the first synclinal basin. We had just said good-bye to the quartz rock of Holden; and on the route of the section to Brewer we find scarcely any ledges, the country being covered to a considerable depth by alluvial deposits, as it is upon the east bank of Penobscot river. At Brewer the rock is very quartzose, dipping from 30°–60° N. 20° W. Across the river in Bangor the rock is similar, dipping from 45°–55° N. 30° W. and N. W. The application of the term talcose to these schists would be more appropriate than to any other ledges on the whole section. Yet the rock here is not really talcose; it has no magnesia in it, as the analysis shows, and an inspection of many of the layers exhibit particles of mica snugly stowed away. Often the strata in Bangor exhibit interesting curvatures. Professor D. T. Smith, of the Theological Seminary, pointed out one such instance to us about two miles west of the city, which was very instructive, showing also the difference between the planes of stratification and foliation (or cleavage.) Such an exhibition is uncommon along the line of the section.

Three miles north-west from Penobscot river, at W. Boynton's house, the schists dip at about the same angle, N. 40° W. At I. Tozier's, a mile further, may be seen an interstratification of clay slate and slaty talcose quartz rock, dipping 65°–70° north-westerly. In the edge of Glenburn the rock is more compact, with thicker layers. In Glenburn and Kenduskeag there are occasional variations from the normal north-westerly dip, but these are supposed to be local matters. The section, thus far, runs on the stage road to Kenduskeag village, crossing the Kenduskeag river in Kenduskeag.

Beyond Kenduskeag village drift deposits obscure the ledges for a great distance. The road passes over a horseback from the village to a cemetery in South Corinth, a distance of three miles. At a school house and cooper's shop nearly four miles from Kenduskeag in South Corinth the first ledge with the south-easterly dip of the west side of the great synclinal axis appears, although it probably begins much sooner. Observations on both sides of the road, which we had not time to make, will settle the exact point where the middle of the basin is. The schists at the school house decompose readily, perhaps containing a carbonate, and dip 80° S. 25° E. Adjacent ledges have a smaller dip. Near East Corinth the rock is more slaty and argillaceous. Opposite J. M. Shaw's house the layers are very much convoluted on a small scale with an average dip of 45° S. 15° E. About a mile and a half north of East

Corinth, appear ledges of bright green schist, argillaceous, and often quite micaceous, precisely identical lithologically with the greater part of the slaty rocks on the East Branch of the Penobscot above Mattawamkeag, which we described last year. Like them, also, these strata are very much contorted, and their average dip is about 55° S. E. These peculiar rocks continue for two or three miles.

In the north-west part of Charleston, at B. Bradley's, a compact schist resembling talcose schist, but really a quartzite occurs, dipping from 70°–75° S. 30° E. The land here is higher than anything passed over west of the Penobscot, and it continues to rise till the summit of the mountain is reached, (M in the figure,) about 800 feet above the ocean. Upon the county map it will be noticed that a range of mountains extends through Charleston and the towns adjacent. This range is the one we are now crossing, and it must all be an anticlinal ridge, marking the line of the most important of all the axial lines specified above. Passing down the north side of this range, (for which we have no name,) we find the opposite dip, making the anticlinal. We are coming into a narrower basin than the one just left, it being only ten miles wide, and it has almost an east and west course. The first observation taken is of a ledge just north of Ricker Hill, in the south-east part of Dover; an argillo-micaceous schist dipping 72° N. 10° W. In this rock the mica is quite abundant and distinct. The dip is similar to this all the way to the Piscataquis river.

The traveller sees at once the superior fertility of the soil in the Piscataquis valley, when compared with that passed over since leaving Bangor. It seems to be due to the character of the rock, and to be confined to this basin of schist. The rock is often calreous and indeed certain layers in Foxcroft are real limestone and have formerly been burnt in kilns for lime. As in so many other instances the character of the rock here determines the quality of the soil in great measure. By calling this superior to that in Penobscot county, we do not mean to underrate the latter—only that this is better. That in Penobscot county is far superior to much that is found along the sea coast and covering granitic and gneissic regions. And most excellent farms are common on the road all the way from Bangor.

The first of the southerly dips noticed is on the Piscataquis river, at the bridge between Dover and Foxcroft, where the strata dip

75° S. 10° E. Proceeding westerly for several miles, we cross the strata obliquely. Near S. C. Pratt's house in West Foxcroft, the slates are considerably argillaceous, dipping 68° S. 10° W. Near the west line of Guilford, on the south side of the Piscataquis river, the schists have somewhat of the character of a fine sandstone, dipping 85° S. 10° W. At the village of Abbot, the rock is firmer and has the appearance of schists. Here it dips 70° southerly.

Clay Slate Formation.

At the crossing of the Piscataquis river in the north part of the town, we come to a different formation, to the clay slate, whose presence is indicated both by the highly argillaceous character of the rock, and the change in the position of the strata. These are both vertical and inclined N. W. 75°, a change in the strike also of 45° degrees. This ledge shows also, a number of the markings of an ancient glacier coming down the Piscataquis river from Blachard—a south-east course. We distinguish these markings from the common drift scratches, because these grooves have been made upon the perpendicular side of the ledge, as well as upon the top, and because they descend the valley, following the course of the stream. Here also are spiral pot-holes, excavated by the eddying rush of the water as it descends; or, such as are called in Oxford county "screw auger" holes. The water falls a considerable at the bridge.

Two miles south of Monson village the strata dip 80° N., and consist of clay slate alternating with argillaceous sandstones. At the village the strata dip 75°N. 20° W., and the rock is a little micaceous, but has a very fresh look, though often paler than most of the formation. We kept on the lookout for fossils, but discovered none in the brief time devoted to the search. We anticipate the discovery of many fossils in this formation in the future, especially of Nereites. In the north part of Monson, we find ledges of a beautiful clay slate dipping generally 85° N. 15° W., and occasionally S. 15° E. or 90°. On all the hills the ledges are exposed beautifully, particularly upon the northern slopes. Wherever seen they are as smooth as a house floor, having been ground down by the ice of the drift period. The ledges are all of a dark bluish cast, and are highly argillaceous. One would think that in this vicinity admirable quarries of roofing slate might be worked. At Spectacle pond, near the north line of Monson, the drift striae run N. 20° W.

In the edge of Shirley the clay slates dip 70° N. 20° W. At Shirley Mills they dip 70° N. 20° W. also, and no other kind of rock is associated with them. We noticed in regard to the drift deposits between Abbot and Moosehead Lake, that they are often quite thick upon the highest summits. It is an accident almost to find a ledge exposed. So smooth are the ledges that one would expect to find slides common on the sides of the steep hills.

The soil is excellent between Piscataquis river and Moosehead lake, some of the vegetation being quite rank. The season is quite late in the spring; but very large crops of the products suited to the climate may be produced, such as hay, oats and buckwheat. The scenery in this region is much superior to that south of the Piscataquis. Here the country is full of large, gently sloping hills and high mountains; while in the distance, the great granite piles east of Moosehead lake and the enormous mountains to the southwest are commonly visible.

Between Shirley Mills and the Forks of the Kennebec, the section runs through a country destitute of carriage roads, and as it was very important that the line should be explored, Mr. Goodale volunteered to explore this part of the section, while the rest of us went around by the way of Brighton and Bingham to meet him at the Forks. We here introduce Mr. Goodale's notes upon this unsettled region.

Goodale's Observations.

While making a continuous section from Mount Desert to the Canada line, it was found necessary to go in a direct line to the town of Shirley, near Moosehead lake. Here the road terminates, and, of course, there is a break in the section including all that territory lying between Shirley and the nearest point on the northwestern road. In order to supply the deficiency caused by this break, I was directed to proceed through the woods in a westerly course to the Forks of the Kennebec, noting all outcroppings of rock, the altitude of the line passed over, and whatever else might be of geological importance. Accordingly, having procured the services of Mr. J. Sturdivant and his son Llewellyn as guides, I started on the 26th of July, from Shirley Mills. The clay slate at the commencement of the journey near the "North road" so called, dips north 75° W. It is dark grey in color, and where not weathered or injured by frost, cleaves readily into fair plates. There was no other exposure of slate till we reached the west branch of the Piscataquis river. Here the dip is N. 82° W. At this point a hurricane two years ago tore up the trees from their roots, twisting them together and forming a formidable barricade for an extent of a mile and a half. The whirling of the tempest had completely obliterated all traces of the wood road, and rendered walking a matter of extreme difficulty. The course of the hurricane appears to have been from south-west to north-east, and reminded me of the whirlwind-traces noticed last spring on the river St. John.

Near the Shirley west line is a ledge of fine roofing slate of great extent and easily exposed. I saw an excavation, evidently an old one, which enabled me to examine the slate quite carefully and form a deliberate opinion in regard to the character of the slate. It is fine grained, bluish-grey, easily fissile into one-eighth inch plates of twelve inches square, even where the frost had exerted some effect upon its texture. Owing to its peculiarly soft character it would make, I think, good school slates. The strata were in this place perpendicular with a strike north 70° east (excl. var.)

At Sandy stream, same rock, same dip and strike

At Alder stream, east branch, the rock is exposed near a slight waterfall, the same in all essential particulars with the last. Near the west branch of Alder stream the rock appeared to have lost its fissile character and to have become more like the compact slates occurring in Scarboro'. Numerous quartz veins run through the slate in an east and west course, but with no metallic deposits. At Moxie outlet the same rock was again seen. It is harder and has entirely lost its distinctive fissile structure. Near Clark's camp, three miles or a little less from the outlet of the pond, clay-slate was noticed dipping north 85° east. This convinced me that I had passed over a synclinal axis. At the Forks of the Kennebec the slate dips north 70° east, and confirms, I think, the opinion formed in regard to the axis.

We were able to reach the comfortable hotel of Mr. Murray at the Forks late in the evening, having walked through a thick forest of second growth and swamps of cedar, part of the way in a hard shower. The distance is variously estimated from eighteen to twenty-two miles. I incline decidedly to the latter opinion. Accompanying this, is a vertical section* of the slate between the

---

* Reduced and incorporated into the general section of the whole route. The description of the figure will show the localities of the route and the dip of the strata; *h* are the clay slates, and *i* the slates west of Moxie Falls. C. H. H.

two points before mentioned, giving, of course, the altitudes as I was able to compute them from a good aneroid barometer.

Moxie Falls. The stream flowing from Moxie pond has a tortuous and troubled course. It makes its way over several miles of the upturned edges of clay-slate strata; here smoothing and polishing them, and again, quarrying deeply into the mass, and even excavating pot-holes in its path. The descent for most of the way from the outlet to the Kennebec is very rapid, owing to the contour of the district and the frequent waterfalls. At a point two miles from its confluence with the main river it makes its greatest descent. This cataract is caused by the abrupt breaking down of the strata, much resembling certain faults. Upon each side of the base of the fall the rocks rise to an altitude of 115 feet and are very precipitous. The water comes down over a jagged ledge, eighty feet in a single leap of foam which gains additional whiteness by its contrast with the blackened slates. It is a cataract second in its singularity and beauty to none of our minor waterfalls, and will well repay the visit of the pleasure-seeker, the tourist or the artist.

The strata of the greyish-black slate dip apparently to the northeast about 75°. G. L. G.

Slaty formation in the Kennebec Valley.

The rocks upon which we come at the Forks of the Kennebec are different apparently from any already passed over on the section, and yet perhaps not more diverse than what might be expected on the different sides of a great anticlinal axis. The dip of the strata would carry these micaceo-argillaceous slates beneath the clay slate formation, and come up on its eastern border. This is not impossible; but we incline to the opinion that the strata here are either inverted, or else rest upon the clay slates unconformably. Although the strata at the Forks have a fossiliferous aspect far more than anything to the southeast, no organic remains have yet been discovered in them; yet the time cannot be far distant when they will be discovered. These slates must belong to the same formation which Mr. Houghton described last year on Moosehead Lake as mica schist, since both formations are similarly situated with respect to the clay slate just passed over, and are succeeded by Oriskany sandstone on the other side. It is the most natural thing in the world to suppose these slates and schists to form a portion of the Silurian system, and more probably the Upper than the Lower Silurian. On Moosehead Lake there is very little difference lithologically between the mica schists at the south end of the lake, and the schistose sandstones at Soccatean Point containing Oriskany fossils. And as Mr. Hodges, Jackson's Assistant, found "madrepores" at the base of Squaw mountain, which are without doubt the common Favosite coral of the Upper Silurian, we have some evidence upon which to found our conjecture of the Silurian age of these slates and schists.

Our observations of the dip at the Forks are scanty. East of the Forks Hotel there is a long bluff of slates interstratified with limestone, dipping 40° S. E. The rocks alternate in great measure, yet occasionally the limestone is three or four feet thick. Dr. Jackson analyzed this limestone and found it to contain Carbonate of lime 50.0, Silica 27.0, Magnesia 9.0, Alumina 8.4, Carbonate of iron 2.8, and Oxides of iron and manganese 2.4; total, 99.6. He recommended that this rock be burnt at a red heat, then pulverized and mixed with fifteen per cent. of clay and ten of manganese, so as to form a hydraulic cement equal in value to that imported from England. We know not whether this anticipation has ever been realized. This limestone is more or less abundant in the ledges on both sides of the river, so much so that it must form a part of the lithological name of the formation, if it shall be found to occur over a wide region. The dip is higher in going east from the bluff, and the formation extends certainly to Moxie Falls.

On the west side of the Kennebec river we found a great many ledges of slate with a high south-easterly dip, as we supposed at first; but we soon discovered a ledge showing this view to be erroneous. We found traces of a gentle dip, say 20°, to the north-east, which we suppose to be the true one, while the very prominent highly-inclined planes are those of cleavage. To illustrate this fact, we introduce here in Fig. 45, a sketch of a small curve of the strata cutting across the cleavage planes. It is a sharp synclinal fold. There are two beds of limestone, the upper two, and the lower one foot thick, found interstratified with slates. Only the latter have planes of cleavage, and these are represented by the finer lines inclined to the right (south-east) at a high angle, while the two beds of limestone rest each upon a bed of slate. Several very crooked veins of white quartz occur in the upper limestone bed, while the layers of limestone are not very much contorted.

The strata in the lower limestone bed are curled up and crowded together very closely, as is attempted to be represented. The

Fig. 45.

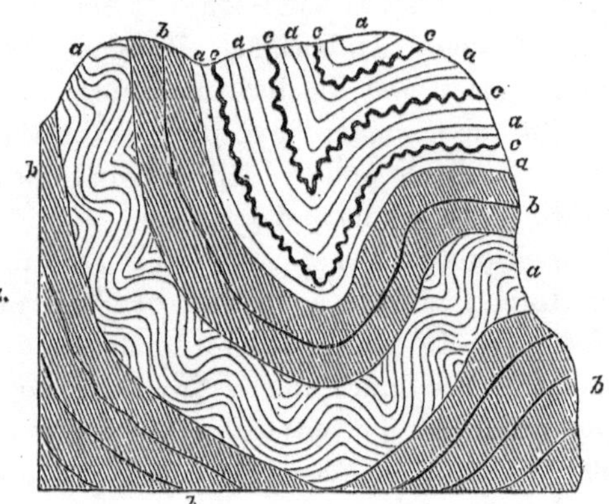

aaaa. Limestone.
bbbb. Slate.
cccc. Veins of quartz.

Planes of cleavage and stratification.

coarser lines, parallel to the limestone beds, indicate the stratified layers. These lines are not imaginary, since they may be seen intersecting the cleavage planes in all the adjacent ledges. This case is an interesting one, and makes it necessary to examine all the strata in this region very carefully. So long as we can find beds of limestone to guide us, we cannot fail to discover a marked distinction between these two kinds of planes.

Nearly two miles west of the Forks, on the Canada Road, may be seen a large mass, perhaps a dike, of coarse syenite, almost a trap. Its location may be seen on the general section, *k*. Most of this is an argillaceous rock, quite compact, and so smoothed down by drift that it was difficult to obtain a specimen, or ascertain satisfactorily whether there were any planes present different from those of cleavage, which were inclined about 70° N. 20° W. Depending upon the observations thus far made, we feel sure that an anticlinal exists in this formation, which is properly represented in the section. This makes this slaty and calcareous formation dip beneath the Devonian rocks to the north-west.

Oriskany Sandstone.

Precisely where this formation begins we cannot tell. A mile and a half east of Johnson's mountain we noticed a slaty rock dipping high to the north, which may be connected with the Oriskany group. This is near the north line of No. 1, R. 5. The rocks in this township are generally concealed by drift.

It will be recollected that this formation was fully described last year, so far as it was possible, and we need not repeat those remarks here. It was stated that the fossiliferous boulders so common in the fields in southern Maine, were principally derived from this rock, and that it belonged to the very oldest of the Devonian rocks. Parlin pond was also represented as the best locality for the Oriskany fossils, which we shall speak of presently.

On Johnson's mountain appears a very tough sandstone. On a spur of Bald mountain, still further west, a slaty rock crops out dipping 50° N. W. Both these mountains are in No. 2, R. 6, and are parts of extensive ranges, of which not the remotest idea is given upon any map of the State or County yet published. Our maps are deficient in respect to the topography of our mountains. Near the west end of Parlin pond we obtained specimens of a trilobite, the *Dalmanites*, we suppose of the same species with the one found last year at Stair Falls on the east branch of the Penobscot, and which we hope will be described and named in the last part of our Report.

The best locality of fossils in this region known to us, is on a knoll west of the Parlin Pond Hotel. But the fossils there are not in a ledge: they lie in the soil, as if quite near the parent ledge. A little digging did not reveal the ledge, however. We followed out the letter of Jackson's description of the locality where he obtained so many specimens, and could find nothing that would correspond with his description of the abundance of the specimens, so well as this. We collected a large number of fossils, both here and at every other available source, and shall hope to speak more fully of them hereafter. The fossils previously collected here belonged to the following genera; *Strophomena, Chonetes, Orthis, Rhynchonella, Rensselæria, Leptocœlia, Spirifera, Modiolopsis, Cyrtodonta, Avicula, Murchisonia, Platyostoma* and *Orthoceras:* and the following species have been already identified; *Strophomena magnifica, Orthis musculosa, Rhynchonella oblata, Rensselæria ovoides, Leptocœlia flabellites, Spirifera arrecta,* and *S. pyxidata.*

On a tributary of Parlin stream in the north part of the township of Parlin Pond, a few fossils were obtained from thick bedded, fine grained, dark colored sandstones of this formation, dipping from

10°–15° northeasterly. The last seen of this formation is just in the edge of Jackman, where a few fucoids and shells were obtained. The dip of the strata is 15° N. 60° E., and that of the cleavage planes much greater in the same direction. The rock is clay slate. This locality is at the clearing called the Jackman Farm, and upon the highest land in the Canada road between the Forks of the Kennebec and Moose river.

A range of mountains runs from Attean pond across the Canada Road just beyond the Jackman Farm, of which Owl's Head is one portion. It also crosses Parlin stream north of the pond, and then trends northeasterly, perhaps to the Kennebec. It is composed of granite at Owl's Head, but not where Parlin stream crosses it, for the sandstone there is beautifully exposed, containing a few fossil shells. The strata dip 20° N. 10° W., at a beautiful cascade of this stream, a couple of miles below the pond. On some of the layers ripple marks were exposed, covering several square yards, and appearing just as perfect as when made in the sand so many thousand years ago. The rocks here reminded us of those coarser sandstones containing fossils on the east branch of the Penobscot at Johnston's camp and Matagamon lake. We felt uncertain about their position relative to the Oriskany belt last year. We are confident these strata are synchronous with the latter, as is proved by the discovery of the *Dalmanites* in the vicinity; and we think that by following Parlin stream up and down, the connection of the two kinds of fossiliferous rocks will be found. Moreover, here will probably be the best place to make a detailed measurement of the Oriskany rocks, and a careful exploration of the whole formation from top to bottom. No one can surely feel averse to such an examination here, although the country is so wild, since the scenery is so grand, the rocks so interesting, and the attentions of Mr. Sylvester at Parlin Pond Hotel so kind and obliging. It would be a toilsome but a delightful task to explore in this vicinity.

Granitic Rocks.

Two or three ledges of granite and protogine appear in the road between the Jackman farm and Moose river. The country is so much covered with drift that ledges are very rarely seen. We do not suppose these rocks extend as far as Moose river, although no other ledges show themselves previously. Still the width of the granitic region must be about three miles, although extending five miles along the road, since the road crosses the belt transversely.

Clay Slates again.

At Moose River settlement, ledges of another clay slate appear, though we somewhat hesitate to call it clay slate, because it is different from the great clay slate formation below Moosehead lake. The cleavage planes are generally different from the planes of stratification, and it is unusually difficult to distinguish them. For this reason previous observers have confounded them. The rock is more properly a shale over much of its area, rather than a slate, and this is the chief difference between the two formations, and may indicate a more recent origin. The first ledge seen is on the north side of Moose river, with strata dipping 28° S. 10° E., and cleavage plains dipping 40° N. 10° W. The argillaceous odor is scarcely perceptible in this ledge. The valley of Moose river is very pretty and is lined with terraces, two or three of which are well marked. The second is the meadow, and is a quarter of a mile wide. It is composed of clay mostly. There is quite a settlement here, and the soil is very good. On the south side of the river the meadow is very short, and the surface for a great distance is covered with large boulders of granite. Many high mountains are in sight from Moose river, which are all unknown to the outside world, being barely named.

Quite a tract of level land is located north of the settlement, say two miles in width. Ledges of slate show themselves occasionally for several miles, all having the cleavage planes inclined to the north-west. Near the north line of Holden is another ledge of the slates, which we supposed at the time to dip 35° N. 20° W., but made the remark in our note book that very likely this dip was merely of cleavage, and that the true dip was 40° S. 20° E. We make all these remarks where any doubt exists, partly that others may decide for themselves this question, and partly to show that the difficulty of obtaining good observations of the dip of strata is often very great. Near the top of a spur of Bald mountain in the south-east corner of Sandy Bay township, the rock is argillaceous, though closely resembling the schists west, and dips 50° N. 20° W. This hill is almost as high as the Boundary mountain.

Mica Schist.

The last formation on the section is a genuine mica schist, more apt to be confounded with talcose schist than clay slate. The eastern portion of it comes up on the west flank of Bald mountain.

The dip everywhere is south-easterly, underlying the clay slate formation. This mica schist is undoubtedly the same formation with the rock upon the upper St. John river, which we called talcose last year. It is also the same as the auriferous schists in the Chaudiere region in Canada. In Sandy Bay, the last township in Maine, only a small portion of the formation comes into view.

In the west part of the township the schists dip 55° S. 30° E. At the Boundary line upon the top of the mountain the dip is 70° south-easterly. Some of the layers are very thin, like slates, and are arenaceous. We went into Canada a mile and found the same rock all along the road. On the American side of the ridge we saw a few small inverted anticlinals, which may possibly indicate that this ridge is an inverted anticlinal; it is certainly not an improbable supposition. It was a little cloudy when we were at the summit, and this may be the reason why we did not notice the fine view described by Jackson. A deserted house is built upon the line between the two countries. The line runs along the height of land between Maine and Canada for many miles, or from the New Hampshire corner to the head of the south-west branch of the river St. John. The line is indicated by strong iron posts inserted at intervals by the Joint Commissioners of the United States and Great Britain under the Treaty of 1842. Inscriptions bearing the names of the Commissioners are found upon every one of the posts. As this line follows the height of land, it is consequently very crooked, and it is marked by a line of second growth trees, since the surveying party cut down a wide road for their purposes, when erecting the iron posts in their proper places.

We found several quartz veins in the schists on the Maine side of the boundary, and cannot doubt that gold might be discovered in this region. In the great valley of Sandy Bay we noticed quite a number of these veins, and think them auriferous. Near Hilton's house the road crosses a branch of the West Branch of the Penobscot river, and the appearances are very favorable here for successful gold washings in the soil. There is very much of the "black sand" here in which the gold always occurs. And the fact that this valley is in the Chaudiere gold region confirms this view. Still, the Chaudiere region is not like one of the great gold fields of the world, although it affords a fair yield in some portions of its area. We know not why the gold should be any more abundant in Canada than in the Maine part of the deposit. Our limited time did not permit us to stop and test the question of the existence of gold in this stream by actual trial.

We have just now queried whether this mica schist formation, of so great extent in Canada, and dipping easterly, may not form a gigantic synclinal axis with the argillo-mica schists of the immense region between Bangor and Dover. The two formations run parallel to each other in Maine for more than a hundred miles, and the formations adjacent to each are the same, and appear also to have a synclinal structure. For instance, as shown in our section, the clay slates of the Moose river region may come up again on the other side of the basin in Shirley and Monson. And in confirmation of this view, the rocks between the two clay slate formations are all newer, being largely of known Devonian age, while the clay slates cannot be newer than Lower Silurian or Upper Cambrian. This suggestion we throw out for all it is worth. It will furnish the clue to the geology of the whole State if it proves to be true.

III. Jottings in Central and Western Maine.

It was expected that Dr. Holmes would present here a sketch of the geology of the Kennebec and Androscoggin regions of the State. This sketch was designed to be a notice, not merely of what he had seen during the past two summers, but the results of his observations for the past thirty years. Circumstances have prevented the execution of this plan in season for this report; and therefore it becomes the duty of the Geologist to mention a few facts which have fallen under his observation, as he has travelled from one field of labor to another in the discharge of his assigned duties.

Piscataquis and Somerset Counties. In journeying from Shirley to Skowhegan, and from Skowhegan to the Forks of the Kennebec, one sees chiefly slates and schists. The results of every observation thus far noted by the Survey, are, that a line drawn from Abbot to Bingham is the boundary between two immense formations, the one to the south-east being an argillo-mica schist, and the one to the north-west being a clay slate. Both of these formations are of great width, and extend very far in a north-east and south-west direction. Several lines of axes will probably be traced through them as the work of exploration goes on; and these results will be both highly interesting and useful. The directions of some of these axial lines in Penobscot county have been noticed already in this report. Few geologists in their surveys have carefully traced out such axial lines. Sir W. E. Logan, however, has done so in Canada East; and these strictly scientific results have proved to be of the utmost importance to those miners who have recently been exploring that remarkable metaliferous country for cupriferous veins. It would not be more strange to find as important mines in this mica schist region in Maine, than it was in Canada half a dozen years since.

These axial lines may be indicated often, in the absence of more definite marks, by the repetition of the same bands of some peculiar rock, as limestone, quartz rock, or schists more or less argillaceous. Between Skowhegan and Shirley we should look for these indications in the form of schistose bands, charged with a greater or less amount of argillaceous matter.

Between Shirley and the edge of Brighton we consider the rock to belong to the clay slate formation, with possibly a mass of granite oozing out on the top of Russell mountain. Three axes have been noticed in crossing the strata on this line. At Shirley Hotel the dip is north-west, but two miles beyond on the Blanchard road, the dip is to the south-east. There is an anticlinal here then. In the deep valley of Blanchard the slate takes a little mica into its composition. At C. S. Whitehouse's estate, on the southern slope of Russell mountain, the dip changes to the north-west again, and the rock consists of alternate beds of clay slate and mica schist. Unless this great mountain is made up of granite, as one would think after seeing the summit rocks glisten in the sun at a great distance, it must be a synclinal pile of strata.

An excellent clay slate formation occupies the greater part of Kingsbury, and in it there appear to be one or two axes, probably an anticlinal, and then a synclinal axis. In the north-west part of the town there is a very fine view of the country towards the east and south-east, in Penobscot county. The fact that the rocks between Shirley and the north edge of Kingsbury are somewhat micaceous, while there are variations in the strata, suggests the inquiry whether the strata in Shirley and Blanchard are not a repetition of the older schistose strata. It is much easier from a cursory survey to raise questions than to answer them.

A fine anticlinal axis is displayed as we pass into Brighton, being upon the edge of the mica schist group. In this town the rocks are all thoroughly micaceous. The dip changes so that the middle of the town is situated upon a geological basin or synclinal axis, the same with that in Dover. Between Brighton and Skowhegan the ledges are very much concealed ; but the rocks are everywhere micaceous, though at Athens they are very argillaceous. Going south from Skowhegan towards Waterville the clay slate reappears, containing Nereites.

Between Skowhegan and the south part of Bingham the same mica schists occur as between Skowhegan and Brighton. We had not the opportunity of noting the variations in the dip. It is a beautiful clay slate, with a high north-westerly dip, that is first met with in Bingham. In Moscow the dip changes to the south-east, hence forming a synclinal. Mr. George L. Goodale discovered some gold by washing, near Carney's Hotel in the north part of the town. Near the same locality was found several years since the boulder of "eurite," or more properly, leopardite, described last year on page 202. Mr. Carney kindly presented the stone to us to be placed in the geological collection at Portland. It is handsomer than any that have been obtained from Charlotte, N. C. The manner in which the black stems could have been produced in the white quartz rock will afford abundant opportunity for speculation. It would be a matter of great interest to find other boulders of this rock, or the ledge from which it has been drifted.

Across the river, in Pleasant Ridge, on the estate of Mr. Moses D. Townsend, we were shown a ridge of good roofing slate. It is precisely similar to that now quarried in Brownville. We see no reason why a good quarry could not be opened here. The facilities for drainage could not be greater, and the road to the nearest railroad station is remarkably level. And possibly the Kennebec river might be employed at some seasons of the year to float the prepared slates to market. In the south part of Caratunk the slates dip south-easterly. The north-west boundary of the clay slate formation must be in the south part of the Forks Plantation, or the north part of Caratunk. Then we find the slates with interstratified calcareous layers, belonging perhaps to an Upper Silurian series. The Moscow slates contain the same fossils as those found at Waterville.

Waterville Fossils. In the Preliminary Report, upon pages 231 and 232, a list is given of all the curious fossil forms which have been found and described from the slates in the Kennebec river at

Waterville. They are called Nereites or Nereograpsus, and Myrianites. Upon page 232 it is stated that "a perfect set of these fossils could be found in no Cabinet in the whole State." We desire to correct this statement, and say that in Waterville College there may now be found not only all the species hitherto described, but also one or more new *genera* of these curious impressions, and these the finest specimens that ever have been obtained from Maine. It is fitting that the Cabinet of this College, so near the place where these earliest inhabitants of the State of Maine flourished, should now contain the most perfect set of their remains.

Prof. Hamlin of Waterville College, at our request, procured a large number of these fossils, and sent us a full suite of them. His efforts in our behalf, have received our warmest thanks, and deserve honorable mention in our report. He says in his letter that, "for eight weeks I gave, on the average, two days per week to collecting. In this time some four hundred specimens were found good enough to be worth taking home. On picking over the whole lot on Monday last, all the really good specimens that could be mustered were those I have sent you, and about twenty-five that I have retained for the College Cabinet." Very few professional men would take as much pains as Professor Hamlin has done to oblige a stranger. In return, the Survey will send shortly a number of the rocks, plants and marine animals collected by them, for the College Cabinet. It was owing entirely to Professor Hamlin's generosity that the Cabinet had an imperfect set of these impressions in 1861, when we examined the collections. At the earliest possible moment these Silurian fossils will be carefully studied, and compared with the impressions of living worms and crustaceans, in the hope that their true character may be satisfactorily determined. Of late their similarity to certain forms of crustacean impressions has been suggested.

Franklin and Oxford Counties. Several items of interest were noticed by us on the way from Leeds Junction to Rangely Lake. In Leeds there appeared to be examples of singular alluvial deposits, at considerable height above the ocean. The rock is mica schist in Leeds and Livermore, though at the Falls it resembles talcose schist. There is a fine quality of granite between the two villages of Jay. East of the depot at North Jay it is quarried extensively. Near East Wilton the schists, almost a slate, reappear, and continue up the valley of Sandy river to Phillips.

Between North Jay and Weld, via Wilton, the following varieties of rock were noticed. Just south of Wilson's pond are ledges of gneiss dipping 30° north-westerly. Mica schist is seen all the rest of the way in Wilton and perhaps in No. 4. Then succeeds a narrow strip of gneiss. In Perkins Plantation, or perhaps in the eastern part of Weld, there is a limited patch of granite. In Weld the prevailing rock is gneiss. We went to No. 6 in search of lead ore, but failed to find the precise locality. The rock on the side of Tumbledown mountain is gneiss, dipping 70° N. 30° W. A large boulder occurs here, which is estimated to be thirty-eight feet long, twenty feet high and twenty feet thick. The scenery of Weld is charming. The village is surrounded by lofty mountains, like an amphitheatre. Some of the peaks consist of naked rock, affording thus a pleasing contrast to the universal green. Some of these mountains are nearly three thousand feet in height.

The rock is mica schist about Farmington, with the exception of a few patches of granite. So it is up the valley of Sandy river. Near the village of Strong the mica schist is well characterized. Some of the strata yield copperas by decomposition, dipping 70° S. E. Layers of limestone are common, interstratified with the schists. Two of them occur back of the store, each several feet wide. On the farm of Samuel Worthly other beds of similar limestone were formerly burned for lime. A narrow band of granite commences about four miles below Phillips, passes through the village on the east side of the river, and continues up the valley about a mile and a half. Fig. 46 shows how some of this granite changes into mica schist. There is no seam or break between the two kinds of rock in the ledge, and no other indication of division than the sudden change from the dark colored schist to the lighter colored granite. This sketch is taken from a ledge thirty feet high, by the side of the road, north-west of the village of Phillips. The granite band is invariably indicated by the presence of large granite boulders which have fallen from the ledges.

Fig. 46.

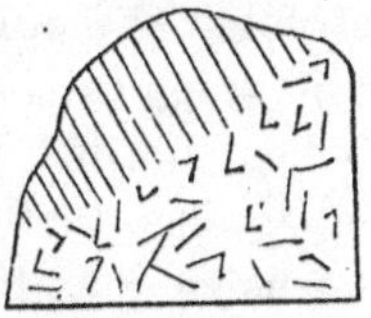

Mica schist and granite.

Two miles north of Phillips are two openings in a bed of white azoic limestone, whose layers dip 35° northwesterly. Lead ore has been obtained from one of them. We saw specimens of black lead, copper and iron pyrites, from the excavation. Mr. Seward Dill informs us that upon the top of a high hill north of Phillips, there is a huge boulder, as "large as a barn," which has been split into two pieces. His son, also, has picked up very interesting fossiliferous boulders, which must have been derived from some undiscovered belt of rock. In Dr. Jackson's third report, page 27, there is a notice of a large boulder of magnetic iron ore, weighing 174 pounds, which was found near the village of Phillips. Last year a similar boulder was found in Salem. Whoever can find the vein from which these boulders were derived, will do good service to the community. Passing on towards Rangely we find the axis of Saddleback mountain to be composed of mica schist, only it has more of a gneissic aspect than the rocks in the valley. At Long Pond a new variety of mica schist succeeds, which has been described in a previous portion of this report.

The geology of Oxford county is peculiar. From the little we have seen of it the following is our impression of the whole. Originally the whole country was occupied by a schistose formation of essentially uniform character. Being in a favorable situation for the action of metamorphic changes, so much granite has been produced and thrust among the strata, that it is now the prevailing rock. The country is made up of mountains, hills and valleys; without any level tract except scanty meadows along the banks of rivers. Now nearly all these mountains and hills are composed of granite, while the lower districts are occupied by mica schist, gneiss, silicious slate or quartz rock, and indurated limestones. We have never heard of any other district of the same extent with such a singular arrangement of azoic rocks. Upon a properly constructed geological map the colors will be seen to conform to the topographical features of the country; and numerous sections can be made which will conform to the old theory of the structure of mountains, that their central and interior portions are composed of intrusive granite. It is hardly necessary to add that an accurate geological map of the western part of the State will present the true arrangement of the Oxford county rocks, besides correcting any false impressions which may have been already received respecting them.

The following changes in the character of the rocks were noticed in travelling between Welokenebacook lake and South Paris. Granite extends for about three miles south of the lake. Then, commencing at the summit of a mountain range, succeeds mica schist, often gneissoid. About eight miles from the lake we saw the first ledge of granite, in the town of Andover. This rock seems to be the prevailing variety all the way to Bryant's pond, except a strip of gneiss about two miles wide in the north part of Rumford. Andover is another village nearly surrounded by an amphitheater of high mountains. Ellis' river runs through the township and lies in a wide valley at the village, with extensive meadows and well formed terraces. The "Corners" village is built upon the third terrace. About Bryant's pond the granite is the tabular variety. Three or four miles before reaching South Paris station may be seen several ledges of obscure mica schist, which probably extend through the whole of the low land bordering upon the Little Androscoggin river.

At the famous Tourmaline locality upon Mt. Mica, we find that mica schist was the original rock, but it has been mostly displaced by granite; while the tourmaline vein is a vein of doubly coarse granite in granite. The term "mount" is hardly appropriate to this swell of land, in the midst of so many higher summits designated hills.

At Paris Hill the rock is granite. It is granite on the road to North Buckfield nearly to the village, then succeeds mica schist. Between North Buckfield and Buckfield village is a very large vein of granite. Mica schist occurs in the west part of the town. The magnetic iron ore however, occurs in granite. Between West Buckfield and Paris the rock is entirely granite. In North Buckfield one sees many patches of impure limestone.

Meteors.

Black heavy masses of metallic iron called meteors have sometimes been seen to fall from the sky. One such example in Maine was in Castine, of which specimens can be found in some collections. Another mass fell in Sidney twenty-five years ago, but the fragments have been lost. Such fragments may sometimes be found in the fields, of whose time of falling there is no record. We present these brief statements in the hope that if any person who reads them may chance to know of the existence of such fragments anywhere in the State, he will convey such information to the Survey. There is no value attached to such specimens other than what interest is connected with them on account of their source. They must be fragments of other worlds, or each entire meteor may be a world by itself. It is an interesting theory that "between

Jupiter and the smallest meteor, (say the size of a walnut,) there is an unbroken series of planets." Whoever, then, finds a meteor and believes in this theory, may congratulate himself upon being the possessor of an entire planet! The geologist desires to examine them, because he finds out what other worlds are made of. No meteor yet discovered has revealed the existence of any chemical element not known upon our own globe; but we know not what may be found in specimens from new localities.

B. GEOLOGY OF THE SCHOODIC REGION.

In the eastern part of Maine are two groups of large lakes, commonly known as the Eastern and the Western Schoodic lakes, although this designation is not found upon the maps. The former form the boundary between Maine and New Brunswick, while the latter are entirely within the limits of Maine. These waters give rise to two large streams, which unite at the north-east corner of Baileyville, and flow into Passamaquoddy bay. This river has the name of the St. Croix, and the same name is commonly given to the eastern branch in connection with it; but the western branch is not named upon the maps. This name of the river was given by the early French settlers. The Indians called the western branch the Kennebasis; the eastern branch the Chepedneck, often spelled Cheputnecticook; and the united waters the Schoodic river, signifying low and swampy ground: thus manifesting a more philosophical view of things than their successors. These Indian names are partly in use now; and in the hope that the designations may become permanent we will adopt names which seem the most appropriate. First, then, we have the name Schoodic to be applied to the lakes in general, with the primary division into eastern and western. Secondly, we would adopt the Indian names for the two branches, the Chepedneck and the Kennebasis. And thirdly, we would use the name St. Croix for the main river below the junction of the two branches. If the names given in the Treaty of 1842 compel the use of St. Croix for the Chepedneck, we can certainly retain every other proposed name. And in describing the geology of this part of the country we will speak first of the country watered by the western Schoodic lakes and the Kennebasis; secondly, of the country bordering upon the St. Croix and Chepedneck rivers; and thirdly, of the country watered by the eastern Schoodic lakes.

I. Geology of the Western Schoodic Waters.

We started to explore this region August 5th, immediately after returning from the Canada line. We went to Calais, and spent a little time in that vicinity before venturing into the forests. The first three of the lakes, Lewey's, Long and Big lakes, we were enabled to explore in Mr. Sawyer's small steamboat, the Gipsey. This was a very convenient boat for our purposes, since it enabled us to perform our task with great ease and rapidity. On account of the shallow water and the poor road between Big and Grand lakes, we could not use it, though it was offered us, beyond the first three lakes. For the rest we used birch canoes. Our guides were S. W. Haycock of Milltown and B. D. Wyatt of Calais, who accompanied us through both the Schoodic regions, and showed themselves to be skillful and efficient pilots. We will not enter upon the details of the history of our expedition, nor describe what we saw in the order of time, but will describe the region systematically. We were obliged to travel twice over the same ground.

The geology of this region is very simple, so far as the rocks can be seen. Three varieties of rock occupy this area, granite, mica schist and calciferous clay slate. The precise boundaries of these formations it was difficult to ascertain, on account of an immense thickness of the unmodified drift deposits.

The principal lakes explored by us west of Big, were Pleasant, Junior, Sysladobsis, Pocumpus, Wawbawsoos, and Witteguerguagum or Grand lake. One of these is not delineated upon the State map, and the rest so incorrectly, that we should despair of making the details intelligible by mere references. These lakes are situated at the corners of three counties, and being in a wild country it is not strange that their topography should be so little known. We had with us three different plans of a portion of this district, and while no two agreed with each other, they were all equally remote from the truth. For the credit of the Washington county map, we will say that its delineations are essentially correct, while the others are very incorrect, and the maps issued since the Washington county one have not even copied the true topography, but have used some inferior authority. We have been compelled to make these comparative estimates of the different maps unwillingly, but necessarily that our readers can follow us. For the same reason particular specifications must be given hereafter.

In the town of Carroll in Penobscot county the principal rock is mica schist. It is the only rock seen upon the east and west road running through the town. At Mr. H. Gates' in the west part of the town is a very fine bed of dark bluish limestone, whose layers dip 45° N. W. The bed is several rods wide and of unknown length. Mr. Gates manufactures from 100 to 300 barrels of quicklime annually out of this bed. It is capable of producing much more, and furnishes lime equal to the best. It can be produced cheaper here than at Rockland, and can successfully compete with that in the market hereabouts. Hints of other beds of limestone in Carroll reached us in both directions, particularly in the north-east. There is said to be a bed on the land of Mr. Ames. A similar bed also may be found on Mr. Coffin's land, near the centre of the town. These limestones correspond better to the beds in Azoic schists than with the Eolian limestone on Penobscot bay.

The mica schist of Carroll extends uninterruptedly as far as Musquash lake in Topsfield, in an easterly direction. In the east part of Carroll there is an anticlinal axis, the limestone being upon the western side. The south-east dip extends to Musquash lake, and into Tallmadge and the Indian township. The north-west dips occur at the saw-mill in Tallmadge, and about three miles from Princeton in the other township. These observations indicate the presence of a synclinal.

A high range of mountains in the south part of Carroll is evidently granitic, and connects as a mountain range with the syenite on Musquash lake, and has been traced into New Brunswick; and the provincial geological map carries this granitic belt entirely through the province to the Gulf of St. Lawrence. A large portion of the Western Schoodic region is granitic, forming a belt several times wider than it appears in Topsfield. We have evidence of the existence of granite on the north side of Pleasant lake, on Mill Privilege lake, Scragly lake, the third Chain lake in No. 4 Hancock county, upon Stone island in the west part of Witteguergaugum or Grand lake, and at the west end of West Musquash lake in Tallmadge. This gives us an elliptical granitic expansion of the great range, of fifteen miles long and ten miles wide. It probably extends to meet the granite of Greenfield, if not the range extending through Dedham, Orland, etc., to Mount Desert. East of the granite the rock is mica schist all the way to Princeton. We saw such ledges on Wawbawsoos lake, and Mr. Kelley of Calais states that he has seen them on the upper Machias lakes.

We will say a word about the more western of those lakes which we visited. Pleasant lake lies in townships 6 and 7 of Washington county, but does not extend into Penobscot county as the State Map would indicate. High hills of granite are upon its north and east sides, and somewhat on the west. The entrance from the south is very rough, the road being filled with logs and fallen trees, besides its natural crookedness. The stream was too shallow to permit our canoes to float into the lake. Logs were common here four feet in diameter. Many large boulders of granite are scattered over the surface.

Pleasant lake empties into Scragly lake by a *thoroughfare*, so called because the streams connecting the different lakes in the Wild Lands are the only thoroughfares of travel between them. There is a remarkably deep hole in this stream at the head of navigation, said by the hunters to be bottomless. It is certainly twenty feet deep, which is remarkable considering the narrowness of the outlet. Between this hole and Scragly lake, about a mile, the banks are very low, and commonly marshy. Rainy brook at the south-east end of Scragly lake is entirely in a swamp, over a mile and a half in length. The thoroughfare between Scragly and Pleasant lakes connects their very extremities, contrary to its appearance on many maps. The general outline of Scragly is given best on the Penobscot County Map, but it should lie partly in No. 6. We ascended a short thoroughfare from the east part of Scragly to Shaw lake, and found no ledges but an immense number of boulders of granite and trap. This is a very pretty lake, but much smaller than Pleasant. It is not represented at all upon the State Map. There are many islands in Scragly and Junior lakes, and a few ledges of granite.

Junior lake is connected by a short thoroughfare with Scragly, and it is six miles in length, and perhaps four miles wide. It is represented correctly upon no published map. Two small lakes are situated near its north end, Duck lake and Mill Privilege lake, which are either omitted or not named upon the maps. No ledges occur either upon Junior lake or any of its small tributary lakes upon the north and west sides, of which there are five, which are incorrectly located or else omitted upon all the maps. Nearer the south end of Junior lake the boulders become small and much water-worn, consisting of an interesting conglomerate, trap, schist and granite. Close by Junior stream they appear to be piled up in a ridge, much like a rampart. Magnificent veins may be seen upon any of these lakes.

Upon Junior stream large boulders of granite occupy the bed so much that it is difficult to manage a heavily loaded canoe among them. Two interesting boulders attracted our attention, as they had been worn into the shape of an hour glass. We suppose the neck of the stone was worn most because the strongest currents chiefly exert their powers at that altitude. As is common to almost every thoroughfare, so here on Junior stream, the upper part is very stony with quick water, and the lower part with very deep water and marshy banks, insomuch that chiefly sedges grew upon them; while the surface of the water abounds in white and yellow water lilies and pond weed. At the mouth of the stream the land is a little higher, a coarse beach separating the thoroughfare from Grand Lake, in which we found boulders of metalliferous trap, pyrites, conglomerate, granite, clay slate and schists.

The character of the shores of both expansions of Grand or Witteguerguagum, Pocumpus and Sysladobsis lakes is uniform and may be described as a whole. The immediate shore is composed of angular blocks of granite, often of mammoth dimensions, with scarcely any soil over them. They are covered with moss, and the trees of the forest shoot down their roots among them with difficulty. The shores rise up gradually to hills and mountains, without a single clearing to give evidence of civilization. But no ledges appear, although their fragments are so common. These boulders often lie in the lakes away from the shores, and may project above the surface. One such in Sysladobsis lake must weigh many hundred tons. Where the summits of the boulders just come to the surface they render the navigation difficult. It is extremely rare to see any rock represented among the boulders upon these lakes, except the angular granitic fragments. Most of them are of the porphyritic variety.

On Sysladobsis lake may be seen the finest views of any of the western Schoodic lakes. That part which lies in No. 5 is correctly represented upon the County map, showing the "Big island" at the south end. That part of the lake which lies in No. 4 has a due north and south course, its northern extremity being only half a mile distant from Bottle lake. The Chain lakes in 4 and 5 are mostly small and swampy, except a single ledge of granite on the upper or third Chain lake, which is the largest of the three.

Leaving Sysladobsis lake for the east, we come first to a long narrow lake five miles in length, and from one mile to one-fourth of a mile in width, called Pocumpus. This is given correctly upon the County Map, but is not laid down at all upon the State Map, there being in its place an enormous body of water called Grand lake. Pocumpus has a north-west and south-east course like several other lakes and coves in the vicinity. Its shores are ragged but not ledgy. From its southern extremity we passed up a remarkably crooked thoroughfare to Wawbawsoos lake. We found a few ledges of mica schist and micaceous quartz rock with vertical strata, and having a north-easterly course, before arriving at the lake. The quartz rock lies to the west of the schist. Wawbawsoos or "Machias" lake is very shallow. Near the outlet there is on the shore a short ridge resembling a small horseback, though composed of much coarser materials. Several years ago we described similar ridges upon lakes in Vermont under the name of *Lake Ramparts*, from their resemblance to the ramparts of fortifications. This one is six feet high and wide, and not less than a quarter of a mile in length. Something similar was noticed at the foot of Junior lake. They are analogous to the sea walls frequently noticed upon the sea shore of Maine, and specially described in our last report.

We explain the formation of the Lake ramparts in this way. They are formed by the ice of winter, and only in shallow ponds, or where the water is shallow near the shore, and the bottom is covered with boulders. The ice of the winter seems to inclose the stones with perhaps some of the gravel of the bottom, and from its well known property of expansion it would by freezing gradually force the rocky fragments towards the shore. In one year the progress would be small; but in each succeeding winter the work would be resumed, until at length the fragments would be driven to the shore; and as the level of the lake is commonly higher in the winter than in the summer, they might be crowded a considerable distance beyond low water mark, and in the course of ages the accumulation might be very large. Thus the manner of their formation is like that suggested for the formation of the sea walls, except that ice is substituted for water. Farmers who build fences on the edge of a wide ditch often find them prostrated or bent over in the spring, probably for the same reason, that the expansion of the water in freezing has pushed them over.

A sand bar almost separates Pocumpus from the western irregular expansion of Grand Lake, (See the County Map for the true delineation.) And the west shore of this part of Grand Lake is low and gravelly, contrary to the general rule. We turn around Coffin Point to enter the largest portion of Grand Lake, and we find upon the rocks here the remains of the coffin which was the reason for giving the name to the point. It was brought here many years ago for the body of a man who was drowned while warping a raft of logs, and left because it was too small. The death of this man was considered a judgment by his associates for his profanity, and the remains of the coffin have made his memory vivid to a succeeding generation.

Grand Lake, called Witteguerguagum, (*a forehead,*) by the Indians, is the largest of the Western Schoodic Lakes, being ten miles long and four wide, but not uniformly of these dimensions. Several deep coves make the shape very irregular. Numerous islands are found in the lake, upon one of which we found a ledge of granite—Stone's Island. No other portion of any shore was found interspersed with ledges, though everywhere rocky. Quite high mountains appear at the end of Whitney or Deep Cove.

We made a short excursion up Ox Brook to Ox Brook Lake, and found boulders of a fine-grained granite and Devonian fossiliferous sandstone. No ledges were apparent. Ox Brook passes through an immense swamp filled with sphagnous plants and their concomitants. Loons and ducks abound in these lakes. Flocks of them were often seen; and the notes of a loon were always the first sound heard in every lake visited. Immense numbers of white fish and togue are caught at the Pocumpus thoroughfare, while the Grand Lake stream is known to amateur piscators as the finest locality in our country to catch the Salmon Trout, (or *Dwafed Salmon,* according to eminent authorities,) in June and September. We saw white sea-gulls on these lakes during the prevalence of storms on the coast.

There are two falls on Grand Lake stream, called respectively Big and Little. At the Dam the rock is an argillo-mica schist dipping 80° S. 60° E. At Big Falls the rock is less argillaceous, with layers of hard sandstone dipping 80° S. 20° E. Drift striæ and grooves cross the stream and valley transversely here, with a course of S. 30° E. It is common also to see small and elegant curvatures in the strata, showing that these rocks have been subjected to great pressure. At Little Falls the dip is north-westerly, so that we have here a synclinal axis, the same with that alluded to recently in the Indian Township. Near Mr. Gould's house on this carry the schists are a little calcareous, but not enough so to yield quick-lime.

A few ledges of mica schist, with a nearly vertical dip, occur upon Big Lake, but there are none upon Long or Lewey's Lake. The land is much better than on Grand Lake, and several large clearings have been made, both upon the islands and main land, and have been cultivated for many years, yielding good crops. At the east end of Big Lake is an Indian village, not represented upon any of the maps, where about two hundred Indians of the Passamaquoddy tribe reside. The Governor of the tribe, Louis Neptune, showed me crystals of quartz which were found in the Indian Township.

It would be highly desirable that a road should be built from Princeton to Greenfield, a distance of nearly forty miles, in order to connect Calais and Bangor with each other partly by rail. This would make between fifty and sixty miles of turnpike road and thirty-five of railroad, instead of the hundred miles of stage road on the Air Line route; so that the journey from Calais to Bangor, or the reverse, could easily be made in a single day, without the fatigue of an all-night's ride in a stage. A railroad would be still better. The easiest route would be to keep on the north side of Big Lake, crossing Grand Lake stream near Mr. Gould's house, passing south of Grand and Pocumpus Lakes, and cross Nickatou Lake where it is so very narrow. This would be comparatively a level route. A part of this route might be by water, from Gould's to Princeton, or from Pocumpus Lake to Princeton, by constructing a canal two and a half miles long, large enough to permit the passage of a small steamboat. This would require the building of a turnpike road between Greenfield and Pocumpus Lake only twenty-seven miles long.

At the head of the Kennebasis river at the foot of Lewey's Lake are dikes of trap. Below the banks are alluvial mostly being capable of yielding good crops. Tomar stream, the eastern boundary of the Indian Township, is a beautiful creek, navigable for four miles above its mouth. About a mile above the St. Croix river may be seen ledges of mica schist, dipping 40° north-westerly.

II. Geology of the St. Croix and Chepedneck Rivers.

Below Devil's Head in Calais, on the west bank of the St. Croix river, the rock is granite and syenite. Above this headland indurated slates appear, forming a very narrow border to the granitic rocks, so narrow that it escaped our notice last year. This border extends, with a single interruption, to Milltown. It is never over a few rods in width. Near the lower steamboat wharf are two dikes of traps cutting through the slates or schists. More of the schists appear at Salmon Falls, but the last traces of them disappear at Milltown. The river runs through a gorge in this rock below and at the Mills. Beautiful specimens of pyrites were shown us which were said to have come from beneath the Union Bridge. If abundant, this deposit may be very valuable.

A band of syenite crosses the St. Croix above Milltown, showing itself for five miles along the river and railroad. At Baring station and on the summit level of the railroad in St. Stephens, N. B., the syenite is rendered beautiful by the occurrence of micaceous nodules. The nodules are black, but the stone is white. These nodules may at one time have been pebbles, and the agency that changed the sedimentary rock into syenite, may not have operated long enough to have obliterated the sedimentary character of the rock. The nodules are not as regular and handsome as the so-called "petrified butternuts" in the concretionary granite of northern Vermont.

There is considerable syenite in the granite about Calais, and it is possible that an accurate knowledge of its distribution might assist greatly in determining the truth of the theory that these rocks were once sedimentary. If the various masses of syenite are disposed like stratified beds in the granite, the theory would be confirmed. The syenite seems to run into the granite like a spur or a bed, on the road east of the principally inhabited part of the city towards Vose lake. On Bog brook in Hardscrabble, in the same rock, is an immense mass of white quartz and chalcedony, in which specks of gold may be found. On the summit of Macwahoc mountain the granite rocks have been beautifully embossed by the drift agency, looking like a great number of large haycocks crowded together. It is not common to find so good an example of these embossed ledges in this part of the State. Small pieces of calcite have been found in the west part of Calais. Small pieces of gypsum have been picked up on the shore near the steamboat wharf—whence it becomes a matter of doubt whether the gypsum was derived from the ledges or was imported. On the New Brunswick side of the St. Croix, proceeding in a north-westerly direction, we saw syenite for the distance of two and a half miles beyond the St. Stephens bridge. Then mica schist succeeds. This we followed for six miles, where are large veins of auriferous quartz and plumbaginous strata on Mr. Bolton's land.

Very near a water station of Lewey's Island Railroad, in New Brunswick, the mica schist commences. At Sprague's falls the schists dip 45° S. 30° E. Nearly a mile beyond on the railroad, the dip is from 40° to 70° S. 65° E. A number of rods further the dip is from 20° to 80° S. E. This ledge is full of small contorted veins of auriferous quartz. Several small pieces of gold were found in this vicinity during our visits. Some of the strata contain pyrites, which has decomposed very much, so as to impart a reddish tint to the ledge. About five miles from the Princeton line, estimating on the railroad track, ledges of a calciferous clay slate occur. They are common all the way to the line, dipping 60° N. W. A single ledge near the line has a talcose aspect. Near the village of Princeton, this clay slate dips 60° N. W. Its decomposition must make fertile soil, as the limestone is very abundant, almost as much so as the slate itself. This formation is not very extensive. We suppose it runs into New Brunswick, and also some distance south-west from Princeton village. It does not extend north of the Kennebasis river.

Near the Princeton line the railroad crosses a horseback, at least five miles long, and running south-westerly. It slopes considerably towards the river as it disappears in the woods, more rapidly than any other horseback we have seen. A mile nearer Princeton another horseback appears, of inferior proportions and somewhat imperfect. Through the generosity of the Superintendent we were enabled to make these observations along the line of the railroad both by a special engine and the use of the regular trains.

Between Sprague's falls and the mouth of the Kennebasis we did not travel on the river, but did so on the whole of the Chepedneck. Near the head of the St. Croix are large falls, called the Grand falls. There is an alluvial island with very good soil at the union of the three rivers, and upon it is located the dirtiest house that we have ever seen, both without and within. A short distance above this island are the Frying Pan rips, where the strata of mica schist dip 50° N. W. For several miles there is a meadow on both sides of the Chepedneck, but narrow. Half a mile above the last rips are another set of rapids, the Tea-kettle rips. Four miles above the confluence are the Chepedneck falls, by which we were obliged to carry our canoes and baggage. The schists are quite green here, with a very few veins of white quartz. They dip 60° N. 30° W. Upon Enoch brook are some extensive flats, called Catamount meadows, which are very productive.

Clark's point is at the mouth of Millsberry brook. There are three farm houses in this vicinity, two of them on the New Brunswick side. We had what is called "dead water," i. e., still, for five miles above Clark's point, and the banks are alluvial and fertile. Near King brook, (N. B.,) are two deserted farm houses. In the south edge of No. 1, R. 2, or the Dyer township, are greenish mica schists precisely identical in appearance with the schists on the upper part of the river St. John. It was this resemblance that first suggested to us the probability of the rocks on these rivers being identical in age with that on the St. John. These rocks dip 75° N. 30° W. There are wide meadows here. In this part of the township is an Indian village. Above the village are the Canose rips, where the water is very strong. In the dead water above, called Loon bay, are more farm houses. Near the mouth of Little Sim Squash brook is the farm of the enterprising Mr. Keene, who has cultivated it for many years. A mile below Keene's there is a change in the dip, to the south-east, so that we have here a synclinal axis. Above Keene's we pass over Hog island and Meeting-house rips, the latter named from the roof-like shape of the large granite boulders in the river. At the latter rapids are ledges of a calcareous ferruginous schist, perhaps a limestone, dipping 48° N. W. It has a metaliferous aspect.

We have noticed along this river and further west boulders of conglomerate, which probably came from a Devonian formation to the north or northwest. Some appear like the red conglomerates of Woodstock, which is probably of the age of the gypsiferous rocks of the Tobique. Their position in the drift would indicate that they came from Maine; and if so from some outlier not yet discovered. Below Big Sim Squash brook in No. 1, R. 3, at the Tunnel rips the dip of the mica schist is about 30°, but irregular, N. 20° W. Some of the strata are curved in the fashion of small anticlinals and are argillaceous. Large boulders of granite and conglomerate are common here. Some of our party had the pleasure of seeing a deer crossing the river here upon a log. Terraces are common along the St. Croix and Chepedneck rivers, but rarely over three in number at any one place, and they are composed of gravel. At the mouth of Big Sim Squash there are two very pretty ones on the British side of the river.

Two miles above Scott's brook the schists dip south-east. We are now upon the great bends in the river, which one would suppose to be indicative of vast meadows; but it is not so. The land is not very high upon the banks, but the meandering course of the river winds among many ledges. At the west end of the first great bend of the river the rock is talco-micaceous schist, dipping 55° S. 20° E.; as we enter Vanceboro' they dip 30° in the opposite direction. Little falls and Rocky rips, both very strong water, lie between Scott's brook and the south part of Vanceboro'. There are two miles distance of dead water with very wide flats adjacent, which yield large quantities of hay to the lumbermen and the farmers of the Lambert lake settlement in No. 1, R. 3, whose location will be sought for in vain upon the maps. This marshy land if covered with water would make a large lake, and it is possible that it may have been formerly the bottom of a lake. An obstruction of a few feet at the Rocky rips would now turn it into a lake again. Above this dead water are the "One mile rips," and one mile still further, or three miles below Chepedneck lake dam are the Elbow rips. The rocks are mica schist the rest of the way to the dam. We were impeded in our course below the Corporation House by a jam of logs a mile long, over which we found the work of carrying very tedious. This was an unlooked for obstacle, as the laws of the State require all such obstructions to be removed after a certain date in early summer. We found the dwarfed salmon again both at the foot of Chepedneck and of Grand lake above. The Corporation House at Chepedneck dam is an unusually fine edifice for such a locality. Three-fourths of a mile south-east from it, on the British side, are ledges of mica schist. Thus on the St. Croix and Chepedneck rivers the rock is uniformly mica schist above Sprague's falls. Over this distance are five axes, two synclinal and three anticlinal. The low land soils are invariably very good. That of the high lands may be compared with that between Bangor and Charleston. This section was explored the last week in August.

III. Geology of the Eastern Schoodic Region.

Immediately after leaving Chepedneck lake dam we travelled over Chepedneck, Grand and North lakes, which belong to the general appellation of Eastern Schoodic lakes, and form a part of the boundary between Maine and New Brunswick. But the lakes when taken separately are not named Schoodic, as will be seen upon the map. The lower one is Chepedneck and the upper (of the two large ones) is called Grand lake. These are the names by which the neighboring settlers know them.

Near Chepedneck dam we picked up a boulder of beautiful red jasper which must have been derived from some ledge in Northern Maine. We did not find a single ledge upon the whole western shore of Chepedneck lake, and had not time to examine the British shore. In this respect this lake is like Grand lake of the western Schoodic region. We suppose the greater part of the western shore of Chepedneck lake is occupied by granite. Immense boulders of this rock strew the shores every where and render the soil barren. A remarkable feature in the form of this lake is that the large coves on the British side are to the eye much larger than the principal lake, which bends greatly to the west. One who has not been informed of this feature, when travelling up this lake will surely wander out of his way two or three times. The scenery is very fine here. We were much troubled by heavy storms and head winds while upon this lake. There are one or two houses at the south end of the American side, and far in the distance on the British shore a few clearings may be seen.

From Chepedneck we carried across to Mud lake, about ninety rods distance. This was much easier than to have followed up the thoroughfare. At the outlet of Mud lake the water falls over ledges of porphyritic granite very beautifully. There are no other ledges on Mud lake which is about two miles long. Passing through a thoroughfare of about a mile's length we are ushered into Grand lake, where we were permitted to see evidences of civilization again, in the distance. After a rather hasty examination of the lake we found no ledges upon it except some bold bluffs of white granite on the west shore, in the south-east part of Weston. There is a horseback upon the west shore of the lake in the south-east part of the town. The prevailing rock in Weston is mica schist, with some large masses of quartz rock. At the north end of Grand lake is the termination of the great horseback running up beyond

Houlton. At this end it gradually dies away in a swamp. The east side of Grand lake is underlaid by granite. Upon the hill between Grand and North lakes may be seen the junction of the granite with a mica schist deposit, apparently of the same age with that on the St. Croix river. On North lake high mountains of granite may be seen upon its southern and eastern sides. On the contrary the northern and western sides are low and flat. We went up the Boundary branch a few miles, and found the country very low, but fertile. It is the commencement of the rich Aroostook country. One or two ledges of mica schist may be seen upon it. Between North and Grand lakes we found several fossiliferous boulders, probably of Lower Helderberg age, which must have been derived from some unknown belt of this rock in Maine, as we have never before seen anything precisely like it in the State.

General Remarks.

By comparing our results with those obtained by others in New Brunswick, we may learn somewhat respecting the continuance of our rocks in the province. And first, we would say that the great belt of granite described by us as extending from Jonesport and Addison to Calais, is almost entirely cut off by the St. Croix river. It breaks out again in Charlotte county, N. B., and extends, with a single interruption, to the river St. John, at the boundary between King and Queen counties. Second, the great belt of mica schist noticed at Columbia last year, undoubtedly connects with that on the St. Croix and Chepedneck rivers, and has been traced through New Brunswick in a north-easterly direction, nearly to the Bay of Chaleur, where it is covered up by red sandstone. A spur of it follows the granite last spoken of beyond the river St. John. It probably underlies a great portion of the New Brunswick coal field.

Third, the granite which we have traced from the third Chain lake through the eastern Schoodic region, (and have suggested may connect with the granite running down to Mt. Desert,) has been traced through the province to the Bay of Chaleur, parallel to the belt of mica schist upon its south-east side. If this range commences at the islands off Penobscot bay, then we shall have here a belt of granite two hundred and ninety miles long and from two to twenty-two miles wide. Fourth, the mica schist is repeated on the north-west side of this long granite belt, and is even longer.

In Maine, in an essentially unaltered form, it extends to the Saco river; and in the province it extends quite to the Bay of Chaleur, a total distance of three hundred and seventy miles; its width varying from nine to forty miles. It would not be strange if the name Cambrian which was applied to both these belts of mica schist in New Brunswick many years ago, and is now generally discarded, should ultimately prove to be their correct appellation. In this connection it is certainly an interesting fact that a long ridge of granite should lie between these two long belts of mica schist.

Fifth. Last year, when descending the river St. John, we noticed a narrow outlier of red conglomerate upon its banks, several miles above Woodstock. We noticed a similar deposit this year near the Furnace of the Woodstock Charcoal Iron Company, which is probably the same deposit, as those outliers are only ten miles distant. In Fig. 47 we give a sketch of the relations of this conglomerate to the underlying argillo - micaceous schists. At the east end of the section runs the river St. John. Upon its bank is the Furnace, represented by a small house on a pile of rubbish. Passing over a level tract we presently come to a few small hillocks, which are composed of red conglomerate, dipping from 45° to 50° a little north of west. Further west we see a hill where a section has been exposed by digging into its side, and we distinctly see the conglomerate resting upon highly inclined strata of clay slate alternating with thin beds of limestone. It is still more slaty as we go west. The strike of the two rocks varies certainly thirty degrees. Here then we have an example of one formation overlying another unconformably, and both dip essentially in the same direction. The positions of these rocks is not merely theoretical, it is an actuality, ascertained with pick and shovel. We introduce the figure to illustrate the geology of other parts of New England, where similar sections have been by some supposed to exist and by others denied. Another interesting fact of a general nature illustrated by this section is, the difference of the dip of the strata in eastern Maine and New Brunswick and western New England. Here the dip is very common to the west and north-west; there the dip is almost universally to the east and

FIG 47.

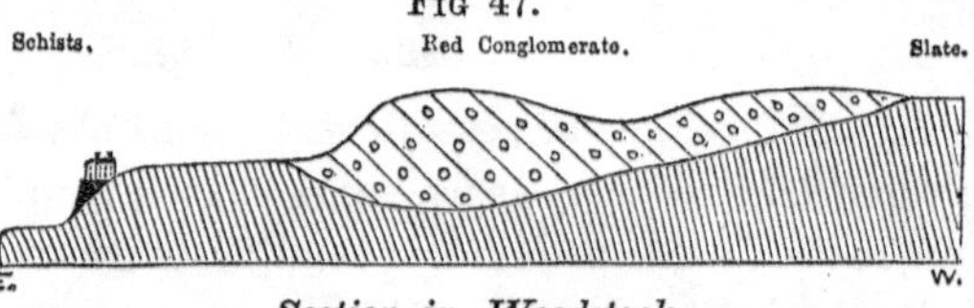

Section in Woodstock.

south-east, and that where sections similar to Fig. 47 have been supposed to exist. This outlier of conglomerate has been traced about a quarter of a mile south-east from the Furnace. It is not over an eighth of a mile in width in Woodstock. Nor were finer layers discovered here like those at the Ferry containing the impressions of rain drops.

In the south part of Woodstock we examined some strings of copper ore, situated in a syenitic calcareous rock. They were formerly mined, and hence had made some noise in the world. The common rock in the vicinity is a micaceous quartzite traversed by dikes—probably of this syenitic rock. These strings are all perpendicular, but none were over half an inch in width. The ore is copper pyrites, and beautiful hand specimens might be obtained here. The mode in which the ore occurs here indicates the manner in which it would be most apt to occur in the adjoining parts of Maine. It would appear in a light grey granite-looking rock, which is a common looking rock, and hence not so apt to be noticed.

In passing down to Frederickton we had a fine opportunity to see the geology of the route, for ledges are very common on the banks of the river St. John. Five different ranges of rock are passed over between Woodstock and Frederickton; first, the mica schist; second, granite; third, mica schist again; fourth, red Devonian sandstone; and fifth, the coal measures. The geology of this province is exceedingly interesting, and the pecuniary benefits that would result from a scientific survey of its territory, would much more than repay all the outlay from the public treasury. There are very few States or provinces on our continent which give such promise of undeveloped mineral wealth as New Brunswick. We trust her authorities will regard the true interests committed to their charge and seek to develop their resources.

NOTE.—The following report of Mr. Goodale, relating to the Botanical and Agricultural character of the western Schoodic region, is inserted after the geological and geographical description of the same district, contrary to our usual order—owing to delay in receiving the manuscript. C. H. H.

Reconnoisance of the Schoodic Valley.

To Ezekiel Holmes, *M. D., Naturalist to the Scientific Survey of Maine:*

Sir :—You will remember that when our parties reached Woodstock, N. B., in 1861, after having examined portions of the Penobscot and St. John rivers, it was proposed that Mr. Hitchcock, with an assistant, continue down towards Eel river, and thence come to the eastern Schoodic lakes. These plans, which were admirably arranged, were completely frustrated by a continuance of stormy weather, and the sudden illness of one of the party. Therefore it became necessary to reserve the examination of the south-eastern portion of Maine, till the ensuing season.

After various divisions of our corps had visited, in succession, Somerset, Aroostook and Knox counties, during the summer of the present year, 1862, Mr. Hitchcock desired me to meet him at Calais on the 6th of August, for the purpose of assisting in an examination of the valley of the St. Croix, or Schoodic, river. Mr. H. arrived in Calais August 7th, and commenced immediately a general survey of the western shore of the river between that city and the town of Princeton. In this and many other portions of our tour, we received very much assistance from Wm. W. Sawyer, Esq., Superintendent of the Calais and Lewey's Island Railroad. Through his kindness we were enabled to make a minute and somewhat protracted survey of the country lying along the line of rail, and were offered every facility for free transportation of our party and their baggage, over the railroad. His politeness was sincerely appreciated by us, and deserves this public acknowledgement.

There are two branches of the Schoodic river, one coming from the chain of lakes lying to the west, and the other from the north. Each of these branches were to be examined by us, and we chose the western for exploration first, because such a course would harmonize better with subsequent plans. Accordingly we arrived at Princeton, on Lewey's lake, on the 14th of August. Princeton, the western terminus of the "Calais and Lewey's Island Railroad," is an enterprising village, in which the main business is lumbering and the shingle trade. It is quite advantageously situated upon the eastern shore of a small lake named for one of the Passamaquoddy tribe of Indians. The fall at the outlet of the lake is such as to afford fair water privileges to several mills, without interfering with river driving in the spring. Lewey's lake is the most south-easterly of the Schoodic chain of lakes, and was, in all respects, the most eligible point from which we could commence our tour. The party was made up of C. H. Hitchcock, the State Geologist, J. Henry Robbins of Calais, and the present writer; with two guides, Benj. D. Wyatt of Princeton, and S. Wallace Haycock of Milltown. These guides, acting as boatmen, accompanied us during the entire tour, and were, in every respect, deserving of the highest praise as skillful, willing and gentlemanly assistants.

Since we should be obliged to retrace our steps over the western lakes after we had reached the one called Sysladobsis, we adopted the plan of examining one side of the lakes or river, on our ascent, reserving the other till our return. This enabled us to make more extended observations than we should otherwise have been permitted to do.

The voyage across Big lake was accomplished in a miniature steamer, belonging to Mr. Wm. W. Sawyer of Calais. He kindly placed it our disposal, and by its aid we hoped that we could examine certain portions of Big lake with more ease than in our canoes. Various circumstances combined to prevent our taking advantage of this assistance, except in transporting our baggage from Lewey's lake to Grand Lake stream.

Premising, then, that a part of our western tour was accomplished in this tiny steamer, and the remainder in birches, it will be necessary to commence the Botanical and Agricultural Report by a brief description of the land around Lewey's lake.

Upon the northern side we have, mostly, swampy land, covered by a thick growth of small Juniper, (Larix Americana,) and, back of this, a scattered forest of slowly growing Pines. The soil, from the lake north through Indian township, as far as the Waite plantation, is mostly made up of disintegrated mica schist, and belted with granite. Under the most favorable circumstances, such as well drained drift-hills and sunny valleys, we find some fair farming land. Very much of the territory is rocky, having exposures of compact schists and slate which might serve to discourage any farmer. It is, on the whole, better fitted for a timber region than for agriculture, and it is stated that the first growth of pine, and saplings generally, was excellent, but subsequently fires have kept down the pines to a size about that of the average of trees in similar districts in Washington county.

Lewey's lake is four or five miles in circumference, perhaps a little less, and is connected by a short rocky thoroughfare, with two small lakes, called "Long" and "Round," or, in the Indian tongue, Pētbekis and Pētquokmus, which are pronounced with a kind of indistinct, gutteral cough. At the point which divides one of these latter bodies of water from Big lake, there is an Indian village of two hundred inhabitants and eighteen houses, or thereabouts. Seldom are more than half the people at home. The settlement is called "Peter Denny's Point." The Indians have a small Catholic church, similar to the one at Oldtown, a fine, new school-house capable of seating twice as many children as there are in town, and a capacious town hall where dances and weddings are conducted. I have been thus particular in noticing this aboriginal settlement because it is omitted from Chase's new map of Maine. There are few points of interest along the northern shore of Big lake. A little way up the lake is a flourishing farm, which confirms what I have previously said, viz: that under a favorable combination of circumstances fair farms can be made in this district. The distance from Grand Lake stream, which is the head of Big lake, to Princeton, is variously estimated at a dozen miles and more, the steamer performing the trip in a little less than three hours, with many stops for examination of the shores. So that the short journey from Calais to the fine salmon trout fishing on Grand Lake stream, can now be accomplished as easily as any trip of equal distance in more thickly settled districts of the country. The land around the head of Big lake is covered with pine of fair growth, except near the water, where the pine gives way to small hardwood, such as Acer rubrum, usually called "*white* maple," notwithstanding the Latin specific name; Populus tremuloides and P. grandidentata or Aspen and Toothed Poplar. The water vegetation is remarkable. The shallow waters along the shores are filled with a large growth of Scirpus palustris or Bullrush. Very many square rods at the mouth of Grand Lake stream are covered with magnificent specimens of this imposing rush, and a stranger would find it difficult to paddle a canoe in the stream at this point, so completely is the mouth blockaded by these thick stalks.

Several undetermined species of Algæ and various other water plants are plentiful at the steamboat-landing. Saguittaria variabilis, *Eng.*, including two strongly marked varieties, S. heterophylla, *Pursh.*, or a form nearly akin to it, also occurs here. Growing with the Scirpus is the "Bayonet Juncus," everywhere common around these lakes. The species, as noticed in Dr. Gray's Manual, is not accredited to any localities north of Tewksbury and Plymouth, Mass., but I have collected good specimens in latitude 46°–40′, and have found it fringing the shores of every large lake in eastern Maine. In many of my specimens the bayonet leaf is *shorter* than the *diffuse* parricle, in others the leaf is much *above* the middle of the culm, and still is shorter than the remainder of the stalk. But they obviously belong to one species, for, with all these differences, these military rushes are not so widely distinct, as we see that many individuals are in the same army.

Grand lake stream which connects Grand lake with the one just noticed, is three miles in length, and has a broken and troubled course among granite boulders and over slaty ledges. In the water of this stream are found multitudes of the well-known "Salmon trout," noticed in the report upon the Fishes of Maine. This dwarfed salmon has strong and quick fins which give the fish great celerity of movement. The tail is disproportionately large, and, of course, aids very much in giving the fish power to stem the strongest currents. The species is an exceedingly interesting one, and deserves more study and attention than has yet been paid to it. Its singular confinement to restricted localities—its likeness to, and its difference from, the Salmon of the Penobscot—the peculiar parasite infesting the scales, (very similar to the Caligus figured in Sill. Jour. 1st series, 70, by Dr. Dana,) all combine to make the study one of much interest to Naturalists.

At or near the mouth of Grand lake stream there is a stock-farm kept by Mr. Gould. Some of the land on the southerly side of the stream produces good hay in abundance, being, for the most part, alluvial soil. By the aid of Mr. Gould and his team we carried our luggage over the rocky road along the side of the stream. The dam at the outlet of Grand lake is capable of resisting a great head of water and is kept in good repair. The sand and gravel plains over a part of the road from Mr. Gould's to the outlet, furnish a good illustration of the open-plain vegetation of eastern Maine. There is almost a total absence of Ferns, no one except the ever-present "Brake," Pteris aquilina, attracting my attention. Occasionally, on similar plains, one meets with Botrychium lunarioides and its perplexing varieties, but such ferns are rare. Comptonia asplenifolia, (Sweet Fern,) Rubus triflorus, Vaccinium Canadense, and dwarfed pines with straggling Junipers, (Juniperus communis) cover the ground. Such is the appearance of a grazing field in which the cattle have not kept the shrubs down. Where grass is allowed to grow unmolested on such plains, we shall find it made up of depauperate specimens of Festuca ovina, Poa amura, Poa pratensis, Danthonia spicata, and occasional tufts of Phleum pratense. Nearer the water's edge, we have generally found, in addition to these enumerated, Juncus bufonius, Alopecurus geniculatus and Juncus effusus. In short, there is little to distinguish such a plain from an exhausted meadow in western Maine or Massachusetts.

The southern shore of Grand lake was next examined. The water in the lake is deep and very clear, enabling one to see the bottom at a great depth. The sand and mud of the bottom is made up, in great part, of disintegrated granite, and consequently presents the characteristic clayey stickiness of such earth. In the mud along the shore there is an abundance of the common Bayonet-rush and Juncus acuminatus. The shores are, in many places, very slightly shelving, so that these rushes extend, in their growth, a considerable distance into the lake. The vegetation of the territory along the southern border of the lake may be said to consist of average pines which had been well culled by the lumbermen, and a considerable sprinkling of maples and beeches. This is especially true of Farm point, so called, where the "hardwood" growth is quite noticeable.

Farm cove, on the southern side of Grand lake, is quite an extensive bay opening to the north-west, and filled with many islands. One of these islands at the mouth of the cove may be taken as a specimen of all the smaller islands in the western Schoodic lakes. It was banked up with boulders of coarse granite but little attrited, on all sides, and covered by a growth of scraggly pines, maples, beeches and birches. The larger islands are more like the main land, and support a good growth of sapling pine, thickly interspersed with hardwood.

The last point around which we passed in Grand lake, was one of considerable interest from a fragment of a lumberman's history connected with it. Upon the extreme point, and in full view of those approaching from the north, is a weather-beaten coffin of which little more than the form is preserved. It was brought up from the settlements to contain, for interment, the body of a river-driver drowned in the lake. The story of his death, as told by a guide, gives additional interest to the melancholy spot.

Rounding this "Coffin point" we had but a short distance to pass over before reaching the strait connecting Grand lake with Pocumpus or "Compass," as it is generally abreviated by lumbermen. The maps from which the new State map was compiled were entirely erroneous as regards this portion of the lake. A more correct one, and yet open to some criticism, is the map of Washington county, by Lee & Marsh. It is difficult to see why this part of our State has been topographically neglected. Very many lumbermen and lumber owners of Calais and Baring have more correct plans than any others we have had the pleasure of examining.

The thoroughfare between Grand lake and Pocumpus lake, is not more than a few rods in length. The current is quite slight, (in August,) and yet enough to completely change the character of the pond-weeds growing in the still water. Potamogeton lucens, *L.*, with ovate leaves floating on the water and making it P. lucens, *L.* var fluitans, was quite abundant in the almost stagnant water of Pocumpus. But in the current of the thoroughfare it became submersed and agrees with P. obrutus, *Wood*, convincing me that Dr. Gray is right in referring the latter plant to P. lucens *L.* var. I hope to place the plants in the hands of Prof. Tuckerman, who formerly published a monograph upon the genus, and who has studied them more protractedly in the fresh state, than any other botanist in America.

The lake thus called variously "Pocumpus," "Pocumsus," and "Compass," is an extensive body of water, oblong in shape, its greatest length lying in a north-west and a south-east direction. It is a lake of much beauty, but does not possess much interest either for the Geologist or Botanist. The banks, as in the other lakes, are mostly walled with granitic boulders, while the white and whitish yellow sand comes well up to the very base of the wall. The woods are mainly pine, with hard-growth intermixed. It was our conclusion, as we examined the soil in this vicinity, that several good farms could be cleared and rendered productive around the lake. Our party also ascended the stream coming from Woboosoos, a lake lying off to the south. This lake has a dam at the outlet, but does not seem capable of high flowage even if the timber on the shores should warrant the expense. Many logs of considerable size were decaying near the outlet, having lain in the mud, apparently, for several seasons. Around the dam the ordinary herbs attained a good size, indicating alluvial fertility. But the land at a little distance from the lake is exceedingly rocky and ledgy. Cladium mariscoides, *Torr.*, ("Twig rush,") grew plentifully along the shore in the soft clay and sand, whereas it usually has been found, in Maine, in Sphagnous swamps. The distance from the head of Woboosoos to waters flowing into the Machias, is very short.

From Woboosoos we returned to Pocompus, and commenced to ascend the arm leading to Sysladobsis lake. In this narrow river-like arm the water is very deep, or at least so deep as to afford a great contrast to that which we had just left. Much of the shore is lined with water lilies, everywhere floating with Nuphar advena and Kalmiana.

Sysladobsis lake we reached by a short carry. This lake is at once recognized by the long island, by no means well represented on the map, parallel to the eastern shore. There was the same appearance of a good pine country all around the lake, as of that we had last been examining. During this day we made a greater distance in our birches than on any day previous, having twice measured Pocumpus and sailed on both sides of Sysladobsis.

The water being so low as to forbid our exploration of the Chain lakes lying off to the west, and the smaller lake at the north, we commenced to return. We arrived at the "Dobsy" carry at ten o'clock A. M., on the 19th of August. As I shall not have occasion to allude to this lake again at present, it will be well to give a brief account of its topographical and physical characteristics. It is a long, twice-curved lake, extending through the length of a whole township, and receiving water from the north-eastern, northern and western tributaries. From the north-east there empties into it a small stream supplied by a pond of very trifling size; on the north it is augmented by water from "Second Dobsy" or Sysladobsis, which is the most western lake of this long chain; on the west it receives a stream from the Chain lakes, which are of small size, in township 4 in the 1st range. There are two large islands in the southern part of the lake, each of which are covered by fair pines. The lake is of greater depth than Pocumpus, and of larger size. From a brief examination of the shores in one portion of the lake we were convinced that a good lumber farm could be cleared if necessary, and it might form a nucleus for quite a settlement, but it is our unbiased opinion that there is so much good land to be obtained nearer remunerative markets that it hardly seems worth while to spend labor in felling and clearing here. This portion of Washington county is much better adapted for purposes of lumbering than for agricultural interests, and good pine is better than half-ripened corn. In again looking at Pocumpus we have only to say, that it is also fitted for lumbering.

We next proceeded to examine a series of lakes lying off to the north of Grand lake. We first reached Junior lake, a beautiful body of water connected by a pretty short but rocky thoroughfare with the lake last noticed. There was the same wall-like look of the shores, the same pine-covered points and long reaches and islands. Beyond the islands are seen what could not be seen on the shores of the two lakes just left—houses and farms. Passing a very singular line of islands, incorrectly represented on all the maps, we approached the northern shore of the lake. A range of granite hills extended for many miles along a line parallel to the northern shore, and upon the slopes of these hills there was to be seen, here and there, a farm of considerable size.

Leaving our birches in a stream on the north-eastern portion of the lake we walked along the bank for a mile, or perhaps less, arriving at a small body of water apparently quite shoal. The bayonet rush extended for some distance into the water, everywhere tinging the shores with its color. This sheet of water is called "Duck pond," and has quite a stock-growing settlement on its northern shore. It is a few miles from the road extending from Topsfield and Lee.

Upon the point at the mouth of Duck pond stream is sand of surprising whiteness, much resembling the sand of the sea-shore. It comes from the decomposing granite of which the hills are made, the granite being remarkably free from hornblende or mica. In fact the blanched, weathered appearance of exposure of this rock, and of the soil which is formed from it, is noticeable at a great distance. While camping among the abundant and pertinaceous musquitoes of this shore, we heard the familiar lowing of cattle in a direction where our maps gave us no indication of any settlement. A protracted search, late into a dark evening, failed to discover any houses, or obtain any answer to our anxious holloas, even while we could distinguish human voices at an apparent distance of half a mile. In the morning we learned, upon further search, that a pond of considerable size stretched between us and a little colony of five houses and flourishing farms. Neither were the houses or the lake and stream represented upon our largest map. This little settlement, thus shut out from the world, consists of the farms of Messrs. McLaughlin, Robinson and Moore, with one or two others, and is called "Mill Privilege." The soil has a true granitic character, and yet is of such a friable nature that the yield of grass is abundant. In the town of Carroll, a few miles from this settlement, it was my privilege to examine somewhat hastily the well-conducted farm of Ezekiel Brown. The soil did not differ materially from that at Mill Privilege, and Mr. Brown was obtaining fine crops from his well-tilled farm. There are indications of more or less crystalline limestone through this section. It crops out, as noticed by the geologist, in Carroll, and probably aids in giving increased fertility to the soil. In concluding this notice of the section which may be called South Carroll, I would say, that there appears to be in it much available, productive farming land. Considerable of the timber is of good quality, but much of the kind has been swept by fire. Pursuing our journey, we arrived at a lake of irregular shape, situated on the east of Junior lake, and in many respects resembling it. We examined, in the vicinity of this lake, two other smaller ones, each reached by a devious path through tangled second-growth and fallen trees. One of these was Pleasant lake, so called, and it is rightly named. We had before seen it from the east, when Mr. Hitchcock and his assistant searched with Mr. Bailey for the native copper on the shores of the lake. It has high granitic banks upon the north and east. The ledges on the eastern shore assume very fantastic forms, such as extended shelves, making long and open-mouthed caves of granite. In one of these caves, or rather under one of these immense roofs, we found several fungi of much interest which will be hereafter noticed. But with the exception of these and some other Cryptogamiæ we saw little worth noticing as regards the Flora of the district. The extended exposures of hornblende granite (syenite) forbid our speaking of the eastern and northern shores of Pleasant lake as eligible farming land. The other lake referred to, was much like the one just described, only it had very large timber on some parts of the shore and lodged at the outlet. In short, the country extending from this lake to Mr. Gould's farm on Grand lake stream must be considered timber land instead of farming land.

The water in these lakes, not excepting Grand lake, was filled with minute particles, much resembling some of the confervæ, which proved to be a Protococcus. The water, drawn from a considerable depth, was also found to contain them, and they did not appear to be more abundant in one part of the lake than in another.

Having returned to Junior lake stream, we dined on the point of land extending out into Grand lake; and while here, found among the boulders on the shore, a mass of Bi Sulphide of Iron of great compactness, and weighing several ounces. It is our opinion that the many Indian traditions, and the stories of early settlers, in regard to great metallic wealth among the hills around the lakes, are referable to the discovery of pieces of Pyrites similar to the one just spoken of. That there may be, and probably is, some iron and copper scattered through the granite, is not difficult to believe, but the many circumstances which must conspire to make a mine of either of these metals profitable will be noticed in the chemical report.

At the mouth of Ox brook, a small stream entering Grand lake not far distant from the point just referred to, and into what is called Whitney cove, we commenced an overland tour in search of "Ox brook lake." The vegetation around the lake was unmarked from any noticed before in the vicinity of similar bodies of water. In a swamp of large extent which we passed over in approaching the lake, I found a Juncus, subsequently, referred with specimens from another locality, to the rare Juncus Stygius, *L.*

The rest of our journey back to Grand lake stream was soon accomplished.

Upon the following day we passed down Big lake, examining the shore untouched as we ascended the chain. I cannot resist the conviction that there is some excellent farming land lying along some parts of the southern shore of Big lake, and there is needed only the encouragement of fair roads to have such clearings undertaken by good settlers.

We reached the town of Princeton late in the evening of 22d August.

The continuation of the survey of the western branch was resumed on the 27th, the intervening time being occupied by explorations in and about Baileyville and Calais. The river below Princeton, as far as its junction with the east branch, falls gently, having only two sets of rapids with the name of "rips." At the Tomah stream where the water is "dead" the vegetation was noticeable on account of its luxuriance. The meadow hay in certain localities along this part of the river was principally made up of Phleum pratense and Calamagrostis Canadensis.

At 3 P. M. we arrived at Young's island which marks the confluence of the two branches of the Schoodic river. Having reached the terminus of our tour on the western river and its lake expansions, it now becomes necessary to take a brief resumè of the results attained by a topographical and botanical study of the waters. The western river extends, from the source of its remotest tributary to the eastern branch a distance of — townships or approximately — miles. It drains portions of *twenty* townships lying in a general east and west direction. The greater part of the course of the river is south-easterly, but pursues a north-easterly direction in the last portion of the way. The larger portion of the townships drained by the western river is average timber land with the usual proportion of inaccessible lumber and hackmetac swamps. The remainder of the territory may be called fair farming land, especially so much of it as borders upon Mill Privilege in Carroll, also the lower part of Six in the first range, that around sections of Big lake and along the remainder of the river as far as "Young's."

In conclusion, the whole "water-shed" of this branch of the river inclines to south-east by east, S. 80° E.

I have the honor to be, sir, your obedient servant,

G. L. GOODALE.

C. GEOLOGY OF THE MORE NORTHERN AND UNSETTLED PORTIONS OF THE STATE.

Treating this topic very briefly, on account of the want of space, we will first speak of the geology and country watered by the large lakes on the upper Androscoggin; secondly, a very few notes respecting Moosehead lake; thirdly, the region of the west branch of the Penobscot and the upper St. John; fourthly, the Alleguash lakes; and lastly, Dr. Holmes will give an account of his observations in Aroostook county the past season. These brief accounts are all fragmentary, often supplementary to last year's researches.

I. Geology of the Upper Androscoggin Waters.

Our examination of this region was only preliminary to a more careful exploration. This part of the State was visited late in the season, and want of time in part prevented much examination. The lakes passed over are Rangely, Cupsuptic, Mooseluckmaguntic, Mollychunkemunk and Welokenebacook.

The rock between Phillips and Long pond in Sandy River plantation, is a primitive looking mica schist, standing on its edges. It changes very suddenly at the north-east corner of Long pond to a micaceous schist of Paleozoic aspect, dipping about 35° southeasterly. Ledges of this newer rock were seen all the way to Rangely lake. It is worthy of remark that this newer rock is upon the west side of a great range of mica schist and granite, of which mount Saddleback is a part. We suspect it to form a part of an important basin, to which belongs the great belt of Oriskany sandstone described last year, and also the newly discovered fossiliferous belt in the north part of New Hampshire. It is a noticeable fact in this connection, that the strike of these schistose rocks in the north-west part of the plantation is 180° different; or in other words, the strike of the newer rocks would carry them directly across the older rocks of Saddleback.

Passing up Saddleback stream—the small brook crossing the road at Indian Rock Hotel in the north-west part of Sandy River plantation—the rock is evidently an altered sandstone, and a large number of pebbles of various sizes are present. The rock is a little argillaceous, and dips 45° southerly, apparently underlying the rocks at Long pond. The planes of cleavage are nearly perpendicular; it being unusually easy to distinguish them here from the strata; for the latter appear like successive ribbons of different shades of color on the sides of the ledges, and upon the surface of some of the layers are ripple marks. It is rare to find more satisfactory distinctions between these two sets of planes. Furthermore, these ripple marks are *right side up*—that is, they have not been overturned, and consequently they prove that the strata containing them have never been overturned. It may be of great importance, by and by, to know this fact. Horizontal jointed planes are also present in the ledges on this brook. All these phenomena are exhibited within a short distance of the hotel in a narrow gorge.

Passing northerly we find a coarse conglomerate very abundant, whose dip is greater than that of the adjoining schists, being successively 75° N. 10° W., 90°, and 75° S. 10° E. Many of the pebbles are distorted in some way, as they are very apt to be in disturbed localities. The pebbles are composed of granite, a schistose rock, sandstone and hyaline quartz. All kinds of them in certain layers are flattened, elongated, and sometimes indented, just as if all them had been somewhat plastic; and when the strata were elevated by the great plicating agency, these layers of pebbles, on account of their yielding nature, were compressed into a smaller bulk than before, the different fragments altering their shapes so as to be accommodated to all the crevices of the mass. Even the granite pebbles have been distorted here, which is not common elsewhere; they appearing the most unyielding. These pebbles are commonly flattened in the direction of the dip, showing the force to have been a pressure simply, without the tension of curvature which seems to have elongated pebbles elsewhere in the direction of the strike.

Last year we called attention to this subject so fully that we will add nothing now, (See page 178, *et seq.*, where is a sketch of elongated pebbles.) It is a new subject in geology upon which but few geologists have yet given their opinions. No geologist doubts that many fossils have been distorted by pressure, and that exerted in the same manner in which we suppose the pebbles have been misshaped; and if fossils can be distorted by pressure, why not pebbles, which must have been somewhat plastic during the process of metamorphism? For instance, Dana in his Manual of Geology, page 109, says: (though he has not expressed any opinion respecting the distortion of pebbles,) "These uplifts of the rocks, besides disturbing the strata themselves, cause distortion also in imbedded fossils,—either (1) a flattening from simple pressure, or, in addition, (2) an obliquity of form, or else (3) a shortening, or (4) an elongation." This language would describe admirably the changes undergone by these pebbles in the same circumstances.

We will only say further in relation to this subject, that this is the finest locality to exhibit these phenomena of any yet observed within the State. It shows the process just as finely as near Newport, R. I., which was described last year as the classic ground. Every feature exhibited near Newport may be seen also in this plantation. Both the altered and unaltered pebbles are present, as if to show the differences by contrast. This locality is in an older rock, and hence may present a single feature not common in R. I., viz., the presence of veins cutting across the pebbles. These veins cut the pebbles in two, and one can break out the dissevered fragments and fit them together again. We did this repeatedly here, in one instance the halves of the pebbles having been separated sixteen inches. The inhabitants of this region have often noticed these distorted pebbles, and have reasoned respecting their origin precisely as we have attempted to do. The proprietor of the Indian House, Mr. Prescott, had often been in the habit of exhibiting them to his guests. We have now discovered four localities of this nature in Maine, viz., at No. 9 of Washington county, Cushing, Mt. Battie, and Sandy River plantation. We were informed that a ledge of granite was contiguous to the conglomerate on its west side, but had not time to visit the place. This might be a case of the metamorphism of conglomerate into granite, similar to that described by my father in the Geology of Vermont, vol. 1, page 40. The conglomerate belt runs across the east arm of Rangely lake into Township No. 3 an unknown distance.

Two or three terraces appear at the mouth of Saddleback stream, as one sails over Rangely lake. On Birch point is a schistose rock, apparently the same with that on Saddleback stream, dipping 50° south-easterly. The same rock was noticed upon Ram island and in the western part of the lake. The boulders on the shores are largely made up of green grits and schists.

It was a matter of surprise to us to see so fine an agricultural region in Rangely. The hills about the lake are all rounding and smooth,—that undulating character always found in fertile upland countries. From our sail-boat not a boulder could be seen anywhere. Good crops can be raised here also. Hay yields two tons to the acre, which is a very good yield with scarce any cultivation. Unless it be an alluvial meadow, it is very safe to say that it is not common in the western part of the State to find land as good as that in Rangely. Up the Kennebago stream, still nearer the Canada line, the land is said to be even better. Here then is a large tract of fertile land, probably equal to that in Aroostook county, whose claims to settlement are urgent. This district has the advantage over Aroostook county in its proximity to the market, and is one whole degree of latitude south of Houlton. It seems strange that such fertile tracts of land as this, with that in Aroostook county and on the west branch of the Penobscot, should have been brought to notice at so late a day, and after so many persons have emigrated westward. In our estimation the northern portions of Maine are more attractive as settling lands than the prairies of the West. We do not know the limits of this fertile tract in Franklin county. We presume it is about equal to the area of twenty townships, each six miles square, or seven hundred and twenty square miles. Upon the lakes west of Rangely the soil is poorer, or like the average further south.

On the road from Rangely lake to Indian rock the ledges are seen at a saw-mill to dip south-easterly. The rock is a talco-micaceous schist, similar to that seen at the Canada line in Somerset county. Indian rock is a ledge of similar character, only more soft and talcose, appearing in general to dip northerly. But it is made up of strata contorted in a wonderful manner. The ledge is full of small anticlinals and synclinals; and it is no wonder that it is visited by the curious, on account of these bent layers. Whether the Indians admired these convolutions is a matter of doubt, as they would be more apt to prize its adaptations to meet their piscatorial wants. It is known that it was a favorite place of resort for them. Opposite this rock is a very nice camp, constructed by gentlemen in Boston, to accommodate amateur sportsmen and piscators who frequent these lakes during the warmer months.

Cupsuptic lake lies partly in Rangely and partly in No. 4, R. 2. We passed across the lower end of it, which was low and marshy. On a small island at the north end of Mooseluckmaguntic lake is a large ledge of granite, the quartz of which is remarkably sharp angled. It is immediately west of Bald mountain, which we suppose also to be composed of the same rock. As there are so many Bald mountains in the State, we will specify concerning this one, that it lies on a narrow neck of land between Rangely and Mooseluckmaguntic lakes. We passed through this region early in October, when the most gorgeous colors clothed the forests; and never in any part of New England have we seen such bright and distinct colors as were exhibited upon these trees.

No other ledges were noticed on Mooseluckmaguntic lake. Its shores are lined with immense angular boulders of granite, and a very few of siliceous slate. In the distance are high mountains. These Androscoggin lakes generally afford grander scenery than any others in the State. Their waters afford several kinds of fish not found elsewhere, and wild animals are common in the forests adjacent; so that there are fine places of resort among them for the student of Natural History. The *Salmo Oquassa*, Girard, or blue back trout, an uncommon variety of dace, and a red-sided sucker, are peculiar to these waters. The togue and pickerel are not found here.

On the west side of Mooseluckmaguntic lake is the thoroughfare to Mollychunkemunk lake, a shallow wide stream very full of boulders. A very fine dam is built about half way across it for the benefit of the lumber owners, and a good corporation house is located here, besides a house erected for the benefit of the amateurs. We had no time to stop here, but leaving one boat behind took another on the next lake. Birch canoes are not used upon these lakes at all, while in the other lakes of the State they are generally esteemed more highly than the heavier row-boats.

Instantly after emerging from the woods skirting Mollychunkemunk lake, a beautiful view bursts upon one. Two large and singularly shaped mountains appear on the right, while to the left are still higher immense granite piles, interesting because in contrast with the sugar-loafed shaped Eskahos, or Eziskahos mountain of many authors. They are all of granite. Eskahos lies between this lake and Wilson's mills. It is bare on one side, looking very much as if it had been cleared by man, as much of its surface is yet green. The shores of this lake resemble those of the previous one. Some parts of it are, however, low and sandy, forming quite a large meadow, which is too sandy to be very fertile. The surroundings are similar on Welokenobacook lake, the next of this chain, and the last one above Umbagog. At the south end of this lake are several ledges of granite, and sandy beaches are not uncommon. There is another dam across the outlet of Welokenebacook lake, and also another hospitable corporation house. These last two lakes frequently pass under the general name of Richardson lake. The granite extends for three miles below the south end of Welokenebacook on the road to Andover, where it is succeeded by gneissoid mica schist.

It is necessary to carry between Welokenebacook and Umbagog lakes, a distance of five or six miles; but we had not time for the undertaking. The rock about Umbagog lake according to Mr. Goodale's notes last year, is mostly granite. Beyond Wilson's mill on the Megalloway river the rock is said to be clay slate.

But we have a threefold evidence of the existence of a fossiliferous belt in this vicinity. First, there is the great Oriskany sandstone belt, known to extend from Aroostook river to Parlin pond, which is better developed at its most south-western limit known to us than elsewhere in its course. This must extend further than Parlin pond in this direction. Second, we find boulders of a peculiar fossiliferous rock in Phillips, different from anything yet seen by us in the State, which must have come from this vicinity, or from the Kennebago region, judging from the common course of the drift striæ in the vicinity. These boulders were sent to us by Seward Dill of Phillips. Third, a gentleman residing in New Hampshire and employed in the topographical survey, informs us of the discovery of fossils in situ near Umbagog lake. We give here extracts from his letter:

"I was at work with a party of our topographical engineers in that region all last summer, from the Umbagog up the Megalloway to its source, thence to the 'Crown monument,' and spent the summer in that vicinity." "Since Dr. Jackson made his geological survey of New Hampshire nothing has been done in that line, and so while prosecuting the work on the topographical survey, I have also given much time to the geology, and more particularly to the mineralogy of our State." "While at work on the line of Maine and New Hampshire, I have frequently made trips into Maine, until I feel very much interested in the geology and mineralogy of your State." "I am glad to know that you have discovered a fossiliferous region in Maine. We put to rest forever, last summer, the theory of many eminent savans that 'no fossil ever existed in New Hampshire.' We have made important discoveries in that direction, and have sent away many excellent fossils obtained last summer, and got orders for more which we cannot fill," etc.

(Signed) JOHN EDWIN MASON.

Manchester, N. H., May 24, 1862.

This is an important discovery, and we shall hope to hear further from Mr. Mason on the subject. It will be remembered by scientific men, that in Vol. I, (N. S.) of Silliman's Journal of Science, Professors H. D. and W. B. Rogers gave some account of supposed fossils discovered near the Notch, referring them to the Clinton group of the Upper Silurian. Some of the supposed genera were Agnostis, Cytherina, Atrypa and Lingula, besides fish scales.

In the fifth volume of the same Journal, it is stated that upon a further examination of these specimens, they are satisfied they are not fossils. This locality is at least fifty miles distant from the new locality described by Mr. Mason. We mention this that no one may confound the two. This new locality is probably the prolongation into New Hampshire of the Oriskany sandstone belt of Maine. Many of its strata are clay slates, such as are described as occupying the whole of the Megalloway above Wilson's Mills, and such as have been shown us from the Kennebago valley further east. And in the final report upon the Geology of New Hampshire, an account is given by Prof. J. D. Whitney, (now State Geologist of California,) of a clay slate region in the very northern portion of the State—*See final Rept. Geol. N. H., page* 68. This rock does not extend beyond New Hampshire into Vermont, unless in a very unaltered state, as along the whole of the north-east part of Vermont the rocks are granite and mica schist.

II. Notes upon the Geology of Moosehead Lake.

We had an opportunity of visiting this lake in the early part of the season, but not of exploring it. We visited one or two localities there, and give our impressions in general respecting the geology of the whole lake. We found Mr. Houghton's observations correct, which he described in the Preliminary Report. Our remarks are designed to be supplementary to his report.

The rock at the south end of the lake is clay slate, part of the great belt described previously. At Greenville the dip is rather north-westerly. Immediately succeding the clay slate is a narrow band of syenite, a rock entirely distinct from the common granitic rocks of the vicinity. It is our impression that the clay slate dips south-easterly immediately adjacent to this syenite. This syenite belt cannot be over two miles in width. At the base of Squaw mountain, an immense range on the south-west side, the rock is mica schist dipping 60° south-easterly. Half way between the mountain and shore the syenite appears again, being on the south side of the schist.

Fully half the length of the lake is occupied by this mica schist, a rock supposed to be fossiliferous, as the *Favosites Gothlandica* (?) has been found in it at the base of Squaw mountain. The dip of the strata is very high to the north. At Mount Kineo we come to a narrow band of siliceous slate. A mountain upon the west side of the lake is evidently the same rock, also Kineo Jr., and the associated peaks to the north-east. Hornstone or flint are other common names for this rock. It is extremely difficult to ascertain the position of the strata of Mt. Kineo. On the west side one would fancy that the strata were nearly horizontal; and a view of the east side appears like an inverted crushed anticlinal axis, the strata at the base of the mountain being nearly vertical. The best way of learning the position of these strata will be to ascertain the position of the adjacent rocks, and infer thence the position of the former. On the beach south-east from the mountain, and upon the south side of the range, is a slaty micaceous sandstone dipping under the mountain, or 75° N. W. Mt. Kineo is almost an island, being connected to the main land only by a sand bar.

In a short excursion to Farm island we found specimens of great interest. The rocks are indurated sandstones dipping 60° S. E. Ripple marks are occasionally found upon the strata. The dip is smaller near the north end of the island; and upon the west side drift deposits have entirely obscured the older rocks. They dip towards Mt. Kineo. The most interesting thing discovered upon Farm island is a fossil plant, allied to the *Fucoides Cauda-galli* of authors, so called from its resemblance when spread out upon the rock to the tail of a rooster. West of New England and in Lower Canada this fossil is the characteristic form of life found in one formation of grits or sandstones; hence receiving the name of Cock-tail grit or Cauda-galli grit. This formation of fifty or sixty feet thickness is situated just above the Oriskany sandstone elsewhere; and so it is here. For Oriskany sandstone fossils have been obtained from the shore north of Farm island, whose strata dip southerly towards this locality. Hence we are enabled to chronicle the discovery of another fossiliferous formation in Maine, similar to those well known elsewhere. Upon our map last year, this formation must occupy the south-eastern border of the belt of country represented as Oriskany sandstone.

The thickness and distribution of the Cauda-Galli grit in Maine cannot of course be even conjectured from the discovery of this sea-weed. Perhaps this may not be the very same species with the one found elsewhere, but it is generically the same; that is, if a separate genus was made of the common species, this new one would be a Cauda-galli from Moosehead lake, and this generic resemblance, we suppose, is sufficient for the identification of the formation. It is smaller than the common form. In our tabular list of the older formations given last year, page 150, this grit, with the two other formations discovered in Maine, will be found given. It is singular that so many of the rocks found fossiliferous in Maine thus far should belong to these three successive belts, viz., the Lower Helderberg, Oriskany and Caudi-galli groups.

At the extremity of the north-west arm of Moosehead lake we found ledges of an argillo-micaceous schist, supposed to belong to the Oriskany standstone, but without fossils. It is inclined 50° S. W. Owing to the supposed superior importance of other districts in the State for geological research, we were compelled to pass by the exploration of Moosehead lake this year, but hope to be permitted to resume its exploration another year. It seems to present subjects of unusual scientific interest, as the glimpse we have had seems to indicate. The first voyage of our survey across its waters proved the existence of a new formation in the State—the Oriskany—and the second expedition across its waters, accidentally stumbled upon another fossiliferous member, the Cock-tail grit. Neither of these voyages were anything more than reconnoissances, and hence we may expect other important scientific discoveries there when its exploration shall be commenced in earnest. Moose river should be explored in connection with it.

A very few more facts respecting the alluvial geology of Moosehead lake will be found given in the section entitled *Surface Geology*.

III. Geology of the west Branch of Penobscot River and of the River St. John.

The first work performed by the Scientific Survey the past season was the exploration of the west branch of the Penobscot river above Chesuncook lake, and the upper portion of the river St. John. We started the day after the ice disappeared from Moosehead lake, while the snow was still lying upon the ground in many places. This early start was necessary in order to avoid the annoyance of the black flies, which are the most troublesome the first part of June. As Mr. Goodale has written some account of the natural features of the country over which we travelled, with some botanical and historical notes of our progress, we will insert his report before speaking of the geology of the route.

To E. Holmes, *M. D., Naturalist to the Scientific Survey:*

Sir:—It gives me pleasure to present, in obedience to instructions received in May of the current year, the following report upon the Physical Geography, Botany and Agricultural capabilities of the valley of the west branch of the Penobscot and the upper main St. John rivers.

Our party of exploration consisted of the five following persons: Mr. Hitchcock, Geologist; Mr. White of the Amherst Scientific School; two guides—James Bowley of Shirley, and George O. Varney of Greenville—with the present writer acting as Botanist. We arrived, with our canoes and luggage, at the north-west arm of Moosehead lake, upon Monday, the 19th of May. The ice had broken up on the day previous, and we had every prospect of a cold, comfortless tour. To add to the cheerlessness of the first day of our journey, we had a drizzly, penetrating rain, which made the morning quite dark and inauspicious for the commencement of our tour. But these and similar discomforts peculiar to a spring campaign, were more easily endured for two good reasons; first, we should probably have abundance of water to enable us to examine the smaller streams with care; secondly, we hoped to have comparative exemption from those torments of summer life in our Maine woods, the Simulium molestum or black-fly, and the Simulium nocivum or midge.

Our plans for the tour may be stated, briefly, as follows: having ascended the Penobscot west branch as far as the St. John waters, we will pass down the river St. John to Seven isles and carry into the Alleguash. Returning from this point we desire to examine the lakes and mountains immediately west of Chamberlain lake; sailing down Caucomgomoc stream to Chesuncook, we will ascend the west branch as far as the Rail portage to Moosehead. We were fortunate in being able to accomplish this undertaking in a shorter time than we had allowed for its completion, and in a comparatively comfortable manner, finishing the tour by our safe return to Moosehead on the 7th of June.

Since our journey began and ended, so far as the exploration of the district is concerned, at Moosehead, it has been thought advisable to defer a detailed description of the lake, to the latter paragraphs of this report. The account of our journey will, therefore, commence with our arrival at the "North-west carry." By this portage, the carrying distance to the Penobscot is one mile and three-quarters; but we were enabled to take advantage of a sluggish and very crooked brook, which afforded our canoes good water for more than half the way. The brook creeps through a dense growth of low trees, which gave, in the mist of the cold day, an unnatural darkness to the water of the stream. Emerging from this black pathway we found the commencement of the carry to lie in a forest of mixed growth, where a small spot had been cleared for lumber operations. At a point ten or twelve rods from the brook, a road crosses the carry, running in a north-westerly direction. It is much wider than the portage path, having been originally intended for a military road. It was cut, by contract, in 1842, and commonly goes by the name of General Wool's Military road. Of course twenty years of disuse have permitted the growth of many quite large trees in the very roadway. The portage is half a mile in length, and terminates at Seboomook-meadows pond. This pond, in early spring, is one hundred and twenty rods in diameter, its size principally depending upon back-water from the Penobscot. Late in summer it is barely a quarter of a mile in circumference. The blue-joint grass around the borders of this pond is said to be very good, and has afforded, in a single season, eighty tons of meadow hay of fine quality. The brook which forms the outlet of this pond was quite deep at the time we visited it, and enabled us to have an easy journey to the waters of the west branch. At the union of this brook with the main river, we first noticed particularly the fertility of the soil. In many respects the vegetation of this district reminds one of the luxuriance of the plants of Aroostook.

West Branch of the Penobscot.

The river, at this point, is the finest we have yet seen in the State. It is full, swift and strong. The banks are covered with hardwood, elms and maples, with here and there some fair pines and spruces. But the immediate vicinity of the river has been well cleared of available timber, and there is now little left of what is called "good sapling pine." The land, even at this very early season, had such a flourishing growth of spring plants, Dicentra, Atragene, Viola and Claytonia, that we could easily imagine ourselves in the maple woods of York or Kennebec county, and many times we looked around for some clearing and farm house on the shore. But although the banks have such a homelike look there is not a cultivated spot nearer than the rail-carry at Moosehead. This part of the county of Somerset deserves much attention from those intending to purchase farms. The land, as I have already remarked, is very good for agricultural purposes, and is much nearer an available market than much of the better grain lands of the west. A person can start on one day from this part of the river and reach Bangor or Skowhegan the next night. Since then, to fertility is added accessibility, we can conscientiously recommend the land along this portion of the west branch, to the careful consideration of farmers.

The point where the dividing line between the towns of Seboomook and Plymouth crosses the river, is marked by a whitened pine stump on the northern bank. When we had reached this bend of the river, we felt the large drops of an approaching thunder shower. We had hardly time to place our luggage under the inverted canoes before the rain came down in torrents and drenched us to the skin. Some distance above this we arrived at an expansion of the river into quite a bay. An island, or peninsula, in this small bay is covered by elms of vigorous growth. Not far beyond this we arrived at Gulliver's falls, by which we were obliged to carry our birches and luggage. At the foot of these falls is a brook five or six feet wide, called by the name of the man who first conducted lumber operations upon the stream, Gulliver's brook. Forty or fifty rods above the falls we come to other rapids which we also carried by. The next five or six miles were over rapid water, running between banks of hardwood growth intermixed with conifers. It was near the end of this rapid water that we saw a smoke some distance from the shore, and this we found proceeded from the camp of the Masterman's, father and son. They received us very hospitably, shared with us a quarter of moosemeat, and gave us much information concerning the upper part of the river. Their camp was located near a brook of good water and in immediate proximity to a cedar swamp. Here the old man John and his son Edward had camped since April, and had been quite successful in hunting.

The next carry after we passed the "Forks," was by Leadbetter's falls. The water rushes over clay slate, dividing by an island in the middle of the river. The soil here begins to be less fertile and the conifers more plenty.

Up to this time we had seen very little high land since leaving

Moosehead. The country through which we had passed had been gently rolling, but with no marked eminences. Near this place we saw a mountain of considerable elevation, and Mr. Hitchcock determined to visit it. The mountain is on the west side of the river, and overlooks a flat pine country lying to the north-east. We were not able to ascend to the summit of the hill on account of the icy covering of the slanting, exposed rocks. Upon this exposure, which very much resembles a quarry for some time abandoned, I was able to find many interesting mosses and lichens. One or two fresh-water algæ were growing in the cold pools of water in the crevices of the rocks, but these cryptogamia were all I could detect upon the comparatively naked schists. The quartz veins running through the schist are noticed by Mr. Hitchcock in his report upon the geological features of this valley. Owing to the coldness of the season we occasionally found it not only expedient but comfortable to walk instead of going in the birches with the morning mist clinging to their sides. While walking in this neighborhood Mr. Hitchcock called my attention to some splendid specimens of the charming Calypso borealis. *Salisb.* This delicate orchid should be cultivated in a cold bed of sphagnum or peat-moss. While continuing our walk after dinner, we were, for the first time, lost in the woods. Wood-roads for hauling lumber, and the tote roads for hauling supplies, were very plenty, intersected very often, and were consequently bewildering. Wandering away from the river for an hour or two, we at last found a brook which we followed down till we struck the river, and we waited for the boatmen. At last the thought occurred to us that they *might have gone ahead!* Impressed with this idea we shouted till we were so hoarse that our voices could not be heard a few rods off, and sat down to await events. While anxiously waiting on the shore to hear the slightest noise in the water, I found plenty of leisure to botanize. A viola was growing among the sphagnum which seemed different from any described in Gray's Manual. It is a well marked variety of viola palustris, or else a species, as yet undescribed.

In the course of half an hour our guides came down the river, having poled up against the current for some distance, thinking that we had passed along. We camped during that night near the southern line of township 5, in 18th range. There is a good deal of high land lying off to the north-east of the river, and this we suppose to be the ridge dividing the west branch of the Penobscot from the Baker lake branch of the St. John. The fungi in the vicinity of this camp are very abundant, and comprise several most interesting species. During the 23d of May we were so fortunate as to reach what our guides supposed to be lake Abacotnetic. It was said to be much like a submerged marsh, but this lake was more like an inundated alder swamp. The flat land on the north-west and north, led us all to suppose that we had reached the St. John's carry. A walk of exploration that same evening convinced us that the portage was much more than two miles in length. Camping half way between the two streams upon a cold eminence where such plants as Braun's fern, and the highland cranberry luxuriated, and where we shivered, we passed the Sabbath. The distance between the waters of the Penobscot and the St. John is fully five miles by this carry. The Penobscot is scarcely three rods in width, while the St. John is even less. There is very little interval land in this vicinity, and not much which can be called good for farming.

St. John Waters.

A whole day was occupied in carrying from one river to the other, and it was not till 5 P. M. that we were afloat upon the Woboostook. This stream is very crooked and its banks are covered with cedars and black spruces. It must have been formerly a good timber country through which it flows. After a couple of hours paddling down stream we reached a portion of the stream which was very winding. We finally made our last vexatious turn and entered upon the waters of a beautiful lake. There was very little high land on the shores of the lake or immediately back, nearly the whole country being flat and covered with pines and spruces. These conifers are mingled with beech and some species of birch on occasional rises of land. The lake is about six miles in circumference, and has a smooth, gravelly bottom. Most of the pebbles are worn pieces of mica schist. On the north-west and west are the only elevations of land to be seen, and these hardly vary the monotonous character of the country. The borders of the lake are covered with loose boulders, and present as uninviting a prospect for a settler as can be found in this district. It is called, Lower St. John pond.

From this sheet of water to Baker lake formed a pleasant journey for one day. The water was low in the river, and subjected us to the discomfort of often walking through the cedars on the shore. Baker lake is a fine body of clear water, rather shallow, and with beatiful shores. Here we found some of the characteristic plants of the St. John district, and this can be said to come properly in the northern belt. The pines on the east of the lake are fine, and seem to have very little spruce intermingled with them. Below the outlet of this lake we saw two beavers sitting demurely on the alder shore. They were not frightened at our approach, and waited for a few minutes before they plunged into the stream.

Rhodora Canadensis and Myrica are abundant and in full bloom. In the neighborhood of an old lumber camp I collected all the introduced plants, and was surprised to see how numerous they were. They had been introduced with the hay of the winter supply. Cerastium viscosum, Rumex acetosella, Poa pratensis, Festuca pratensis, Elymus, Leucanthemum vulgare not in fl., Chenopodium album, Taraxacum Densleonis, Antennaria margaraticea, Gnaphalium uliginosum.

Having reached south-west or boundary branch we endeavored to ascend it, but the water was too low to admit of it. Here we found the shore tinged in some places with the rosy hue of Primula Mistassinica, and in others purple with Viola palustris. Ribes rubrum is very abundant, and an allied species, perhaps a variety of R. rubrum, occurs with it. The rivers at their confluence form the upper main St. John. The shores in some places are high banks of gravel, probably terraces; in other parts of the river they are low and rocky, covered with small trees of second growth. The water washes along a shallow shore most of the way, even where the terraces are highest, and over rocks and stones so firmly bedded in the sand or clay as to resemble pavement. This peculiarity of the main river can be noticed even as far down as Frederickton, and permits the use of horse boats during a large part of the open year. The boat is allowed by a long rope fastened to one side, to swing out into the stream, while the smooth, paved pathway affords a capital chance for the horses to walk. By the resolution of forces the boat keeps the middle of the stream, or deep water, and thus an easy mode of navigation is secured. The river is in many of its turns obstructed by boulders of granite and schist, and in others by the troublesome gravel beds which are often of large extent and resemble "sand-bars" at the mouths of ocean harbors. Huronian tansy, Tofieldia, Oxytropis and Astragalus, are plentiful along the banks and in the woods immediately back from the river. They are also found on the tributaries of the main river.

This description of the St. John applies to the whole river as far as the Seven Islands farm. This farm, owned by Holman Cary of Houlton, afforded us our first night's shelter in a house since the commencement of our tour. We reached the house on the evening of the 29th of May. The farm is of immense extent, and a good deal of it is under fair cultivation. The annual crop of oats and hay is sufficient to support a large drove of cattle and afford supplies to the up-river lumbermen. The farm is under the superintendence of Mr. Currier of Eastern Aroostook, and in his hands has proved, it is thought, an exceedingly profitable undertaking. He is a good farmer, prudent and enterprising. Seven Isles is now quite a settlement, having several houses in the neighborhood, and numbering in all twenty or twenty-five souls. It is distant from Quebec only a ride of a day and a half on horseback, and the same from the nearest railway station. One goes to St. Jean, Port Joli, thence a very good road till within twenty miles of Seven Isles.

We carried from this place to the Alleguash river, eastward thirteen miles, by the assistance of Mr. Holmes. He brought our luggage over the rough road in a creditable manner, and deserved our thanks for it, and a higher price than he demanded. Having now left the river St. John, let me give a brief resumé of its geographical features, and the contour of the valley. The river is shaped not unlike the inverted figure five, ç having the commencement of the curve in latitude 45° 56′ N.; thence bending around the State of Maine, receiving water from its tributaries, St. Francis and Madawaska, from a point as far north as latitude 47° 47′. It now pursues a southerly course as far as Woodstock, where it turns to the east, and finally empties into the Bay of Fundy at the city of St. John. It is only with its upper part that we have now to deal. The water flowing into the river comes from a water-shed or valley bounded on the south by a range of hills, separating it from that of the west branch of the Penobscot. Westerly it is limited by the Chaudiere line of hills and the mountains along the eastern shore of the St. Lawrence.

In this basin, the highest land is probably in Canada, at least such is the opinion of Mr. Greenleaf, (Atlas, map No. 3,) and it is confirmed by our own observations.

Alleguash Waters.

The Alleguash flows into the St. John, and is one of its largest tributaries, deserving to be treated of as a separate river, since it is one of the largest in our State, and of considerable importance in lumbering operations. This river was visited by us last year, and a pretty full account of it is found in the Report upon the Physical Geography and Agricultural Capabilities of the Wild Lands visited in 1861. Therefore I shall pass rapidly on as far as Chamberlain lake, merely stating that we found few features of botanical interest additional to those brought out in the report of last year. I collected in the woods, near Alleguash rapids, a large number of specimens of Calypso borealis, like those already alluded to, and in the same place where I found the delicate bulbs of the plant in the autumn of 1861. We reached Churchill lake upon the 2d day of June, and were pained to find an extensive fire raging in the woods at the south of the lake. It is the occurrence of such calamities as this, that renders the value of our immense tracts of woodlands so changeable, and it is this which causes proprietors of woodlands to shut out settlers and discourage the building of roads. From an accidental fire, the value of a township may be reduced thousands of dollars in a single week. This fire seemed to have had its origin at a point near the thoroughfare between Churchill and Eagle lakes, and was slowly working round on the eastern shore. Finding that all the exertions in our power would avail nothing in arresting its progress, we made a hurried examination of Spider lake and pushed on to Chamberlain farm. A single word will suffice concerning Spider lake. It is a small body of water on the north-east side of Churchill, connected with it by a narrow rocky stream. The lake is about two miles in length, and on account of the absence of timber and the abundance of musquitoes, was peculiarly uninviting to us all.

We reached Chamberlain farmhouse at noon on the 3d of June, having journeyed, in all, from Moosehead lake, three hundred and one miles. Our party here divided, Mr. White, and G. O. Varney acting as guide, departing over the lake to Caucomgomoc, and thence to Chesuncook, and back to the world; the other party, consisting of Mr. Hitchcock, J. J. Bowley as guide, and the writer, passing southerly towards Mud pond carry. We were very kindly treated, as, indeed, we were last year, by the inmates of the Chamberlain farmhouse, and were able to obtain fresh supplies of many necessary articles. We reached Mud pond, which is distant from the farm about eight or ten miles, during the same afternoon, and commenced carrying across the portage. It is not an easy task to carry on one's back a heavy load across a muddy, miry portage, of two or three miles in length. From Mud pond we had a fine view of Mt. Katahdin, the exposure being, of course, to the northwest. The land around Mud pond and the adjoining lake Umbazooksus is low and covered with mixed conifers. A good deal of fair lumber still remains to be cut from this district, although it has already been pretty well thinned.

June 4th. The black flies have troubled us sadly ever since the first of June, or the day when the wild cherry trees blossomed. We have had them at all hours from the rising to the setting of the sun. The approach of the musquito is sufficient warning of itself, but the black fly comes noiselessly and gives no intimation of his presence till he is ready to fly away. The insect as figured in Harris' Insects of Massachusetts, is a little smaller than we have usually found them. Since they are common in the woods of Maine from June to August, it becomes the duty of one who urges settlers to take up the lands described in this report, to call attention to the remedy for this nuisance. At Chamberlain farm I was told that the workmen were not much annoyed by them after the first few days. In many cases the application of some unctuous substance to the skin is found to give an absolute exemption from the pest. But after a large clearing is made, when several houses are quite near together, the black fly becomes rare. So that this drawback which has kept so many from entering upon the occupation of new land, is really not so formidable as it at first seems. It may be thought indiscreet in me to say anything about this discomfort to settlers, but I am sure that "although a survey may gain a temporary popularity by the exaggeration of certain facts and suppression of others," it is better to state all the facts, *pro* and *contra,* as they really are found.

Penobscot Waters again.

The stream flowing from the lake Umbazooksus to Chesuncook is quite crooked, and, near its mouth, runs through a widely extended meadow. This interval land, and other land in the vicinity, appears to be desirable for farming; in fact, at Chesuncook lake, a few miles further down, we found fields under a good degree of cultivation. Chesuncook is a long, narrow, and exceedingly beautiful lake. Its water flows into the west branch of the Penobscot, meeting that from Chamberlain, which we have just left, at the town of Medway or Nicatou. We now had reached the west branch of the Penobscot again and commenced to ascend the river. There are at Chesuncook several settlers, among whom may be mentioned Messrs. Walker, Ansell, Bridges, Smith and Folsom. The land is good, and not very far from a market, being twenty miles from Moosehead carry.

The west branch for four miles above Chesuncook lake, is dead water, and we saw much fine interval land and many islands. We recognized the vegetation we had left at Seboomook meadows; it had now sprung into the full leaf. This is one characteristic of the summer in the Aroostook belt, the ice once out of the river and the frost out of the ground, all vegetation grows with extreme rapidity. The river is quite strong and swift a good deal of the way from the lake to the Carry farm. We reached the Carry farm on the 5th of June, and were drawn across the railroad by the aid of Mr. Young. The land between the west branch and Moosehead lake is rather low and flat, the greatest elevation not exceeding forty or fifty feet, and for most of the way the "height of land" is not so great as this. This ridge divides the waters of the Penobscot and Kennebec rivers.

Moosehead Lake.

Having now reached Moosehead lake, it becomes my duty to describe briefly its physical character and the vegetation of the shores. The latter can be told in a single word or two, pines and other cone-bearing trees. Where the few clearings have been made, there are good farms and quite productive. Moosehead lake is an expansion of Moose river. This stream comes from the west, passing through a chain of many ponds and empties into this lake, having drained a district of twenty-three townships. The lake is very irregular in its shape, being twice as long as it is wide, and having five unequal arms. The river drivers who have warped logs across this water say they have found that a greater length of warping line was required in the vicinity of Kineo, a mountain in the centre of the lake than upon any other lake in Maine. It is the occurrence of such deep places as this which renders Moosehead such an unfailing resort for fishermen. The lake trout and togue of this lake are considered equal to any in New England.

The water of Moosehead finds an outlet through the Kennebec river; thus flowing out in nearly the same direction and at almost the same place as it entered.

The other features of the lake have been already fully described in the able report of last year of Mr. J. C. Houghton, who visited it late in the summer of 1861.

I regret, sir, that the work of describing in a familiar way, the physical geography, botany and agricultural capabilities of this vast district, did not fall into abler hands than mine.

Our journey was finished by our return to Greenville upon Saturday, June 7th, having accomplished in our birch canoes the entire distance of four hundred and twenty-two miles, in twenty-three days.

I am, sir, most respectfully,

Your obedient servant,

GEORGE L. GOODALE.

Portland, Oct. 31st, 1862.

Geology of the Route.

The rock on the west branch of the Penobscot, from the head of Chesuncook lake to the north-east carry into Moosehead lake, probably belongs to the Oriskany group. At the north end of Chesuncook, the slates dip from 10° to 15° north-westerly, while the cleavage planes are nearly perpendicular. We suppose that the high dips of the strata over the whole of this lake given last year by Mr. Houghton, are those of cleavage; he was certainly in error respecting the dip at the head of the lake. In these ledges at the head of the lake are a few fossils, which are largely changed into nodular masses of pyrites. The organic structure is very often apparent in them. At Pine Stream falls the dip of the strata seem to be about 25° north-westerly. The rock is a clay slate, with cleavage planes dipping 74° N. 20° W.

A few drift striæ appear at Pine Stream falls, running due south, and cross the river transversely. But near the mouth of Rag Muff stream are true glacier striæ, running down the river N. 15° W., the stoss side being on the south. The cleavage planes (and perhaps the strata also) dip 65° N. W. at Rag Muff. Between Pine stream and Rag Muff, rock exposures are common, all of a clay slate, with about the same position of the cleavage planes. Between Rag Muff and the north-east carry into Moosehead or

Seboomook lake, no ledges are seen, the banks being alluvial, and the meadows extensive and fertile. The excellent character ascribed to the country between Chesuncook lake and the North Branch mountain for agricultural purposes, by Mr. Geo. L. Goodale, we can fully endorse—and wonder that it has for so long a time escaped the notice of settlers.

Between the Carry farm and the mouth of Seboomook Meadow brook in the township of Seboomook, we have not explored the west branch, but know that many ledges occur upon this part of its course, since there are falls in the river. On the east side of Seboomook Meadow pond, are a few ledges of a micaceous appearing schist, like one at the north-west arm of Moosehead lake, and also on the north-west carry. No ledges were seen upon the west branch of the Penobscot till we had proceeded about two and a half miles into Plymouth (or Boyd.) This rock is an indurated schist, of a slightly hornblendic appearance. It forms a low ridge running across the valley transversly, and reminds one of the indurated belts of schist frequently found in a talcose region. The dip is 85° N. W. At Gulliver's falls other ledges show themselves, with strata dipping 56° S. 60° W. The rock is a mica schist, passing gradually into the variety just described. These *falls* are very strong rapids, which it was necessary to carry by, and they are near the mouth of Gulliver brook.

In Pittston the tough variety of schist reappears, running apparently N. 65° E. with vertical strata. In this township the west branch of the Penobscot divides at the "Forks" into the south-west and north branches. About a mile below the Forks the mica schist dips 60° N. W. Upon the ledges are striæ running S. 28° E., or in the direction of the valley. Possibly they are glacial, and are connected with those at the mouth of Rag Muff. More mica schist ledges appear at the Forks. Ledges must be very abundant at the "Canada falls" upon the south-west branch, three or four miles above the Forks, as the river falls many (some say 90) feet perpendicularly, but we had no time to explore them.

In the edge of township No. 3, at the mouth of Lane brook, are many boulders of quartz containing carbonate of iron and other minerals. Boulders of mica schist contain pyrites. These fragments reminded us of an auriferous region, and consequently our long suffering tin pan was brought into requisition to wash for gold. But not finding any of the precious metal after a couple of trials, we became discouraged, and left all the gold behind for more pains-taking explorers.

About Leadbetter falls the rock is clay slate, with the planes of cleavage and stratification remarkably distinct from each other.

Fig. 48.

View of the strata at Leadbetter Falls.

Fig. 48 represents the falls with the adjacent ledges of slate. On the right is an island, very near the north shore, where the strata dip 30° N. 30° W., but the cleavage planes dip 72° S. 30° E. The strata are represented by the coarser and the cleavage planes by the finer lines. Upon the south side of the river the position of the cleavage planes is the same as on the island, but the strata dip 30° S. 30° E., forming thus an anticlinal axis. This is a very instructive example. The fall of water is too great to allow the canoes to ascend the river here, hence it was necessary for us to "carry by." At the further end of the portage is a large horseback, which terminates here in a ledge larger than the ridge itself. We traced this horseback up the river for three miles, and found it was not parallel with the course of the river. Just above Leadbetter falls the strata are much contorted, and appear like an inverted anticlinal. The rock is more micaceous also. Still further on, the dip changes to 20° S. 20° E., making a synclinal in addition to the anticlinal in Fig. 48. This belt of clay slate is very narrow, and is hardly to be considered as belonging to a formation distinct from the mica schist on both sides of it.

In No. 4, R. 18, are two interesting mountains, crowding closely upon the river, one of which is called North Branch mountain, and the other upon the east side of the river has no name. They must be over 1000 feet high. The rock is a little argillaceous, but largely micaceous ; being very much like the prevailing rock about Bangor. About 800 feet above the river the dip is 15° N. 30° W., having the same strike with the slates at Leadbetter falls. Near the north base of the mountain is a large excavation in the strata appearing very much like a quarry. The soil was probably removed by slides, and by the agency of frosts and gravity very much of the rock has been exposed to view. A very large curve in the strata is thus developed. The strike and dip vary very much at the different portions of this anticlinal curve, but it is needless to state all the particulars. The curve is much larger than the one represented in Fig. 48; and the variation of the strike to a more nearly east and west course seems to have been produced by the crowding of the strata by the plicating agency. The rocks here are abundantly traversed by quartz veins, both large and small, and minerals similar to those found in the quartz boulders at Lane brook abound. These veins may be auriferous. Similar rocks are found above these mountains on the river, and indeed so far as its source in Abacotnetic lake.

North of these mountains, in Nos. 4 and 5, R. 18, the valley spreads out much wider, and often there occur large tracts of low alluvial land, or meadows, which when cultivated will make excellent grass lands, being overflowed annually by freshets. In No. 5, ledges of mica schist appear, showing both cleavage and stratified planes, the former dipping 67° S. E., and the latter dipping 40° S. Here also are more glacial markings. As we walked much of the way in the tote road we did not examine all the ledges cropping out on the shores, but have no reason to suppose them to be different from the common schists.

Names of Rivers and Mountains.

The north branch in Nos. 4 and 5 divides into the middle branch, (sometimes called Dole brook,) the north-west and north-east branches. When near Abacotnetic lake, if we should try to inform the public of our exact location it would require many words, for we were travelling up the north-east branch of the north branch of the west branch of the Penobscot river. Evidently a change in the nomenclature of the different branches of this river would be desirable. Although objection has been made to the almost *unpronounceable* Indian names of the rivers and lakes of northern Maine, we cannot see how the branch method of giving names is an improvement. Certainly it is not in this case.

As this subject of names has been broached, we desire to say a few words respecting the inappropriateness of many names used to express geographical relations in Maine. Names may be inappropriate either from repetition or uncouthness. No one would wish to give to a child a name of which the youth would be ashamed in after years; nor can we suppose that a community would desire such uncouth names to remain attached to beautiful natural objects in their vicinity as always provoke a smile upon the lips of strangers. Many romantic localities in the State might become famous as places of resort by summer visitors did not the names repel them. Nor can inelegant names ever appear in poetry. Such designations as Hogback mountain, Tumbledown Dick, Bull hill, Goose-eye mountain, Potatoe hill, Quaggy Joe, Jockey Cap, Singepole, Ben Barrow's hill and Hedghog mountain, are certainly inelegant, both in poetry and in prose.

One would be surprised to learn how often the same name is repeated in Maine. A few moments examination of the map discloses the existence of two Hogbacks, three Tumbledown Dicks, two Spencer mountains, four or five Pleasant mountains, the same number of Black hills and mountains, four Owl's heads, three sets of Seboois lakes, five Grand lakes, six Grand falls, four sets of Chain lakes, five Alder brooks, six Round ponds, nine Mud ponds, and nine Bald mountains. The nine * Bald mountain peaks in the State are generally quite high, and of grand proportions, which renders the constant use of the name all the more undesirable. There are two townships also having the name of Bald mountain. Like the autumnal tints of the forests, the names of colors applied to mountains is quite varied. There are three Blue mountains, a Red, two Green, and several Black mountains.

It is much easier to state the difficulty occasioned by the redundant use of respectable or inelegant names than to suggest a practical remedy. As popular language has sanctioned their use, it is only the popular will that can change them. In the new region of northern Maine there are many mountains particularly not yet named. Such objects may at this late day receive appropriate names, but very few of the others can easily be changed. Whenever a party of citizens may desire a change to be made in the names of beautiful objects in their vicinity, it is easy to fix the

* Lest our statement should appear extravagant, we will give the locations of all the Bald mountains mentioned: 1. In the Bald mountain township between Moose river and the Canada line. 2. In township No. 3 near Parlin pond. 3. In the Bald mountain township south of East Moxie. 4. In Clifton. 5. In No. 10 of Hancock county. 6. Upon the Androscoggin lakes. 7. In Newry. 8. In Camden. 9. Bald hills in No. 2, R. 4, of Aroostook county. There may be more which we have not noticed.

new name by appointing a day, when in the presence of a crowd the new appellation may be applied formally with appropriate ceremonies. Then if the community think well of the project, the old name will soon be forgotten.

We have been much gratified at the reception which a few suggestions of this nature in our last report have received. We suggested the name of Mount Pomola for one of the Katahdin summits, and Mount Matagamon for a precipitous summit on the west side of Matagamon or Grand lake on the east branch of the Penobscot. These names have very kindly been placed by Mr. Chace upon the State map without any solicitation. We have no doubt that whatever finds a place there, will be permanent.

With a view to geographical improvement, we cannot resist the temptation to suggest one more change of name, in addition to those spoken of in a previous part of the report on the Schoodic waters, (which are more properly a restoration of the old appellations than the suggesting of new ones.) It would simplify the nomenclature of the branches of the Penobscot river to give a new name either to the east or the west branch. The west branch is the largest and longest, and therefore should retain the name of Penobscot in full. An appropriate name for the east branch might be derived from the largest lake through which it passes; viz: Matagamon; and therefore we would suggest for it the name of Matagamon river, (pronounced Mat-tag′-a-mon.) The word in the original Indian dialect signifies *grand*. We should have then a lake, mountain and river of the same name, which would easily be associated together in the minds of all, students, citizens and visitors.

By a strange coincidence, the name Pomola was applied many years ago by Mr. Young, a botanist, to the same peak to which we thought it appropriate last year. The earlier suggestion is of itself alone worthy of adoption; and the coincidence of suggestion would show that a necessity existed for the application of the name of this Indian Deity to one of the peaks which he was supposed to inhabit.

Geology of the St. John Waters.

We suppose the rocks upon the St. John waters, as far as the mouth of Black river, are the same with those already described upon the upper portion of the Penobscot, viz: micaceous schists with occasional talcose variations. Last year a portion of this mass of schists was donominated talcose; but now, for weighty reasons, we prefer the term micaceous.

Putting together the various items of information within our reach, we suppose this schistose formation extends from Long lake, on the line between Canada and New Brunswick on the river St. Francis, to Bellows' Falls in Vermont. We have ourselves traced it continuously for at least 300 miles out of the 340 of its continuance, and have the observations of Sir William E. Logan and his assistants for the filling in of the gap of 40 miles between the north-east corner of Vermont and the south-eastern portion of Maine. The Canadian Survey have also traced this formation 250 miles further to the north-east, to Gaspe, on the gulf of the St. Lawrence, so that we have a formation here 590 miles long, and that known, not by theory, but by actual observation. We think, however, that the disposition of that portion lying in Maine does not confirm the suggestion that the fossiliferous upper Silurian limestones and slates of Memphremagog lake are traceable continuously to rocks of the same age upon lake Temiscouata. Such a belt must cross a high range of mountains transversly; while the close proximity of the micaceous schists of the vicinity of Connecticut river east of Memphremagog and at Megantic lake to the Maine schists, must render certain the existence of one belt lying to the eastward of the more southern limestones, and perhaps separating them into two belts. Those on Temiscouata lake, however, are of the same age, and may be a repetition of the Memphremagog basin upon the opposite side of a complex anticlinal axis.*

On the Woboostoock stream near its sources are four large ponds, known as St. John ponds. We carried across from the north-east branch of the north branch of the Penobscot river, so as to strike the Woboostoock midway between the two most northern of these ponds, in No. 5, R. 17. Descending the Woboostoock stream, we find no ledges until we pass the last St. John pond, in No. 6, R. 17. Boulders of mica schist and quartz are common on the shores of the stream and the pond, except for two

* Sir William suggests, after hearing our statement of the difficulty, that possibly the Silurian limestones do not enter Maine at all, but that the Memphremagog or Dudswell series are connected with the Temiscouata rocks north of the river St. John; thus lying altogether in Canada.

or three miles south of the pond, where the banks are low and alluvial. Excellent grass lands might be cultivated here with very little labor after the first clearing had been effected. The rock below St. John pond is a clay slate, dipping 68° S. E. Mica schists appear shortly in abundance in descending the stream, which appear to be the predominant rock. In No. 7, R. 16, clay slate ledges appear, dipping 75° N. W.

For two or three miles previous to reaching Baker lake, we passed through an immense amount of low, swampy meadow land, with scarcely a perceptible flow to the current. Immediately adjacent to Baker lake is a tongue of higher land crossing the valley. Hence we came to the conclusion that formerly there was a lake covering this low land separate from Baker lake, though communicating with it by a "Thoroughfare." The rise of water twenty feet by an obstruction at the outlet of the existing lake, would produce the same effect. The hills about these lakes are very low.

No ledges are to be found upon Baker lake. The shores are lined with coarse boulders of mica schist and quartz. Mr. White thinks there is a "Lake Rampart" four feet wide and half the length of the lake on its eastern shore. Woolastaquaquam stream, the outlet of Baker lake, is much larger than the Woboostoock, and falls very much for half a dozen miles below the lake. Near the lake are schistose strata, dipping 45° N. W. Similar micaceous schists crop out near the junction of the Woolastaquaquam stream with the south-west branch of the Wallastook or river St. John, dipping 75° S. E. In this vicinity we prospected a little for gold, but discovered no more of the precious metal than at the mouth of Lane brook. We went up the south-west branch a couple of miles, or as far as the unusually low stage of the water would permit, but discovered no ledges. Boulders of a very coarse conglomerate—the same as those described last year upon this river lower down—are common here also. The source of them is not to be found on the river St. John, but further west. Sir W. E. Logan describes a coarse conglomerate upon [Little] Black river, a short distance upon the Maine side of the boundary line, which is probably the source of these enormous boulders. Judging from the strike of the rocks in this part of the State, this conglomerate belt must lie very nearly along the boundary line from lake Pohenagamook to lake Ishaeganalshegeck. From thence it must continue along the Canadian side of the border, but we do not think its south-western extension has ever been explored.

The rocks all the rest of the way from the junction of the two branches to township No. 11, the limit of our last year's explorations, are mica schist, the same as that described between No. 11 and the mouth of Little Black river. In the north-east corner of No. 9, R. 17, the dip is about 75° south-easterly. In the south part of No. 11 the dip is 8° higher. Terraces are generally very scarce upon the river St. John above the mouth of Little Black river, but very handsome ones are often seen for two or three miles of its course below the mouth of the north-west branch. The striæ run north-easterly down the river, very much in the fashion of a glacier. In No. 11 we noticed a boulder of the coarse conglomerate, 24 feet in diameter. Although we arrived at the Seven Islands' farm on the 30th of May, we passed two large snow-drifts within ten miles of the buildings. This fact shows the lateness of the spring in this part of the State, yet the crops had all been planted when we arrived at the farm. There are probably few years when snow cannot be found on the ground in this region in every month of the year except July, August and September.

Geology of the Alleguash Waters.

Our route now lay from the Seven Islands' farm to Chamberlain lake, or the same route we travelled over the previous year. Therefore we will not generalize concerning this region, and will now notice only some corrections and additions to our last year's report. Between the rivers St. John and Alleguash the rock seems to be entirely mica schist. Three miles east of the St. John the schists dip 80° S. E. At the line between ranges 14 and 13, is a band of clay slate with perpendicular strata. But half a mile to the east the mica schist appears again, though highly argillaceous in its character, with the strike N. 55° E.

At the carry below Churchill dam the argillo-micaceous schist dips 60° N. 20° W.; hence we do not feel satisfied that all this region should be ranked as clay slate. Trap and silicious slate were found upon a small island near the McCatherty farm, in Churchill lake; and on the east shore of the lake a very coarse conglomerate occurs, dipping 12° N. 5° E., whose constiuents are mostly pebbles of silicious slate. This is evidently a newer rock than the schists and slates heretofore described. Near the south end of Churchill, similar conglomerates occur, dipping, say 50° N. W. An expedition from here to Spider lake revealed nothing of interest. There are no ledges upon the lake or its thoroughfare.

We were enabled to confirm the truth of all our observations made last year upon Heron lake. Especially we found the singular trappean conglomerate, with its immense constituent boulders, one of which is nine feet long. The conglomerate composed of the largest boulders lies adjacent to trap ledges; and the further removed from the genuine trap the conglomerate is, the more distinctly the sedimentary character is displayed, and the smaller become the pebbles. On one of the smaller islands these pebbles are arranged in lines of stratification, with the course N. 15° E. We have no longer any doubt that this trappean conglomerate is truly sedimentary in its origin. Near the dam at the north-west part of Chamberlain lake, are a few ledges of trap.

No ledges were seen by us between Chamberlain and Chesuncook lakes. The physical character of the country has been already described.

In these explorations in northern Maine, we were accompanied by Mr. Oliver White of Richmond, Ind. So much interested did he become in the geology of this part of the State, that at his own expense he undertook the exploration of the route from Chamberlain to Chesuncook lake, by the way of Alleguash and Cauquomgomoc lakes, a route never before travelled by any scientific man. His zeal was the more to be commended since the abundance of venomous insects rendered exposure to their attacks almost insufferable. He wrote out an interesting account of his discoveries, which we herewith present under a deep sense of obligation.

Alleguash and Cauquomgomoc Lakes.

To C. H. Hitchcock, *State Geologist:*

Sir:—The subjoined is my report of observations made in accordance with your directions, on the geology of Alleguash and Cauquomgomoc lakes and the vicinity.

After parting from you at Chamberlain Lake farm, June 3d, we rowed directly for the upper end of the lake, and entering upon Alleguash river soon began to discover ledges of the underlying formations. The first ledge seen, two and a half miles up, was a micaceous clay slate, dipping north-west 45°. For nearly three miles these slate ledges are almost continuous, dipping in the same general north-west direction, producing a series of very difficult rapids, and in three places a perpendicular descent of from two to six feet. A few rods above the first falls the strike was north-east and south-west, the dip being 90°, or perpendicular. At the second falls the dip is 75° north-west. Half way up from Chamberlain lake, we came to the Grand falls, where the water descends twenty feet over a ledge having a dip of 80° north 30° west, the strike of course being at right angles, or south 30° east. To the east of the falls the drift has been washed away, exposing a surface beautifully levelled and striated, the scratches having a direction of north 50° west. In addition to what nature has done toward furnishing a reservoir for the reception of logs, the enterprising lumbermen of this vicinity have increased the height of the falls, so as to make quite a large sheet of water above them.

The rock here is a more indurated and less perfectly cleavable variety than what we had already met with. One mile above the falls the strata are very distinct, with a dip of 30° north 10° west. This rock has a large per cent. of calcareous matter in it. The last of these ledges before reaching Alleguash lake has a dip of 80° north 25° west. All the exposed portions of the underlying formations on this river lie within and occupy nearly the whole distance between the south and west lines of township No. 8, R. 13. The river runs this distance directly across the strike of the strata, hence the origin of the falls and rapids.

Entering upon Alleguash lake we soon discovered that the whole of the eastern and southern shore was girdled by the outcropping strata—a light-colored, very micaceous variety of clay slate, exhibiting in some parts a calcareous character; yet hardly sufficient to come under the head of calcareous slates. Almost the first thing that strikes one as peculiar, is the perfect distinctness of both stratification and cleavage, so much so that no one could mistake one for the other. Just south of the outlet the dip is 85° north-west. A little further on, and forming the southern shore of this cove, is perhaps the finest exhibition of the smoothing striating effect of the drift movement easily to be found. It consists of a surface shelving down under the water at an angle of 12°, 500 feet long and over 50 feet wide, as smooth and regular as a floor, and covered with fine scratches running north 50° west.

Farther south, and near the centre of the eastern shore, ther e an anticlinal axis, having less than one rod of the summit taken away, the strata on the one side dipping 12° north 25° west, and on the other 12° south 25° east. The second island is composed of clay slate, dipping south 30° east, and forming with the third isle, which is to the east and has a dip of 85° north 30° west, a synclinal axis. The cleavage and strata here are at right angles. Several other islands, together with the outcrops along the south-eastern shore, are composed of the same clay slate; and what is peculiar to this part of the lake, is that the strata are tilted so as to be nearly perpendicular.

Upon reaching the extremity of the lake in this direction, we proceeded up on the other side, and found an entire change in the character of the rocks. They are an unstratified trap conglomerate, forming extensive ledges all along the south-west shore. At first it was an unmistakable trap conglomerate—trap boulders and a trap matrix, but further on the cementing material disappeared, the boulders growing larger and assuming that concretionary or nodular character spoken of by you on page 410 of your Preliminary Report as occurring on Heron lake. The largest boulder we measured here was five feet long. From the fact that genuine conglomerate is found in such close proximity and merging into this latter rock, there can be no doubt as to its true conglomeratic sedimentary character.

On either side of a valley running in upon this part of the lake is a range of mountains presenting marked differences in appearance,—that to the east is lower, and has a rounded, graceful outline, showing the soft, easily disintegrating clay slate of which it is composed, while those on the west side have a more broken, rugged contour. The upper part of these latter are very precipitous, so much so as frequently to prevent the growth of vegetation, and leave the rocks exposed, which looked very much like trap. This surmise was afterwards confirmed in passing along the portage skirting the base of this range by the great quantity of angular trap boulders we found. The mountains all along the west side of the lake and extending back from it, were, I have no doubt, trap mountains, as they exhibited the same shouldered wedge-shaped sharpness of outline.

By mistake, we found our way to Ellis brook pond, a small and very shallow sheet of water one mile south of Alleguash lake. The strata are exposed in a great number of places, both around the shore and on islands. The rocks are the same as those of the eastern shore of Alleguash lake, having a general north-west dip. We made but few observations here, of which one on the south shore may stand for all. Here the dip was 75° north-west. In some places there seemed to be a tendency to foliation in the thicker and more indurated layers.

In approaching the stream that enters the Cauquomgomocsis lake on the portage from Alleguash lake, we passed over two well-defined terraces, running parallel to the stream so far as we could see. Entering Cauquomgomocsis, it was evident that the east side would be the most favorable for geologizing, as it is the one next the mountains, and upon which the strata cropped out. Just east of the entrance the first ledge proved to be clay slate containing considerable mica. The dip in this and the next two ledges is 50° north-west. Near one of these ledges is an embossed rock with striae having a direction of north 10° west, which is more nearly north than any we have met with in this region—a change, no doubt, caused by a local deflection in the direction of the triturating force. The most southern ledge on the east side has a dip of 80° north 30° east.

A little way to the left of the outlet is the last ledge, consisting of an indurated variety of micaceous clay slate, and having a south-east dip of 80°. There is here a very beautiful instance of local contortion, forming what in carpentery is called an O. G. moulding, showing that while in a soft, plastic condition, undergoing metamorphism, it had been subjected to lateral pressure. It is the same in miniature as what has been demonstrated to have taken place in the Housatonic and Alleghany mountains. The same phenomenon is presented in the strata a few rods above Leadbetter falls on the west branch of the Penobscot, and on a somewhat larger scale than here.

On the west side of this lake we observed no ledges; and although we did not examine particularly, yet we presume from the low, flat character of the shore and adjoining land, that the rocks do not come to the surface. In the south end, a short distance from shore, is a solitary trap boulder some thirty feet long by twenty feet wide and ten feet high. Now taking the directions of those striae as an index, there can be no doubt that this boulder has once formed a part of the trap mountains heretofore spoken of, and had been transported three or four miles by some great iceberg or glacier of the pre-historic period, and deposited where it now rests.

The rocks of Cauquomgomocsis lake differ from those of Alleguash lake in the obscurity of the stratification; the layers are almost entirely obliterated by the cleavage, or what appears to be more properly such; so that in my notes I find these observations as to the bearings of the strata marked as questionable, with the exception of the last on the contorted strata, which I took as proof of its stratified character. Now this latter has a south-east dip, yet the last observation before this is on the *strike* of the other, and one which I had marked as questionably north-west. All of which would go towards strengthening the suspicion as to the real character of the supposed stratification.

We found no ledges from this lake down to Cauquomgomoc. But when we arrived at the lake, and turned to the left along the north-east shore, we encountered a very curious and somewhat puzzling series of formations. The first six ledges and embossed rocks, occupying a space of less than the eigth of a mile, were composed of a compact, fine-grained conglomerate, broken up by numerous joints, but having no recognizable stratification. It had been somewhat metamorphosed, and had a decided calcareous character. Probably the cementing material is carbonate of lime. Then there is a change to clay slate, having a north-west dip of 75°. Just beyond this is another ledge of an undoubted argillaceous rock, yet massive, and without any apparent stratification or lamination. The next is true clay slate with a dip of 75° north 25° west. With this the slate terminates and is succeeded by a species of conglomerate of uncertain stratification. It appears to be made up of large boulders of varying composition, altered so as to make a rock of almost homogeneous texture. The most conspicuous component is a brownish-red sandstone. Yet the whole rock has the peculiar argillaceous odor.

Beyond this again we came to another set of clay slate strata having a westerly dip of 75°. These strata were very much weathered—filled with long rifts and holes—an effect of its uneven decomposition. Thus within less than one mile and a half, there were four changes in the character of the rock:

1st—A fine-grained conglomerate;

2d—Clay slate;

3d—Coarse-grained conglomerate rock;

4th—Clay slate, much weathered;

Presenting an interstratification of two large beds of conglomerate. We found this same thing in miniature upon arriving at the termination of these rocks, where there is a succession of interstratified slates and conglomerate, ranging in thickness from six inches to two feet, varying in dip from west to north 60° west, the angle being near 75°. The latter were probably the thinning out of the thicker beds below, or the forces and material which produced them must have become exhausted. From here we rowed directly west across the lake to where we had observed the rocks were exposed, and encountered a large island, which together with the neighboring shore, was made up of a greenish colored, very hard, brittle rock, having a strong argillaceous odor, without any traces of stratification, and traversed by a net-work of joints. It looked so much like an altered rock that at first I hesitated whether to call it trap or a metamorphosed schist. But upon further investigation and a comparison of it with labelled specimens, I have classed it with the former, and call it a dioritic variety of greenstone. To the south-east of this, clay slate appears again, skirting the shore and forming embossed ledges. The first determination of the dip we made was on an embossed rock a few rods south-east of the trap island ; it was 75° north-west, and we found by other measurements farther along, that the dip varied little from this either in quantity or direction. Upon the rock first measured there are striæ having a direction of north 15° west. At another place they have a direction of north 10° west.

Cauquomgomoc is a pretty little sheet of water set amid some of the most beautiful scenery we have yet, since the commencement of the trip, had the pleasure of viewing. It is surrounded on three sides by mountains and highlands, in some places coming up to the water's edge, and at others retreating so as to leave a margin of comparatively level land. They begin just to the north of the outlet, with a squat, rounded range, and running around to the north and north-west, growing continually more varied and interesting where lies the culminating point of the scene. The particular feature of beauty just here is the receding perspective of ranges, each one overtoping those in front, until the last seems to mingle with and become lost in the ocean of blue beyond. The clear, sparkling waters in front of you, with the receding series of pine-clad highlands beyond, and over all a bright sun adding its thousand touches of beauty, make a scene which once beheld is not easily forgotten. The south-eastern shore is flat, and the water all along this part of the lake is quite shoal. It was the frequent remark of the guide while here, who was an old lumberman, as he observed the tops of the pines shooting up above the other trees, that "there was yet a good chance to lumber here," notwithstanding there had been so much already taken away.

After leaving the lake we came to a long stretch of swift water. Near the upper part of this, and a short distance below the outlet, there occurs a series of ledges, extending across the river, over which the water rushes, falling within the distance of four rods, twelve or fourteen feet. At the upper part of the falls is clay slate, a less micaceous variety than that seen above, having a dip of 45° north 35° west, the strike of course being at right angles to it, or nearly north-east and south-west. Then comes an interstratification of several, not very thick, beds of fine-grained conglomerate, exactly identical in appearance and structure with what we had already seen on the lake. These beds are irregular in thickness, and seem to pass into the slate by insensible degrees, so that there is no definite line of demarcation between them. The lower portion of the falls is made up of a distinctly stratified, blue arenaceous rock, containing quite a per cent. of carbonate of lime. Some five or six feet of the terminal portion of this strata becomes of a red color, produced by a partial disintegration and peroxidizing of the iron, thus exhibiting more perfectly its arenaceous character. The dip of this last bed is greater than that above, and I suspect that further investigation would show that the former is superimposed unconformably upon the latter. This was the last place where we observed the characteristic formations of east Cauquomgomoc. From here to Black pond the underlying rock is clay slate, as exhibited in two or three ledges in which the strata are perpendicular, the strike being north-east and south-west.

There is nothing of particular interest about the geology of Black pond. The north-east shore is lined with ledges of clay slate, upon which we made the following observations: First after entering, dip 80° north 35° west. Near this is an embossed rock, with striae having a direction of north 10° east. The ledge midway has a dip of 75° north 35° east. A short distance beyond this again the dip is 45° north 30° east, and the cleavage planes are very perfect. I counted four or five other ledges having essentially the same dip. We saw no outcrops on the south-east shore, and from its flat character we inferred there were none.

From here down to Chesuncook lake we encountered two considerable falls and a few ledges composed of trap identical with that of western Cauquomgomoc. We discovered no schists in connection with these rocks. At Chesuncook lake we entered upon a field already explored by yourself, hence the necessity of my reporting further had ceased.

Allow me, sir, in conclusion to say, that these observations were conducted under a degree of personal discomfort which prevented such an accurate and detailed examination of some of the more interesting and obscure points as I could have wished, yet what has been done is herewith submitted.

With the greatest respect,

Yours, &c.,

O. WHITE.

To PROF. C. H. HITCHCOCK, *Geologist of the Scientific Survey of Maine:*

SIR:—In accordance with an agreement that during an exploration of a portion of Aroostook county in pursuit of such of the vertebrated animals as I might find, any incidental observations I might make in geology should be communicated to you, I herewith submit the following for your consideration.

I left Bangor in company with S. D. Besse of Winthrop, on the evening of 29th of May.

My route from Bangor was by railroad and steamboat to Mattawamkeag, thence up the Aroostook road to Patten, which was made a station or central point of action.

Quartz Boulders.

While passing along the northern part of No. 2 (Benedicta) and southern part of No. 3, my attention was drawn to the abundance of quartz boulders of a larger size than that rock generally exhibits, or, at least, much larger than any I have met with in Maine. These boulders have evidently been swept from some localities further north, (perhaps Chase's mountain,) where they were associated with schist of a talcose type, patches of which still adhere to them. From what exact locality they did start, or whether, like the quartz and talcose schists in other parts of the Union, they indicate the presence of gold, I am not able to say, as I made no examination of them in reference to that or any other specific character.

Roofing Slate.

As it required some little time, after arriving at Patten, to prepare for a tour in the forest, what leisure I had was spent in excursions in that vicinity. The rock formation in this neighborhood, as you have stated in a former report, I found to be slate. In some localities I found it to exhibit good qualities for roofing slate.

On the premises of Hon. Ira Fish, about a mile and a half from the village, and on the north bank of the Mill stream, this slate crops out in the form of a bluff, of moderate height, from which we obtained excellent specimens. From a cursory examination, as far as the surrounding forest would allow, I am led to the conclusion that a good quarry might be opened here with a prospect of its yielding a large supply of this useful material of very fine quality. It cleaves readily, giving a smooth even surface, and possesses the requisite tenacity to allow of its being dressed and pierced, or punched in the usual manner.

Conglomerate Boulders.

A large proportion of the boulders found around the village of Patten are conglomerate. None of this rock is found here in place except in one locality. This was in the bed of the stream, near the lower grist-mill. One of the abutments of the bridge, which crosses the stream there, is built upon it. The extent of it is not manifest, as it soon dips below the bank and is hidden deeply in the earth.

Granite Bluff.

But little granite is seen after you pass above Lincoln. I had an opportunity through the politeness of Mr. Haines of Patten, to visit Island Falls on the Mattawamkeag, about ten miles northeasterly from Patten. About two miles east of the falls, is a remarkable bluff of coarse crystalline granite, rising up abruptly to a height of at least three hundred feet. In some places it forms perpendicular precipices presenting all the rude, broken, craggy piles and confused heaping up of angular fragments, usual in such formations. It is of rather coarse texture, and therefore would require considerable labor, skill and care in dressing and preparing it for building purposes. This is probably the most northern locality where granite is found in place in this section of the State. I subsequently had an opportunity of tracing this formation, in a north-easterly direction from the bluff, to the westerly shore of Meduxnekeag lake in New Limerick. On this route it is found occasionally cropping out, and exhibiting the same characteristics as is shown in the bluff near Island Falls. Although, as before remarked, this variety is not so easily wrought as the more stratified, or gneissoid varieties, it will nevertheless become, in time, valuable to the surrounding country, affording an inexhaustible supply of this durable material which you are aware is not so generally distributed in this part of the State as it is in the western and middle counties.

I may remark here that I had an opportunity at Mr. Sewall's to examine specimens of the roofing slate which occurs above this on the banks of the Mattawamkeag in No. 4, Range 4. This slate, which you mention in last year's Report, (page 319,) is of excellent quality, inexhaustible in quantity, and can be obtained in enormously large sheets with the greatest ease. While at the falls I was very kindly entertained by Mr. D. Sewall, an early settler at this place, to whom and to his intelligent family I would here express my obligations for assistance rendered. His son and Mr. Porter of Lowell, who was on his way to No. 4, accompanied me to the bluff and essentially aided me in the examination.

Siliceous Slate.

Having completed our preparations, we took our course for the Seboois country, taking the upper road, so called, through No. 5. The slate formation, as you pass northerly and westerly from Patten, may be seen occasionally cropping out, exhibiting its usual characteristics, until you come into No. 5, Range 6. In this township, on the farm of a Mr. Smith, and about a mile north of the road, is a remarkable formation of siliceous slate, which crops out on the margin of a small stream, forming a bank some twenty or thirty feet in height, and extending westerly over several acres.

The strike, or direction of the strata of this formation, is northeasterly, the dip nearly perpendicular. The strata, or layers of this variety of slate, vary in thickness from an inch to two or three feet, exhibiting, in their cleavage, a clean smooth surface. There are also joints, or cross seams at different points or distances, thereby forming slabs of different thicknesses, and from a foot to ten or fifteen feet in length. The faces of these joints, or cross seams, exhibit angles of about 120 and 60°, to its opposite plane—

thus giving to the sides, or perpendicular surfaces of the slabs, or blocks, as they lie in the quarry, a trapezoidal form. The texture of this rock is compact and solid, very little decomposition being exhibited from the effects of long exposure to the weather. As yet no particular use has been made of this rock, but from its proximity to the road—the ease with which it may be quarried—its location in the vicinity of a thrifty, growing village, like Patten, it cannot fail to be soon brought into practical use as a building material, or for posts and sleepers for fences or other fixtures.

Mineral Spring.

We were informed by Mr. Smith, that about a mile from this ledge there is a mineral spring, which, from his account, I think must be of a chalybeate character, and probably well charged with iron, judging from the abundant deposit which is made from its waters. We had not time to visit it then, and make a note of it now to call attention to the fact of its existence, that it may form the subject of future examination should circumstances allow.

Horsebacks.

After passing through No. 5 we come upon the supply road which leads from the settlement to Grand lake, and thence to Chamberlain lake, and the settlers become more scarce and the road more rough as you progress. After passing Mr. Rich's clearing, you find no more clearings or settlers for several miles, until you come to Shin Pond. Here we found a young man by the name of Crommet, who had made a large clearing and was busily engaged in getting in his crops. We found no rocks in place here. A boulder or two of hornstone, made up the amount of geological indications. After passing a few miles beyond this locality you find the growth indicating a change of quality in the soil. It becomes thinner and of a more gravelly character, and you soon find the road passing along on one of those singular formations called in common parlance "*horsebacks.*"

These horsebacks arrested your attention last year, and were the subject of remark in your report. They are probably more abundant and more varied, as it regards direction and extent, in Maine, than in any other part of New England. I leave it to you to search out the cause and theory of their formation.

They are certainly deserving special attention and research, and on account of their numbers and extent, (some of them passing over a stretch of fifty or more miles,) they certainly would demand no brief space of time and care in their investigation.

Whatever may be their extent, or their direction, they all have a similarity of character and accompaniments. They are, in fact, embankments of gravel, sand and rolled pebbles, of different heights and widths—a stream on one side, and often a bog or a morass in some parts of their extent on the other.

For one thing they are peculiarly fitted. They afford capital roading by the dry, well-drained and compact bed they furnish on which to travel. What the real extent of this one is, we could not well ascertain, on account of the forest covering the country on every side—from appearances, however, as indicated by growth, &c.—it rises some distance north from where the road strikes it, probably from a spur of Haybrook mountains—running southerly to this point, then curving westerly extends three or four miles to the Seboois river which passes through it—thence, a short distance from the west bank, it curves northerly continuing eight or ten miles, until it strikes the south margin of second Seboois lake where it drops off giving place to the lake, but again rises on its northern shore and is lost in the forest beyond.

Six miles further west, directly in the rear of the "Seboois House," is another one. This also rises somewhere in the forest northwesterly from this house, and when near to it curves around easterly, passing again into the forest; how far, I had no means of ascertaining. At the turn of the curve is a singular gap, or break, sufficiently wide to allow the road to pass through without any change of grade. This "horseback" is similar in its characteristic features to all the others. If all of the formations of this kind, that could be found in this State, were accurately traced out and mapped, they would present a singular and interesting representation of what was, probably, one stage of the Lacustrine era of the surface of Maine in the remote ages of the past; and might afford a satisfactory explanation of many changes which have since taken place on this part of the face of the earth.

The Gorge of the Seboois or Godfrey's Falls.

At the point where the supply road mentioned above, strikes the Seboois river, a rough, but strong bridge has been thrown across the water; and a little above it, has been built a dam by the lumbermen, for the purpose of commanding the water at their pleasure, in order to facilitate the floating down of their logs in the spring and early part of summer.

The slate rock, which had not been seen in place since leaving No. 5, here shows itself again above the surface.

Just below the dam, the Seboois, which has pursued rather a sluggish course above, the current rendered more so by the dam itself, begins to move more briskly. This is occasioned by the commencement of a slope, or greater incline of the bed of the river. This slope is confined to the bed, but does not affect the bank, which, on the west side, is a somewhat level plain or plateau, and on the east, hills of pretty high elevation.

This slope continues a regular but pretty fast descending grade for three or four miles, and thus forms a magnificent gorge, through which the waters, crowded on either side by the precipitous banks, which rise, during the last half of the course, nearly or quite three hundred feet in height, finally rushes with immense velocity and power. The bed of this gorge is the slate rock. The direction of the river, in this place, is the same as that of the strata of the rock formation over which it passes, both pursuing the same "strike" the whole distance until it comes to the last pitch, more properly called Godfrey's falls, where it curves to the east, thus cutting the rock strata more at a right angle and terminates the rapids by a leap, and a plunge into the basin below of nearly or quite fifty feet in height.

While the water passes over the slate in the direction of its strike, it wears it away apparently uniformly, thus forming a regularly descending plane obstructed only by the points and knobs, and irregular jutting up of the edges and fragmentary portions of the strata; but when the current is turned across the strata, its effect is to break it off in tabular masses, and thus changes it into a succession of cataracts and cascades over the rough and unequal steps formed by rending away those masses from their parent bed. The river, after this, meeting with comparatively small obstructions only, passes along more quietly until it unites with the Penobscot a little above Hunt's farm, in No. 3.

At the foot of these falls commences the long "carry," or portage of the Seboois, three miles in length, well known to lumbermen and voyageurs in this section. At first, the boats and luggage, after being taken out of the water, had to be "*toted*" up the steep bank, 300 feet above the river, in order to reach the plateau above—then commenced the "carry" on the western bank across the plain, to the dam, where the boats and cargo were again embarked for their upward voyage. This plain was once covered with a heavy growth of enormous pines, but these have long since been destroyed by fires, and a stinted and scattering growth of birches, poplars and blueberry bushes, taken their place. Happily the road in from Patten has precluded the necessity of getting supplies on and over this laborious route, and the portage is now used principally by river drivers on their descending voyage in the spring.

The southern extremity of this gorge, and Godfrey's falls, were very accurately delineated by the artist who accompanied Dr. Jackson in his first survey. A great portion of the eastern bank is a steep, mountainous declivity of slate rock, occasionally interspersed with trees and shrubs, and in some places presenting nearly perpendicular cliffs.

This slate rock is so soft, and so easily abraded and frittered away, that it affords no record of the operation of water long ages ago, as does granite in like situations ; and hence no discovery could here be made of any marks or testimony as to the height of the river in former times. It is however highly probable that, at some early period, the bed of the river was high above its present site, and there was a mighty cataract at the southern extremity of what is now the present gorge, where it took its leap down to the basin below the falls ; and that, by the ceaseless and rapid attrition of the waters, the bed rock has been worn down to its present slope.

May not the finer particles of this formation, ground to an almost impalpable dust by the action of the current, and suspended in the descending floods which poured into the ancient lakes,* far down the present Penobscot, have been gradually deposited therein, and contributed to the accumulations which constitute the valuable clay banks now so useful to the people who reside in their vicinity? This theory may seem extravagant to some, but, from observations and examinations of the effect of water in such cases, we are persuaded the idea is not altogether a geological fantasy. As corroborative testimony that there was once a time when the principal fall was at what is now Godfrey's falls, may be

* The Penobscot river in its whole length, like some other rivers in Maine, was once, undoubtedly a connected series, or chain of lakes.

mentioned the fact, that Shin brook, which rises in Shin pond, a few miles east of the dam, and about on the same level, and running westerly across the range of the slate strata, finally makes a plunge in a cascade of more than seventy feet, before it can mingle with the Seboois waters below, to which river it is a tributary.

Hay brook Farm and Seboois Farm.

Two large farms were some years ago cleared up and established in this vicinity, which afford both a criterion by which to judge of the quality and productiveness of the soil, and also very convenient accommodations and houses of entertainment to any who may be led by business or pleasure into this section.

A few miles before you come to the Seboois river, as you come in from Patten, you leave the Horseback by a road leading northwesterly, which brings you to Frye's farm or Hay brook farm. This is a large interval in the bend of the Nutupsemic stream, commonly called "Hay brook." (Almost every small stream where the wild grasses may be cut is called by lumbermen Hay brook.)

At this time the farm was under the care of Mr. Silas Coburn, whose family consisted of his wife and son. The soil and land adjacent is made up principally of drift from the mountains at the north-east. There are several terraces or steps. The lowest, lying on the margin of the stream, is of finer deposit and contains several beds of grey clay of strong tenacity, and of excellent quality for bricks or coarse pottery. It is frequently overflowed, and too low for cultivating, but very productive in grass. The next higher step is of coarser material and contains gravel and pebbles, but is easily cultivated. Still further back on more elevated land are found boulders of conglomerate and occasionally fragments of Oriskany sandstone containing spiriferae and other fossils.

In good seasons, one hundred tons of hay are cut on this farm, which, when lumbering business is brisk, finds a ready market as a supply for the teams during the winter. There were kept here, last winter, thirty-four head of cattle which are now pastured in the meadows and forest lands on Hay brook above the farm.

Passing onward westerly, by a very decent road over the Seboois bridge, and six miles from it you come to the "Seboois House" farm. Here is a large farm cleared several years ago by Mr. Jos. Twitchell, then of Oldtown, who built a large tavern house and commodious stables. He moved his family in and kept a lumberman's hotel here for four years, to the great comfort and acceptance of woodsmen, and, we doubt not profit, to himself.

This place is on the line of the sandstone strata of Stair falls and the foot of Montagamon or Grand lake of the East branch, and we expected to find it in place here. No rock formation of any kind, however cropped out here, but the prevailing stone or boulders though not large, were sandstone. Of this stone the foundations and underpinning of the house were built. Mr. Twitchell had also built a lime kiln, when he was putting up his buildings, and availed himself of the "Helderberg limestone," some six miles north-easterly, of which we shall speak hereafter, burnt very good lime with which he laid up his chimneys and plastered his rooms. He was piloted to this locality of limestone by one of the Indians who was with Dr. Jackson who discovered it on his tour up the Seboois into the Aroostook, and the discovery was of great service to him in his enterprise of building so extensively so far from the facilities and conveniences of civilized life.

The surface of this farm is rolling, the soil a good sandy loam and productive. We have spoken of the horseback which here shows itself a short distance west of the house, curving round to the east, which is evidently distinct and independent from the one through which the Seboois passes, six miles further east. The farm and fixtures now belong to Amos M. Roberts, Esq., of Bangor, who makes it productive in hay and other supplies for his lumber operations in the forest above. We found it in charge of Mr. Nahum Stackpole, formerly of Augusta, who, though keeping bachelor's hall, nevertheless entertained us very comfortably, and assisted us in our operations at very reasonable charges.

Scragly Lake.

Wishing to obtain specimens of some of the fish and *reptilia* of Scragly lake, I obtained the services of Mr. William Staples, who with Mr. Knox, engaged in this neighborhood hunting, and had a canoe in its waters, to accompany and guide us thither. This lake lies about five miles north-westerly from the Seboois House, and is reached by travelling over a very good supply road.

Its waters are one of the sources of the west branch of the Seboöis. On reaching the foot of the lake we found the *inevitable* dam, which the enterprise of the Penobscot lumbermen have built at the foot of almost every lake (and they are not a few) all over the wild lands—or at least, where there is a pine log to be floated on its waters. We saw no rock in place on our way, but occasionally passed moderate sized boulders of a grey compact sandstone. We found the lake to be about four miles in length, and its shores deeply indented with coves and creeks and its waters, for the most part, very deep. There are several islands in it, and on a part of its eastern shores, and on the islands ledges crop out, some of them forming bluffs of considerable height. After furnishing ourselves with such specimens of the fish and reptiles we came in pursuit of, we landed on the shores and islands to obtain specimens of its geology for your inspection.

They are of three varieties. On the eastern shore, as you pass up the lake, are found ledges of unstratified syenitic rock—next further north, on one of the islands, is argillaceous slate full of seams and joints, the strike or range north-east, dip perpendicular, and next north of this silicious slate having the same range and dip as the other.

I thought it probable that the sandstone might be found in place here as indicated by boulders which we had passed below, but saw nothing of the kind. I was afterwards told by a person who had spent considerable time, a few years ago in the neighborhood of the lake, "prospecting for timber," that on a highland called "Owl's head," not far east of the lake, he found rock that "*contained shells of various kinds.*" This was probably the sandstone in question.

Seboois Lakes.

Returning from Scragly lake to the Seboois House, we commenced preparations for an exploration up the Seboois lakes, and accordingly made arrangements with Staples, who had canoes in those waters also, and who was well acquainted with the haunts of some of the wild animals we were desirous of obtaining, to act as guide and woodsman for us. As his canoe was at the Seboois dam, and from the position of the lakes above, and the direction of them from Seboois House, I thought it might be a saving of time to dispatch Messrs. Staples and Besse to the dam for the canoe while Knox and myself took a "bee line" for the foot of the first lake, and wait until they came up the river to that place. In this calculation however, I was foiled, by the fact being made known to us that the invariable accompaniment of a horseback, viz., a bog or morass on one side or the other, in some part of its extent, was, in this case, actually located between the house and the lake in question—that the horseback, as we have before stated, after crossing the Seboois, turned northerly and stretched along for many miles having the Seboois river on its east side and an extensive bog on the west, and that there was no better way to get to the lake by land than to proceed to the dam and then take a "*tote*" road up the horseback to the point proposed. So we chartered a pair of horses to take our "*dunnage*" to the dam, where we left most of it and then turned them up stream along the route proposed, to the foot of first Seboois lake or "White horse lake" as the hunters call it. Our object for taking the land route thus far, was to examine the deposit of Helderberg limestone above mentioned, which is stated to be near the foot of first lake. As we approached the lake, say a mile and a half or two miles from it, large boulders of conglomerate and larger boulders of Helderberg began to show themselves, indicating that the sites of these rocks in place were not far off. We did not succeed in finding the exact locality of either at this time, but the following fact we did ascertain from the testimony of the boulders on the road, viz:—that the Helderberg locality was to the right of us in the forest between the road and the river, while the source and site of the conglomerates was farther up and even above the upper lake. This story was told us by the boulders themselves by the fact that we soon passed the Helderbergs and saw no more of them, while the conglomerates continued, scattered along the margins of the thoroughfares between the lakes, often impeding the channel; sure evidence that we had passed by the source of the Helderbergs, and not yet reached the parent bed of the conglomerates. The Seboois lakes are three in number, and are very fine and pleasantly located sheets of water, but not very accurately delineated on the maps. They are connected by short thoroughfares. The first lake, or "White Horse" lake, is the smallest; Second lake, or Snow-shoe lake, is next in size, while Third or Grand lake is much larger than either, stretching diagonally across the township. They are located on the height of land between the Aroostook and Penobscot rivers. They are all, as is also Scragly lake, well stored with several species of trout,—with pickerel and several other species of fish. In the summer and autumn trout are caught in abundance, and in the winter the deep lake trout or *togues* are obtained readily by the hunters and lumbermen whenever they fish for them. In Scragly lake I obtained a species of Batrachian (called by the Indians "two legged trouts") found in but few other localities in the State, the distinctive characteristics of which require further investigation before being assigned to its true place in Herpetology.

At the outlet of Second lake, and also at the outlet of Grand lake, the lumbermen have established their dams with all the "privileges and appurtenances" of gates and sluices "thereunto belonging," and we found that the channel of the thoroughfare into Grand lake had been improved by having the boulders cleared away and thrown into piles, so as to allow boats and rafts of timber to float free from their obstructions. Among these boulders, in company with the conglomerates, we began to find abundance of trap rocks, but we found no rocks of any kind in place on the route until we arrived at what is called the Narrows, in the upper lake. These narrows are formed by the jutting in of points of land on the easterly and westerly sides, thus narrowing the passage of the water so as to make them to appear almost like two lakes. A few small islands and shoals are located here. On one of the islands, and the point of land near the narrows on easterly side, the usual clay slate appears, having the north-east range or strike, and nearly perpendicular dip. At the narrows, the islands and easterly shore are composed of trap rock, and quite an extensive shoal, which appears in dry times, is floored over with it.

Beyond and north-easterly of the narrows the rock on the shore is calcareous slate, and is undoubtedly the locality of "argillaceous limestone" mentioned by Dr. Jackson in his report of his exploration through the lake in 1837. After making an examination of these localities, and procuring suits of specimens of each rock for your inspection, we proceeded up the lake in pursuit of one of the objects of our mission, viz., to procure a good specimen of the black bear (*Ursus Americanus*) which Staples assured us was "*at home*" in that region. We landed on the north-westerly shore of the upper section of the lake about middle of the afternoon, where we *raised a smoke* to ward off the black flies, which now began to swarm in the woods and margins of the lake, attacking us with intense hunger and ferocity. Staples and Besse immediately started off into the forest on a bear trail, and before night had one in limbo, which Besse dispatched by sending a bullet through his devoted head

About sunset, they made their appearance bearing bruin between them slung on a pole. He was a fine large, long-legged, lank-sided ranger. As we intended to have his skin stuffed for preservation in the cabinet, we proceeded to take it off in shape for that purpose, which we did through "much tribulation." While both of our hands were employed in skinning the bear, the black flies improved the opportunity to *skin* us, and by the time we had finished, in point of suffering and the entire condition of his skin, the bear was by far the best off.

We camped here during the night, and in the morning turned our faces Patten-ward, and prepared to wend our way to the "*world outside the woods.*"

My next route of observation was up the Aroostook road through Masardis, Ashland, Presque Isle, Fort Fairfield, thence to Houlton, Linneus, and from Linneus across through No. 5 to Island Falls and to Patten.

Marls.

In addition to the extensive bed of marl discovered by you, and described in your report of last year, I am able, from personal examination, to add that several of the ponds in Fort Fairfield have large deposits of it, and that several of the bogs or low grounds adjacent to these waters afford it very abundantly. The pond near mills belonging to Wm. A. Sampson, also a pond in the northwesterly part of the township are floored over with it; and so abundant is it, that in dry times the shores and flats laid bare by the drouth look as if covered with snow. I have forwarded specimens for your examination and analysis. Probably, judging from its external appearance, the carbonate of lime largely predominates over the clay or aluminous portion of it. If so, besides its value as a fertilizer, it might possibly, with very little manipulations, be prepared for use in many purposes of the arts.

Why may it not become a substitute for whiting, and serve for making putty—also for an addition to the several pigments in the manufacture of oil cloths, for which purpose hundreds of tons of foreign whiting are now used in the State?

I respectfully suggest that experiments to ascertain its capability for such economical uses be instituted, and the results, *pro* or *con*, be made known.

Limestones.

You have mentioned many localities of limestone in Aroostook county. The upper half of the county is what may be called, in one sense, a limestone country, on account of the frequency with which this stone is found associated with other rock. Still, but few localities, comparatively speaking, have been found where quarries have been successfully opened for the burning of the rock in kilns for quick-lime.

On the slope of the hill or rise of the second terrace from the river, on the farm of Mr. Phips in Plymouth, (opposite Fort Fairfield,) are found fragments of limestone containing a good deal of calcareous spar. In company with J. B. Trafton, Esq., Dr. Decker and Mr. Phips, I examined the spot with a view of finding the exact locality from which these fragments came. We did not ascertain it precisely, but it is evidently near the brow of the slope, and will probably be found by removing a few feet of soil.

In the south-eastern part of the town (Fort Fairfield), Deacon Fowler opened a quarry on his farm a few years since, where he obtained a very good quality of lime. The stone is of the stratified, compact blue variety, and improves in quality the deeper he gets into the quarry. He had just finished burning a kiln of one hundred casks while I was there, which met with ready sale at remunerating prices.

In Linneus, near the house of P. P. Burleigh, Esq., I observed a locality of stratified limestone, which evidently contains a large proportion of silicious matter. Specimens of this are also forwarded for your examination, with a query whether it has the requisite ingredients for making hydraulic lime?

Magnetic Iron.

In the north-western section of Linneus, iron ore is found, combined with the slate, and thus presenting a stratified arrangement. This slate varies in its impregnation of iron. Some of it undoubtedly contains a large percentage; and in one or two of the localities specimens were obtained that are strongly magnetic. I have forwarded specimens of all varieties for your examination. Should it be found sufficiently rich for smelting, any amount of it could be obtained to supply the furnace, while lime for a flux is abundant in the neighborhood, and an inexhaustible supply of charcoal could be made from the adjacent forest.

Moulding Sand.

The sand used in foundries for moulding must be of peculiar material and texture, or consistence. It should be uniformly fine as to its grains, that it may not give too rough a surface to the castings. It should not have a sufficiency of clay or aluminous particles to make it sticky, and yet enough to render it compact and comparatively solid, when wet and packed or tamped into the moulds. It should also not be so absorbent of water as to take it too long a time to dry, after being moistened. It is difficult to find deposits of sand possessing all these requisites. We were informed by Hon. Shepard Cary, that he had found a good deposit of this kind of moulding sand near his iron foundry in Houlton, which affords him a supply of an excellent quality and which he uses altogether in his iron works.

Helderberg Limestones and Marbles.

Among other objects of this expedition, I was requested to trace out what I could of the localities and boundaries of the lower Helderberg marbles, or Limestone formations, that occur in this section of the State, and report to you. I have done in regard to it what the shortness of the time and the lack of some facilities allowed me. The more and further I searched into this branch of our geological formations, the more impressed I became of the ultimate value they will be to this section, and indeed to the whole State, and of the importance of longer time being devoted exclusively to their study and examination. A belt, or formation of rock, which, as I found, stretches in a continuous direction across not less than five townships, occasionally cropping out, and at each locality of its appearance exhibiting surroundings and accompaniments each of different character, could not be thoroughly explored and all its characteristics ascertained in the three or four weeks allotted to this section, and that time interrupted by a search for objects pertaining to other branches of Natural History.

On page 394 of your first report, in speaking of the geology of the Wassattiquoik while on your way to Katahdin, you observe that "on the Wassattiquoik, near its mouth, we found ledges of a bluish quartz rock very evenly stratified. * * * Above them, on the bank, the boulders and large masses of limestone similar to those seen at Whetstone falls are so numerous that we believe the rock to be in place close by, certainly less than half a mile, if indeed we did not find it in place."

Your conjectures were right. Had you turned and gone up the north branch of the Wassattiquoik a little way into township 4, in the 9th range, you would have found the site from which the boulders you saw started. It is the first locality, or cropping out of this belt of the lower Helderberg formation, east of Mt. Katahdin.*

I was not able to give this locality a personal examination, but obtained reliable description of its location from a person† who had visited the spot, clambered over the bluff it formed on the bank of the stream, and who showed me specimens of the rock identical in their composition and structure with the rock which I visited last year in Murch's lake, in the next township north-east of this, (No. 5, R. 8.)

Considering its geological position and surroundings this locality is one of peculiar interest, situated as it is almost at the base of Katahdin, with its granite battlements guarding it on the west and south—the trap rocks of the Lunksoos range on the north, and the quartz rock of the Maine Wassattiquoik on the east. I leave it to you and other geologists to decide the seniority of age and priority of occupation of these several formations, and to explain by what arrangements of nature this rock, so full of the remains of organic life, was placed in almost juxtaposition with such azoic neighbors. The one, full of tangible proofs of an age teeming with aquatic animal and vegetable life, and exhibiting through its structure the outward forms and shapes of former living tenants of an ocean in which they existed, and from which they drew their sustenance. The others, the very reverse of this—hard, crystalline in feature—silent as to any definite condition of the past—giving no sign of any association with life at any period—their clearest manifestations being those of an escape from heat of great intensity, and of convulsive earthquakes which have shaken and shivered the neighboring mountains and scattered their rough and angular fragments on every side. Whatever may be the theoretic speculations on this subject, one thing is certain. When the advance of settlement up the Penobscot shall bring mankind in greater numbers into this section, and the accumulations of thrift and industry shall enable

* This is undoubtedly the belt of rock from which the boulders of fine statuary marble discovered in 1861, were derived. C. H. H.

† Mr. David Malcolm of Patten.

them to erect mills and houses and public buildings, they will here find no dearth of most durable material for the same—no scarcity of granite and lime and marble to meet all the demands and purposes that may be ever required for architectural strength, endurance and beauty.*

The general direction of the strata is north-easterly. The extent of the formation I am not able to give. It becomes covered by the soil, and is hidden from view. Pursuing the general course of the strike, which leads you in a direction across the township diagonally, it again turns up at the Tunnel rocks in Murch's or Horseshoe lake in the next township, No. 5, R. 8. As a pretty full description of this locality has been given in last year's report, it will not be necessary to say more here in regard to it. It is well, however, to note it, as being the next link in the chain of these Helderberg formations, the existence of which this survey has been instrumental in discovering.

The next show of it, on this line of strike, is that discovered by Dr. Jackson, at the foot of the first Seboois lake, an extract from whose description you gave in your first report (page 413). On his authority it is stated to be in township No. 7. I did not arrive at the rock in place when at that lake, but judging from the range of the boulders and other observations, I think, instead of being in No. 7, it is in upper, or north-east part of No. 6 of the 7th range.†

Dr. Jackson also describes a locality of this rock on Peaked mountain, in No. 4 of R. 7. I have not seen this, but if it is identical with the rock in question, it must belong to another belt, as it is east of the range of the belt we are describing.

Continuing our course, we next find a splendid locality of it, cropping out near the north-east corner of lot 16 in No. 7, R. 6. I explored this ledge some years ago. It breaks up from a comparatively level plain, forming an abrupt, precipitous ledge, on one side fifteen or twenty feet in height. Its true location had been

---

*At Whetstone Falls a few miles below, on the Penobscot, is a splendid water power with a good site for buildings. Had the State reserved the fee of the soil in itself, and given proper encouragement to settlers, there would long since have been a thriving village here. E. H.

†It is very difficult, if not impossible, in a dense forest and in the absence of a correct plan based upon an actual survey, to give the true geographical position of any rock. In this particular we realized the truth of the remark of Sir William Logan, Principal of the Canadian Geological Survey, in which he declares, "accurate topography is the foundation of accurate geology."

lost for several years, and some who had sought for it were unable to find it, until last autumn, when from directions given them, Messrs. Baston and Chase of Rockabema, succeeded in again discovering it, a description of which he gave in a letter to me published in your report (page 320). I look upon this ledge as a very valuable one. Specimens from it were put into the hands of a marble worker, who found that it received a good polish—worked free and made good corners, and was compact and even or uniform of structure. Its proximity to the Aroostook road, and the ease with which it can be quarried, render it a feasible and valuable source from which to obtain marble or lime, to meet the wants of a growing community.

The next indication of this formation occurs in a line of the course hitherto pursued from No. 4, on or near the northern line of No. 8, R. 5. Boulders of Helderberg rock are found here, but the true spot of their original site has not yet been ascertained, and future exploration will be needed in that place to make it certain. Here ended my hurried, and of course imperfect search for this species of rock formations in this part of the State. They are deserving a longer and more careful scrutiny, which shall develop more fully both their geological and economical characteristics. I consider these formations, or beds, to be exceedingly interesting, not only on account of the intrinsic value of such rocks, in and of themselves, as affording a source from which to obtain marble for monumental or ornamental purposes, or excellent lime for cements or agricultural applications, but also for the geological teachings and testimonials they give of the period far back in the ages, when this portion of Maine was submerged 'neath the ocean, and crinoid and coral, and sea-fern, and mollusk, flourished on its shores and in its deep soundings, as they now do in the tropical seas of the south. Interesting too, for the story they tell of the singular changes that have taken place in the condition of the materials which compose them—of the hardening into stone of the soft ooze, while full of animal and vegetable life, embracing, and still exhibiting their organic remains as clearly and distinctly as when they flourished in it in the vigor of actual life—for the unmistaken evidences they give of the mighty upheaving of this ancient bed of the sea, and its disruption into mountain masses in obedience to the laws and commands of Him

"Who thundered and the ocean fled."

Very respectfully and truly yours,

E. HOLMES.

D. SURFACE GEOLOGY.

In our Preliminary Report we went largely into details upon the geology of the Alluvial Period, or Surface Geology. Our definitions of the various forms of the superficial deposits, as well as the theories of their accumulation were there given so fully, that it is now incumbent upon us merely to state whatever new facts have been brought to our notice during the past year, what new illustrations discovered, and whether any light has been thrown upon perplexing points. Already we have incidentally alluded to certain phases of alluvial or drift action in our descriptions of the country or of the older formations. Such remarks will not be repeated, nor will all the details of our observations be presented—only the most striking points.

A few very large boulders were noticed. One of granite is in the water near the south shore of Sysladobsis lake. A boulder of gneiss near Weld weighs by calculation more than 1,000 tons. Another split in two is in Phillips. One of conglomerate twenty-four feet long lies in the river St. John in number eleven. Near South Paris there is a large pear-shaped boulder standing by the side of a tree, and we should say it must be at least thirty feet high, as it is more than half as tall as the tree. There should be a sketch taken of this wanderer with the tree by its side.

It was remarked as a fundamental principle in the science of Surface Geology, that boulders are transported from ledges in right lines, in the direction in which the drift agency operated. Hence it is found that the course of the striae corresponds with that taken by the fragments. Conversely it is true that if we find boulders of some particular rock scattered over the surface, we can always discover the ledges from which they were derived, especially if the course of the striae in the vicinity is known. Now there are several important varieties of boulders in Maine whose source is unknown to us. Therefore we will mention them, in the hope that some one will be able to trace them to their sources. First there are the boulders of white statuary marble on the east branch of the Penobscot river, (Matagamon,) between Medway and the Grand Falls. Second, there are the valuable boulders of magnetic iron ore in Phillips and Salem. Thirdly are the not less important fossiliferous boulders in Phillips. Fourthly are numerous boulders of red and grey conglomerates near the eastern border of the State. They may have been derived from a formation containing gypsum. A few of their localities noticed are the following:—at Tallmadge; No. 7 next to Carroll; very large ones on Wawbawsoos lake; on Junior lake; near the Meeting-house Rips on the Chepedneck river; at the thoroughfare of North lake on the New Brunswick boundary, and in township 28 on the Bangor and Calais Air Line Road.

In the granitic regions of south-eastern Maine large boulders of granite are very common, and it is sometimes the case that the land is barren because it is strewed with them. Such examples may be seen in Hancock and Washington counties, on the western Schoodic lakes, and in No. 7 next Carroll. The rock in the latter case is mica schist, and the cause of the great number of fragments more difficult to explain.

Additional Courses of Drift Striae in Maine.

In the Preliminary Report a long list of the courses of Drift Striae in different parts of the State was given. Our observations of the striae were not as numerous the past season; but we present all that we have. They are compass courses. The inferences drawn from our previous list could also be drawn from this table:

Pine Stream Falls on the Penobscot, N. and S.
Shirley, one mile south of hotel, N. 20° W.
Shirley, three miles south of hotel, N. 10° W.
Monson, north part, N. 20° W.
Blanchard, south part, on mountain, N. 10° W.
Bingham, S. E. part, N. 20° W.
Bingham, near Kennebec river, N. 20° W.
No. 2, north-west of the Forks of the Kennebec, N. 15° W.
Moose river village, N. 50° W.
Near Canada line on Canada road, N. 45° W.
North Charleston, N. 20° W.
Rangely west line, N. 45° W.
Mooseluckmeguntic lake, north end, N. 20° W.
Wayne, N. and S.
South Thomaston, Owl's Head, N. 20° W.
St. George, west part, N. 10° W.
Cushing, south end, N. 3° W.
Cushing, north part, N. and S.
Warren, east line, N. and S.
Warren, one mile N. E. from village, N. 10° W.
Thomaston, south of St. George river, N. 8° W., and N. 10° E., a mile apart.
Thomaston, near village, N. 20° W.
Camden, Simonton's Corners, N. 3° E.

Camden, one mile S. W. from the village, N. 10° E.

Camden, S. E. from village, rocks embossed—force from the direction of Mount Battie.

Camden, at Lily pond, N. 10° E., and N. W. from Rockport the same.

Camden, Hosmer pond, N. 10° W.

Camden, on height of land west of Hosmer pond, N. and S.

Camden, near Ingraham's Corners, N. and S.

Camden, mouth of Megunticook river, N. and S.

Camden, Negro Island, N. 10° W.

Lincolnville, west part, N. 20° W.

Appleton, N. and S., and N. 10° W., two miles apart.

Belmont, N. 30° W.

Trenton, N. 10° W.

Washington county, No. 7, N. 10° W.

Near Witteguerguagum lake, at Big Falls, N. 30° W.

Tallmadge, east part, N. 12° W.

Grand Falls, Alleguash river above Chamberlain lake, N. 50° W.

Alleguash lake, N. 50° W.

Cauquomgomoc lake, N. 15° W., and N. 10° W.

Cauquomgomocsis lake, N. 10° W.

Black pond, N. 10° E.

Traces of Ancient Glaciers.

An example of an ancient glacier in Maine has already been described in our first report on the river St. John. The evidence of its existence has been confirmed the past season, by the discovery of other markings above the lake of the Seven Islands. The most convincing proof was found upon the sides of the boundary or the south-west branch of the river proper. Several examples of striae running down the valley in a north-east direction were there seen. This course is at right angles to the common direction of the drift striae in the valley of the St. John.

Upon three other streams also have we discovered glacial markings. First upon the Penobscot. These were seen at the mouth of Rag Muff stream where the striae run N. 15° W., the force having come from the S. 15° E., or in direct opposition to the drift force, which came from the north and proceeded southerly; in Pittston, where the course is N. 28° W.; and upon the north-east branch. The evidence in the last two cases does not rest upon the direction of the striae, so much as upon the fact that the striating force must have slid down the valley, following all the turns and windings of the river.

Secondly upon the Piscataquis river. In Blanchard the valley is remarkably deep. Upon the north side near the village, there is on the side of the road a ledge with a smooth striated perpendicular side. The course of the striae is N. 20° W. The distinguishing glacial feature here is the wall, or perpendicular side of the ledge, smoothed parallel with the course of the valley. The direction coincides with that of the drift in the vicinity. Similar striae and grooves coming from the northwest, and worn upon the perpendicular face of the ledges, may be seen in Abbot, where the Monson road crosses the Piscataquis.

Third, on Sandy river. The great bending of the striae in the valley of this river mentioned in our last report, page 262, must be an example of glacial markings. Confirmatory traces may be seen on the road to Rangely near the head of the river; for the striae there run down hill most perceptibly for a great distance. This is never the case with drift striae. The descent here must be very great, rather more than the proper average for glacial slopes. No one can for a moment suppose that an iceberg can slide down hill—it must always be the true glacial ice that accommodates itself to the slopes and windings of vallies.

In the valley of Ellis' river below Andover, the sides of many ledges are perpendicular, and resemble the walls which have been mentioned as characteristic of glacial markings. We had not time to examine them carefully. We cannot doubt the existence of many glacial markings in the numerous vallies of the western portions of the State. The rocks, however, disintegrate so easily that they may not preserve the markings very well.

We do not understand Dr. DeLaski to mean byhis great Penobscot glacier exactly what we do in distinguishing certain glacial markings from the drift proper. He regards the drift markings as made by one great glacier, extending over the whole of the northern portion of the continent. He supposes that all the striae upon the rocks were made by glaciers, when the whole continent was much more elevated than it is at present. The view we have adopted and explained in detail, supposes a combination of glacial and iceberg agencies—the traces of the latter being those most commonly seen now upon the rocks. When many of our valleys were filled with rivers of ice, the tops of the higher mountains must have been covered with perpetual snow and the scenery have been strikingly similar to that now exhibited among the glaciers of the Alps in Europe. Even with the striated and embossed rocks in full view, it is difficult for one to imagine that such mighty changes have been effected in the face of our country.

Trains of Boulders.

A train of boulders was last year described in our report. Since then Dr. True of Bethel has published further remarks upon that example in the Proceedings of the Portland Society of Natural History, page 92.

We have had an opportunity the past season of inspecting the train of boulders in Wayne, formerly alluded to. The case is not so distinct and impressive as we could have wished. The boulders are arranged on a line, starting very near the "Devil's Cave," and continue to North Monmouth, a distance of two miles or more. The boulders are all of syenite, and their source can plainly be traced to one hill. The course of the train, as well as of the striae in the neighborhood, is due north and south.

We learn on good authority that there is another good example of a train of boulders ten miles long in Madrid.

But the finest example of the kind we have seen or heard of in Maine is at the Forks of the Kennebec. Mr. Murray of the Forks Hotel took us east of the rivers about half a mile, where one end of the train appears. There are several kinds of rock in the train, but all the pieces are crowded together as much as they would be in a dilapidated stone wall. There were no spaces between the stones large enough to permit the growth of vegetation. The train is about three rods wide, and is known to be at least half a mile long. How much longer it is, no one knows, and we had not the time to ascertain. Its course is N. E. and S. W., or about that of the Kennebec river, on whose southern bank it is situated. This does not correspond with the course of the striae in the vicinity.

Sea Walls and Lake Ramparts.

Three more sea walls were noticed the past season, but all of them are small and of little consequence. Near Owl's Head, in South Thomaston, there is a large pile of clean coarse gravel one-eighth of a mile long, which must have been accumulated in the same way as the sea wall in Tremont. It is exposed to the full force of the ocean's waves. Another wall is at Mosquito Harbor in St. George. It is of less length than the first, and the road passes over it. The third example is near the Herring Gut Light House in Cushing.

The nature of the lake ramparts, with the theory of their supposed origin, has been given under the description of the phenomenon at Wawbawsoos lake. Other examples have been alluded to at Baker lake and Junior lake.

Dr. DeLaski of Vinalhaven, having been invited to give some description of the glacial phenomena about Penobscot bay, has kindly consented, and has furnished us with the following statements:

Ancient Glacial Action in the Southern part of Maine.

To Mr. George L. Goodale:

Dear Sir :—I herewith comply with your request to furnish for the ensuing Report of the Scientific Survey of the State of Maine, an account of my examination of the boulder evidence of the Penobscot bay.

The very limited space suggested by you, will permit me to do nothing more than merely notice a few of the facts which bear upon the question regarding the nature of boulder action in the locality named. I prepared a series of Articles during the past summer for the "Rockland Gazette," on the "Ancient Great Glacier of the Penobscot Bay," in which I went over in detail, the phenomena I had observed, and the grounds of the principal theories which have been brought forward to solve the mystery connected with the drift.

From careful personal examination of the surface of the islands and borders of the great Fiord of the southern coast of Maine, I have been forced to the conclusion that a glacier once occupied that margin of the state, of a magnitude sufficient to cover the highest hills of the region, and to extend far into the interior towards the north. From a glance at the correct county maps of the locality, we observe that the general trend of the islands, headlands, streams, lakes, harbors, creeks, coves, &c., is north-south, suggesting some law of formation. In these directions, I make no allowance for magnetic variation, which is considerable in the Penobscot bay. There are indeed departures from this rule as in the east-west direction of the great thoroughfare separating the Fox islands, where the natural boundary between the two towns was set up at an infinitely earlier period than that of the boulder age; for the irruption of the trap of North Haven broke through the granite and Taconic slates in a line corresponding to this trend, as you and I agreed, I believe.

We also find the hills not rounded and rough, but having an elongated appearance, and a trend also north-south, as if their sides had been subjected to a gigantic system of sculpturing, on the design that these, too, should be directed towards the south. And furthermore, these hills, even where they attain an elevation of one and two thousand feet, as those of Camden and Mount Desert, present gradual slopes to the north and bold fronts to the south; and if of granite, they are broken down more or less into step-like precipices of east-west parallels, the debris of which has not been accumulated as tali, but has been transported south a little distance, often more and more comminuted as we advance.

The formation of the coast is syenitic granite, bordered here and there with a margin of trap or of Taconic slates, highly altered in cases, and often converted into cherty flints as on Isle au Haut—and furnishes from the general barrenness of the surface, a good opportunity to study the boulder phenomena. And this surface is everywhere ridged into furrows, often very deep and in the usual direction of the valleys, &c., and present the finest examples of embossed rocks as described by Charles H. Hitchcock, in his Elements of Geology. This is so remakably the case that one might in the foggiest weather, easily point out north, south, &c., by looking at these rocks; for they represent in miniature, the hills and mountains of the coast as I have described them. Transverse indentations are everywhere common—*lunoid furrows,* I have called them—from an inch in length to four and five feet, having their horns pointing towards the north-east and north-west, and their steep walls facing the south. *These furrows in all cases, are sufficient to tell the cardinal points of the compass as one passes along over them.*

Everywhere, too, the boulder striæ may be found on the south sides of these hills at their bases, and on their sides when dipping at steep or lesser angles towards the east or west, in as finely developed examples as are found on their northern slopes. It is a fact beyond controversion, that the boulder phenomena in the Penobscot bay are *sui generis* in character, and owe their existence to one agent and the same period.

I have found these boulder striæ four hundred feet high on the side of Isle au Haut hill—which is five hundred feet above the sea—and on the southern brow of Megunticook, overlooking a precipice two or three hundred feet, and twelve hundred feet above Camden harbor. Mount Battie, south of that mountain, the nearest the village of any of those hills, and composed of your quartzose conglomerate, is everywhere scored and scratched, and has a very abrupt southern face. Vast masses of rock have been torn from it in this direction, and lie around its base. One large boulder here about forty feet long, must weigh not less than six hundred tons.

There is a series of terraces in Vinalhaven as you remember, seven hundred yards long, rising one above another, the last wall of which forms the highest margin of a dell running nearly due north–south unbroken for four hundred yards, and from twenty to thirty feet deep, and fifty yards wide. This is a trough cut out of the solid granite—a gigantic and splendid specimen of Nature's sculpturing with her rude stone chisels—all she needed in those days, when she had a vast duration before her to prepare a barren country with fruitful soils for the expectant worker, man. Towards the northern extremity of this rim, which is one hundred and fifty feet above the sea, there stands a high rock overlooking the village, apparently in its native bed, presenting a vertical wall towards the south twenty feet high above the soil, and twenty-four broad. No blasting by art, however carefully conducted, could perform a better operation. If this rock be a boulder, as you and I doubted, it must weigh upwards of a thousand tons. But many thousand tons from the south of it are utterly removed. Going a little further north, we reach one of the highest hills in the town, of granite, two hundred and fifty feet. To the north we look away down upon a tide "river," now a mile long, but once three, before the land obtained its present height; and earlier still, very much longer. Looking around towards the east and south, we have glanced over a spacious salt meadow, a densely wooded valley, and a large salt water pond. This depression must have been cut out of a comparatively level crust. From incessant examination of the subject during the last few years, I have seen nothing to induce me to believe that the granite had been materially changed from a horizontal position before the boulder period, as those north–south depressions might suggest. But what was really the depth of the denudation, one can only vaguely conjecture; but I have no doubt but that it has been many hundred feet.

The island of Mt. Desert exhibits the boulder phenomena in a more wonderful degree than those places I have mentioned. I presume you have thoroughly explored the locality. You see the southern brows of those lofty granitic hills everywhere crushed and broken into fearful precipices; whereas their sides turned to the north, present plains of greater breadth, and dip at vastly less angles down towards the level country beyond. The great granitic boulders lie at their southern feet, and those specifically the same but of less magnitude, and transported the farthest off, and are more worn and rounded. We have here as elsewhere in the Penobscot bay, the evidence that it was the special business of the great denuding agent to cover the barren surface with soils, and that those soils are the result of local detritus—gravels, clays and sands crushed and ground out of the detached rocks.

On the Taconic slates beyond these mountains towards Ellsworth, we have the debris of the Taconic formation. Still beyond through Dedham, we have a granitic formation, and see the granitic boulders in the most wonderful profusion and of great magnitude. They were derived from the hills a little way towards the north. The same peculiarity may be said of North Haven above Vinalhaven. On that island, principally a trap region, you see trap boulders and rubbish. In the northern part of Vinalhaven where the Taconic slates are highly altered, you see boulders of the same character; on the granite below, granitic rocks; and still further beyond, where the syenite has apparently been altered—or the cooling crust originally took the form of hornblende—the ruins of hornblendic rocks are found.

Around one of the quarries to the west of Carver's harbor, the ground is literally covered with boulders, some of which are enormous. After repeated attempts, I could not make out more than five per cent. of foreign rocks among them. Many of these turned out of their beds, exhibit the polishing and scratching of the common floor rock of the island. Furthermore, if carefully turned over, we find some of them left just where they had last been employed in scratching the ledges, the parallel scratches of the boulder being placed parallel to those of the rock beneath. Of these foreign boulders we often have little or no grounds to imagine the origin. We have specimens of red and blue granite, trap, gneiss, mica schists, clay slates, and fossiliferous sandstones from the Katahdin region. We can well suppose them to have been dispersed by icebergs, or borne as freight to these localities, by slowly moving glaciers.

Let me ask, then, how could such rending asunder of mighty masses of rock and the general phenomena I have described, be the result of the action of icebergs in their passage south over a *sinking* continent? The conviction can scarcely escape the mind of the observer, that at least very many of these enormous masses must have been detached from their original beds before the country went down into the sea; for icebergs of supposed power sufficient to quarry them, would require a very great depth of water, the pressure of which would assist the rocky bottom in resisting fracture. And no known currents have power to drive icebergs against submarine hills with such force as to separate large masses of rock from them. And if their motion might be supposed to have often been accelerated by violent winds, which they could not materially have been, as their principal bulk was below the reach of such imaginary aids—this supposition would be against the argument that those floating bodies have produced the parallel boulder striae on the rocky floor of the country.

I lay it down, therefore, as a self-evident conclusion—that, icebergs driven by any known currents, could never have ascended such long and steep planes as those which the lofty hills of Camden and Mount Desert present to the north; for in the attempt, their bulk would often have been shattered and lessened, and their freight of boulder materials, frequently unlike the formations over which they were passing, deposited where the bergs had foundered.

That, the idea of a *sinking* country in this case, conforming in its process of submergence to the passage of such ice mountains, is wholly inadequate and untenable in theory.

That, icebergs floating in a liquid whose density was but a little greater than that of their own composition, could not have broken down the southern brows of the lofty hills of the coast.

That, these hills could not have been crushed as Mr. Hugh Miller suggests in a supposed case mentioned in the "Cruise of the Betsey," by bergs turning *backwards* in their journey south, upon those hills, and operating against them as submarine battering-rams; for in that case, the bergs would have given away before the granite would have yielded.

That, icebergs could not have originated the striæ at the southern bases of the highest hills of the coast.

That icebergs could not have materially denuded and regularly scratched the east and west sides of those hills—and especially those standing alone; for they would have been pushed with the current like any other floating body, *around* those hills, instead of *over* their steep sides.

That, the fact of modern icebergs being often driven out of their usual course by getting into counter currents, as in the case of the one which during the past summer grounded off the harbor of St. John, Newfoundland—a novelty the inhabitants had not seen before—in the boulder period, according to their supposed infinite abundance, must have been of very common occurrence; but nowhere do we see diagonal markings on our ledges at all adequate to the supposition that they were thus made by the irregular course of icebergs. In fact, secondary scratches are not found in the Penobscot bay, so far as I am aware of, nor any irregular markings on the rocks that could be interpreted as the result of the grounding and vibratory motions of icebergs.

That, the enormous quantity of boulder materials does not favor the iceberg theory.

That, icebergs could not have denuded the surface rocks and originated the striae, because in the direction north whence the agent came, the base of the highlands of Maine over which it must have crossed, is not less than a thousand feet above the sea.

That, these peculiarities of the boulder phenomena could not have been performed through the agency of diluvial waves from the extreme north when the country stood at or near its present level above the sea; for we find the islands off the coast twelve and fifteen miles, with nearly five hundred feet of water north of them over the river and ocean silt and submerged boulder materials, denuded and scratched precisely as those within the bay. The whole force of the ocean would have opposed such currents.

That, if it be presumed these islands were then a part of those above as "main land," and the country consequently higher, the locality would have been very much *colder* than at present.

That, the theory of diluvial waves would involve a heterogeneous mixture of boulder materials altogether different from the common deposits—materials of granite, hornblende, trap, slates, &c., would lie scattered alike over formations more or less unlike them.

That, Polar floods, if any ever occurred, could never physically speaking, have been projected so far from the localities of their origin, as the southern part of Maine.

My conclusions, therefore, from the facts which I have enumerated, are, that a glacier once filled the basin between the Camden hills on the west, and those of Mount Desert on the east, forty miles wide—extended to a great distance north, involving several hills beside those mentioned, of a thousand feet high, and certainly not less than three thousand feet thick.

And, it has *suggested* itself to me, that glacial action of the coast of Maine, has utterly removed the tertiary deposits from her surface; for if they are presumed to be yet under water, certainly a country like the present, could not be said to be a *tertiary* one.

If these hasty sheets will be acceptable to you, you are very welcome to them. Very truly yours,

JOHN DeLASKI.

December, 1862.

Horsebacks.

We are able to add several more horsebacks to the seventeen enumerated in the preliminary report. The first is one brought to our notice by the *Maine Farmer*, whose account of it we quote:

"There is one of these horsebacks in the northern part of Somerset county, which we have not seen described, and therefore adds another one to the list. It is situated partly in number 2, range 2, west of the Kennebec, and partly in Jerusalem township in Franklin county. Its general course is N. W. and S. E., although it is interrupted by many short zigzag turns. It is nearly five miles in length, and from twenty to sixty feet in height. Occasionally there are sharp pitches or depressions through its course, and here there seem to be a predominance of boulders ranging in size from a hen's egg to a two-quart measure, while on ascending the horseback from these gullies, coarse sand or gravel seems to be the formation. On either side of this horseback, for the whole distance, is a peaty swamp, in places covered with a black growth, and at other places—where the growth has been burnt—showing a stream with a mucky bottom, forming at one place a pond of considerable size. At several places on the bog, there are considerable quantities of cranberries, but they are small in size and inferior in quality. The people in the vicinity regard this horseback as formed on purpose for a road; and it would seem that such were the designs of Providence, for it is the only place for the entire five miles that it would be possible to build a road for public travel."

Near Princeton are two horsebacks. Both are on the railroad, one in Princeton and the other in Baileyville. The latter is the largest, being five miles in length. It has a large slope towards the Kennebasis river, which is unusual. Both cross the river valley with a N. E. and S. W. course.

The only other horseback seen having a slope is in the southeast part of Bingham. It slopes at an angle of 2° 30′, and shows itself for half a mile near a carriage road. It lies on the west side of the valley, with somewhat of a south-easterly course, and runs into moraine terraces.

On the west shore of the Eastern Schoodic Grand lake is a very well marked horseback, though not very long. We cannot vouch from personal examination that it is over half a mile in length. It is in Weston.

In the north part of Weston we were permitted to see the southern termination of the great horseback extending from Houlton to Weston. It suddenly curves to the east, and in less than half a mile's distance terminates in a swamp, gradually dying away. It is probable that the horseback in the south part of Weston was formed by the same general causes which produced the large ones, and we shall expect to learn that the two are connected together by other links, now concealed in the forest on the low land.

In Linneus, north of the post office, there is a very crooked horseback nearly a mile long. Its southern end is in a small pond near the village. In the north part of its course, it is accompanied by moraine terraces. It lies west of the great horseback of Houlton. In Houlton we remarked that the material of which the horseback was composed was stratified black gravel; while the mounds which we have considered to be moraine terraces on the east side are composed of unmodified loamy material, with frequent fragments of slate. It would hence appear that the two classes of deposits must have been formed by different agencies, although they may have operated at the same time.

A fine horseback at Leadbetter falls on the Penobscot, terminating in a ledge, has already been spoken of in **C.**

By all odds the largest horseback we have seen in Maine is what is called the "Whale's back" in Aurora. The air line road passes over it for three and a quarter miles in a south-easterly direction. The horseback then continues on in the woods to an unknown extent. We passed over it too early in the morning to estimate its altitude and width with any precision.

Later in the morning we caught a glimpse of another large horseback in township 28, but do not know its length. This is distinct from the one described previously in the corners of Beddington, 29 and 22.

The stage road passes over an interesting horseback between Kenduskeag and Corinth. The road first strikes it in the west part of the village of Kenduskeag, and continues upon it for three miles to a cemetery in South Corinth. It appears to extend somewhat further in both directions. Its general direction is north-westerly; but there are changes and curves in it, whose precise nature may be ascertained by noticing upon the map of Penobscot county the course of the stage road. This ridge is wide and not so high in proportion to its width as is most common. It is of the whale back type, like the example in Aurora. We estimate its altitude from twenty to fifty feet; and its width from six to fifteen rods. It starts from the lee side of a large but low hill, and the north-west end is higher than the south-eastern. A cut through it reveals a section of gravel, precisely like the ideal sketch of a horseback, on page 273 of our first report.

North-west from Parlin pond there is a curving horseback three fourths of a mile long. Our impression is that it is parallel with the shore of the pond.

But the most remarkable example of a curving horseback has been described by Dr. Holmes in his notes upon northern Maine; where is one of these ridges bent around in the form of a horse shoe. He also describes another one in the vicinity of great interest. These ridges are on the Seboois waters in Nos. 6 and 7.

In examining the surface geology of Kennebec river, we noticed three ridges above Bingham which we are inclined to refer to this class of deposits, rather than fragments of high terraces, to which they are closely related. The longest one is in the Forks plantation, between F. E. Shepard's and J. Steward's. The shortest is in Caratunk, between E. Pierce's and G. F. Chase's. The third is in Moscow, west of J. P. Emerson's. They all border upon the river, upon its eastern bank, forming a high ridge between the road and the river. They are designated as long hills upon the map of the county (Somerset.) The longest must be at least a mile and a half in length, and the shortest not over a half a mile.

Hunters have informed us of other horsebacks in the wild lands; but of them all there occurs to us now the location of only one, viz: upon the more southern of the St. John ponds, north-west from Moosehead lake. Without doubt there are many in the State yet to be brought to light.

This makes an addition of seventeen horsebacks brought to light since the publication of the first report. In other words the number of known horsebacks in Maine has been doubled the past season.

We hope that geologists will pay more attention hereafter to the investigation of those curious gravel ridges. We doubt not tha they will be found common over the northern border of the United States. We were surprised and delighted to see two beautiful examples of them in Lower Canada the present winter. One is in St. Flavien near Quebec, composed of coarse materials. The other lay partly in Acton and partly in Wickham, and is composed of finer materials. A section of it is precisely like our ideal section of the Maine ridges. Both these Canadian examples are situated in the flat country adjoining the river St. Lawrence, and both run N. E. and S. W. Both of them, also, are several miles long. Their general characters agree perfectly with those of the Maine examples.

We do not yet feel satisfied about the true theory of the formation of horsebacks. We could not but be surprised, however, when on Moosehead lake, to see how very similar they are to such islands as Sandbar, Snake and Hogback. A section of the first is like that of the horsebacks. These islands are very long and narrow, being composed of coarse and fine gravel. It would not be strange if some of the horsebacks were formed like Sandbar island, which seems to have been deposited by currents, either with or without the assistance of ice. Sandbar island must have been formed in a past period, when Moosehead lake stood at the level of the terraces which may often be met with at the mouths of its tributary streams. Snake island would appear to be one that is forming at the present level of the lake.

Sea Beaches.

A single example of the more elevated stratified gravelly banks of assorted gravel and sand, which have been referred by us to the action of the waves of an old ocean, was pointed out last year. We would suggest a few others this year, without having had time to observe carefully their altitudes or relations to the surrounding country. One of them is upon the hill south of Pollard's hotel in Masardis. Another is near the height of land between Weld and Wilton. This one is more or less connected with very high terraces. Another is high up on Beech hill, of the Saddleback range, in Franklin county. There is a great amount of detritus collected about the small ponds at the head of Sandy river, perhaps referrible to moraine terraces. Lastly, it seems as if there must be some ancient beaches among the numerous sandy hills in Wayne and Leeds, far above all existing streams. In Leeds one of these sandy accumulations has been torn asunder by the wind, and the sand is being blown south-easterly, much to the detriment of the cultivated fields adjacent. A potatoe patch was covered up in this way to the depth of thirty feet. These hills of moving sand are called *Dunes* or *Downs.*

We have not been able the past year to make any further observations upon the very interesting fossiliferous marine clays that skirt the sea shore and the sides of the principal rivers for a considerable distance inland.

Terraces.

Nor have we been able to observe or map many of the terraces lining so many of the beautiful rivers and lakes of Maine. Kennebec river, particularly above Skowhegan, exhibits these phenomena very finely. We were able to map them carefully for about thirty miles of the way below the Forks, and might have published a map of them here, but preferred to defer its publication until we should be able to give a map of them along the course of the whole river. We will, however, state a few general facts concerning the surface geology of this river, beginning at its source.

The Kennebec river rises in Moosehead lake, not at the extreme southern angle, as one would naturally suppose, but from the south-west side several miles above Greenville. It rushes out of the lake a large river from the very first. Until it reaches the Forks, it is an exceedingly rapid stream, falling hundreds of feet. This section of the route is probably deficient in terraces; yet we have not explored it, as there are no roads along the shores, and navigation with canoes is impossible. The immediate banks we understand are rocky, the river passing through a gorge.

The Forks of the Kennebec derive their name from the junction of Dead and Kennebec rivers. Terraces are very abundant here.

It is commonly the case that the junction of streams gives rise to the formation of a greater number of terraces, than will be found away from the confluence. But the elevation of the highest one is no more than it is elsewhere along the valley. It is just so here. Fig. 49 is a section of the terraces of the Kennebec, crossing the valley just above the mouth of Dead river. The altitudes

FIG. 49.

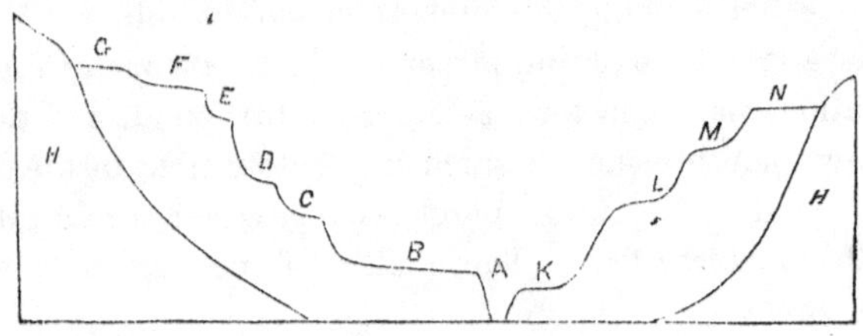

Section at the Forks.

A. Kennebec river.
B. Meadow, west side, 20 feet above the river.
C. Second terrace, 35 feet do.
D. Third terrace, 40 feet do.
E. Fourth terrace, 65 feet do.
F. Fifth terrace, 70 feet do.
G. Sixth terrace, 80 feet above the river.
H. H. Solid rocks.
K. First terrace on the east side, 15 feet do.
L. Second terrace, 40 feet do.
M. Third terrace, 58 feet do.
N. Fourth terrace, 70 feet do.

were taken with an Aneroid Barometer, and are laid off upon a scale. Only the outline of the surface is given. We have not attempted to show the order of superposition of the terraces in relation to one another, but only of the whole to the solid rocks beneath. The highest terraces upon both sides of the river in this case appear to have the same altitude. This is not commonly the case. The lower terraces are loamy, and the higher ones are composed of fine and coarse gravel.

A better illustration of the most common relations of terraces is given in Fig. 50, measured at J. Carney's hotel in Moscow. The materials are very coarse, the most so upon the east side, but the forms of the terraces are perfect. The materials of *E* and *F* are as coarse as common unmodified drift. Two very fundamental facts are illustrated by this section, which must be taken into account in the formation of all our theories respecting their origin; first, the heights of the highest terraces on the opposide sides of the valley do not correspond; nor, secondly, do the heights of any of the intermediate terraces agree. Our theory of their origin has been given in our Preliminary Report.

The valley of the Kennebec is very beautiful. Below Bingham it expands, and at Solon and further down the terraces are so wide

FIG. 50.

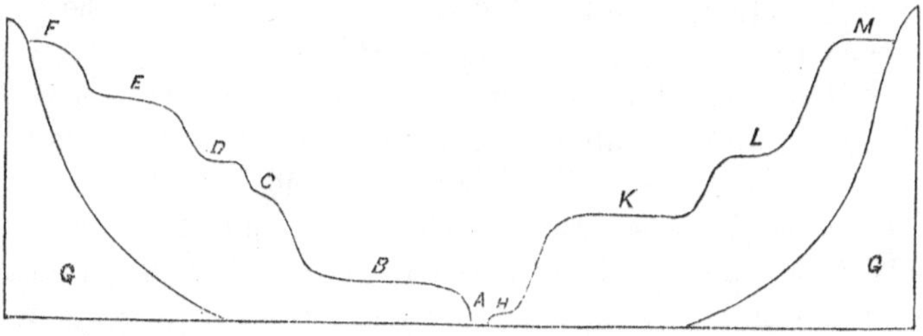

Section at Carney's.

A. Kennebec river.
B. First terrace, east side, 25 feet.
C. Second terrace, 80 feet above the river, and only 25 feet wide at the top.
D. Third terrace, 110 feet.
E. Fourth terrace, 136 feet.
F. Fifth terrace, 180 feet.
G. G. Solid rock.
H. Bank of the river on the west side, only three feet high, hardly worthy of being called a terrace.
K. First terrace, 73 feet.
L. Second terrace, 113 feet.
M. Third terrace, 193 feet.

that several days' work will be required to map them with accuracy. The great plain in Solon is a terrace. At Skowhegan, and perhaps further up, the terraces are more interesting, because they contain marine relics, showing a combination of fluviatile and oceanic agencies in their origin. A similar terrace extends up the Wasseronset river north of Skowhegan. We would compare the terraces about Skowhegan with the delta terraces of the Winaoski river near Burlington, Vt., which we have described in detail in the Final Report upon the Geology of Vermont.

Below Skowhegan the valley of the Kennebec is less beautiful, because it is so wide that it cannot always be taken in at one view. Occasionally the terraces are well developed, as at Waterville and Augusta. A careful map of them through the whole course of the river would be an object of great interest, and having its practical bearings also. We can produce at any time our map of the terraces between Bingham and the Forks.

We were interested to find a few terraces at the mouths of many of the streams on Moosehead lake. These clearly indicate that the lake has been successively at higher levels than at present. The low land at Kineo is entirely made land ; and one can easily detect four different levels. These seem to have been quietly deposited in the lee of Mt. Kineo. The heavy north winds drive around the mountain immense waves full of detritus, which subside when protected and deposit their load. It is this force which has made the foundation land for the hotel. We have hints of a former outlet of Moosehead lake into the Piscataquis river instead of the Kennebec, which we would gladly be permitted to trace out. A knowledge of the facts might be of practical value to lumbermen.

We have noticed interesting terraces at several points on the Androscoggin, particularly at Brunswick and Lewiston; at Andover on Ellis river; on Saddleback stream; on Moose river; on Sandy river; and on the Penobscot. But our limits forbid us to describe them.

Siliceous Marl.

We will now present an interesting letter from Professor Bailey of Frederickton, N. B., upon the siliceous marls, or diatomaceous earths of Maine.

PROF. C. H. HITCHCOCK:

Dear Sir:—You have requested me to prepare for publication a notice of the microscopic Flora of the State of Maine, with a few remarks on the value of "diatomaceous earths" in an economical point of view, and also as throwing light upon certain disputed questions of geological history.

It might at first seem to be of little value to prepare for a report on the agriculture and geology of a country, an account of animals and plants so exceedingly minute as to be in the great majority of cases absolutely *invisible* to the unassisted eye. Of what practical value, we are apt to ask, can the fossil skeletons of beings so minute that fifteen thousand millions may be included in the space of a cubic inch be to the farmer or to the geologist? Is there any way in which the former can employ them in promoting the fertility of the soil, or can they aid the latter in deciding upon the geological age of deposits and formations, from which all other traces of organic life have disappeared? I shall endeavor to show that to both, but especially to the latter, the study of their numbers, distribution, and specific characters, is a subject of the greatest interest, as well from a practical as from a theoretical point of view.

Without entering into the details of the structure and mode of growth of the microscopic Algae, it may be sufficient to say, that notwithstanding their minute size, so incredible are their numbers and so rapid their multiplication, that their accumulating skeletons are even now exerting an important influence in changing the physical features of the globe, filling up ponds and lakes, changing the beds of rivers, and forming shoals or deltas at their mouths. It has been shown by the researches of Roper and other English microscopists that in the mud of the Thames, if we exclude the coarse sand, nearly one-fourth of the finer part of the residuum is entirely composed of the silicious shells of diatoms, while Ehrenberg has shown similar results from an examination of the waters of the Elbe and Scheldt. In our own country the Columbia river has been shown by Col. Frémont to have cut its way through a deposit entirely composed of diatomaceous shells, five hundred feet in thickness! The first question which I am requested to answer in regard to these deposits is whether or not they may be made of economic value?

That they *are* of value has been most satisfactorily proved by the researches of many authors. As early as 1850 the great abundance of the remains of microscopic organisms, fresh and marine, in the low rice-grounds of our Southern States, and the probable influence of their presence in promoting the fertility of those lands, was pointed out by my father. The same observer has also shown that the mud of New Haven Harbor, which has been successfully applied as a fertilizing agent, is largely composed of silicious exuviæ, containing by analysis 58.63 per cent. of silica. To the same cause is assigned by Ehrenberg the wonderful fertility of the lands annually inundated by the Nile and Ganges, while an examination of the tidal muds in any of our large rivers would probably lead to like results. I have myself found marine diatomaceæ in great abundance in the St. John river, at considerable distances from its mouth, while an examination of the ice when melting in the spring, also showed a vast number of fresh water forms. All the cases just mentioned, however, are of recent deposition; will the sub-peat deposits, which can alone be gathered in any quantity also prove beneficial in their application to the soil? I am not aware that any experiments have been made to test this point, but as the deposits are numerous and easily obtained, it would be well worth the trouble to make the trial. It is a significant fact, and one frequently observed, that those plants, especially the cereals, which require a large amount of silicious matter to give the requisite strength to their stems, and thus to promote their healthy growth, have about their roots, in a living state, numbers of these very organisms. Another singular fact affording additional evidence of the utility of such bodies is their invariable presence in great numbers in all the imported guanos. Indeed so true is this, that the poorer qualities of guano have been shown to be the very ones in which the least number of the microscopic algae can be found.

I therefore conclude that there can be but little doubt of the real efficacy of such deposits, no matter whether fresh-water or marine, and that the beneficial action which they exert upon the soil is to supply a large amount of silica in precisely that condition from which it may most readily be taken up by growing plants. As many of these organisms are said to contain a trace of iron in their composition, probably this element may also exert a beneficial effect.

Of the total amount of such deposits in the State of Maine, little is known. Even where their presence has been shown, the extent and thickness of the outcropping beds has been seldom measured. There can be little doubt, however, that a closer examination would prove the quantity to be considerable. The localities now known where diatoms exist in a fossil condition in this State are the following:—Bluehill Pond (discovered by Dr. C. T. Jackson); Brownfield; Newfield; Round Lake, Calais; Chalk Pond, Beddington; Adley Pond, Phillips; Bangor; Chalk Pond, Waterford.

The first of these deposits, that from Bluehill, has long been known and the species contained in it already described. As the four next deposits are but recently known, I will say a few words as to their characters and the species they contain.

Of the four specimens sent me for examination, that from the Chalk pond, Beddington, is by far the purest and the richest. It is as white as chalk, (for which it has probably been mistaken, judging from the name of the locality,) is almost free from foreign matter, and as you have informed me, covers some twenty acres, being of unknown depth, but certainly several feet. The species contained in it are the following:

Eunotia triodon and diodon.
Eunotia pentodon.
Eunotia serra.
Eunotia falx. = E. hemicyclus. Ehr.
Navicula firma.
Navicula ovalis.
Navicula viridis.
Navicula serians.
Gomphonema capitatum.
Cocconema lanceolatum.
Orthosira aurichalcea.
Melosira distans.
Himantidium arcus.
Himantidium bidens.
Himantidium undulatum.
Himantidium gracile.

Navicula rhomboides and Sporangium.
Navicula rhynchocephala.
Pinnularia nobilis.
Pinnularia major.
Pinnularia Tabellaria.
Pinnularia Stauroneiformis.
Stauroneis Baileyi.
Stauroneis phœnicenteron.
Gomphonema acuminatum.
Cymbella cuspidata.
Surirella biseriata.
Tabellaria flocculosa.
Cyclotella Kutzingiana. (Rare.)
Odontidium. (Large.)
Odontidium Tabellaria or
Fragilaria undata.
Encyonema?

In addition to the above a number of other species occur, but not in a condition to be satisfactorily determined. Among these I may enumerate *Navicula affinis,?* two varieties of *Nitschia,* (one of which I suppose to be identical with a similar form detected by A. M. Edwards, Esq., of New York, in a deposit from Bemis lake, N. H., the other undescribed,) and some curious varieties of the genus Himantidium. One of the latter is about as long as *H. arcus,* but with an undulate outline swelling in the centre of the dorsal region into a rounded cone, and also enlarged at the extremities which turn upwards. Only a single pustule was detected. The varieties of this genus occurring here are very numerous and interesting, especially those of "H. undulatum." I shall endeavor to speak more fully of these and other doubtful forms upon some other occasion, such descriptions being out of place here.

The deposit at Adley pond, Phillips, is much less pure than the preceding, containing numerous pieces of sharp quartzose sand. It is more like the common sub-peat deposits of the eastern States than the above. It contains the usual variety of species, less in number, however, than those above enumerated. It is especially marked by great variety in the genus Eunotia, which is comparatively rare in the Beddington earth. Adley pond is, I believe 25 acres in extent, the deposit cropping out upon its edge.

The Brownfield earth is quite impure and not particularly interesting. It is from below peat, and is apparently in a state of decomposition, many of the forms being fragmentary. The earth from a pond in Calais is still less interesting. It consists of indurated lumps, often stained with iron, quite impure, and the forms much broken.

Of the Bangor earth I know nothing, except that such a deposit exists in that vicinity. It has been supposed to contain copper.*

*Mr. A. E. Verrill of Cambridge, states that upon the south side of Chalk pond in Waterford, there is another of these deposits a foot thick. C. H. H.

I have heretofore spoken of the value of the microscopic Algae simply with reference to their employment as fertilizing agents. Although I regard this as by far the most valuable use to which their fossil remains may be put, I should not omit to state, that there are other and by no means unimportant purposes to which they may also be applied. Of these the most important are their employment as polishing powders, in the so-called "metallic lustres" and in the manufacture of porcelain. The substance called Tripoli is almost invariably composed of nothing but the silicious skeletons of the diatomaceæ. A polishing powder has been prepared under this name from the Bluehill deposit already mentioned.

I come now to the consideration of another and exceedingly interesting portion of my subject, but one on which little at present can be said, viz: the uses of diatomaceous earths in determining the unsettled questions of geological history. It will readily be conceived, after reflection upon the figures already given, that organisms which exist in such countless numbers, and which multiply with such inconceivable rapidity should exert a vast influence in promoting geological changes, but have we any key in the characters and species of the formations themselves whereby to determine the nature and the duration of the causes which produced them? May they like other fossils be used in determining the relative age of the several beds in which they occur? From their simple structure, whereby they are better adapted than most organic beings to resist the influence of physical changes, and from the fact that but few, if any, fossil species are known, which have not also been found in a living state, it has been until recently supposed that no such use *could* be made of them. Deposits from various localities were known to be of undoubtedly different geological age, but no means seemed apparent by which that age could in all cases be definitely settled. A more extended study of the distribution of species now bids fair to furnish the wished for key, although at present we can only state what is to be hoped for rather than what may be considered as certainly known.

Some eighteen months ago, while examining the fossil and recent forms from a rich locality in the vicinity of Providence, R. I., I was struck by a remarkable fact, of which I have nowhere seen mention made, although a distinguished Philadelphia microscopist has since informed me that he had also observed a similar fact. The fossil sub-peat deposit referred to contains the usual variety of fresh water lacustrine forms for the most part the same as those already given in the Maine deposits. Upon preparing a gathering, however, from the living algæ of the same pond, I at once noticed the great abundance and large size of the pustules of Nitschia of which not one fragment even could be detected in the deposit below. Here then was certainly an instance in which one genus at least, and that a large and well marked one was entirely absent from a fossil deposit, notwithstanding the fact that it is now and has been for unknown years, growing in countless numbers in the self-same pond, and gradually helping to form by its deposition a new deposit, not three feet removed from the former! Nor is this a single instance. Not *one* of the common *sub*-peat deposits of the country, such as those above mentioned, will be found to contain *any* specimens of the genera "Nitschia" and "Synedra." One deposit only, according to Dr. F. W. Lewis, and that *over*-lying the peat is known to contain these genera. I have already mentioned a Nitschia as occurring in the Beddington earth, and Mr. Edwards of New York, has also alluded to one found in the Bemis lake, (N. H.) earth, but these are all isolated specimens, few in number, and differing specifically from those which swarm in countless numbers in the same localities to-day. A similar fact has been noticed by my father in regard to the great "infusorial deposits" of Oregon and California. The forms obtained from the fresh-water tertiary districts of those States were found to be wholly unlike the recent infusoria from the Columbia river, and other existing streams, and probably the same fact may be observed in all parts of the country. As yet, however, this subject has received but little attention. The great post-pliocene epoch to which most of these deposits belong, has hitherto remained an unknown ground, with little to mark its character beyond the confusion and tumult of the great drift period. Probably the "infusorial earths" were deposited before as well as after, and perhaps during that period. The diatomaceæ in which were found imbedded the bones of the Mastodon in Orange county, New York, in 1843, are exactly those which characterize the greater part of the common sub-peat deposits, already referred to. The species which characterize that deposit may therefore be taken as the type of all similar deposits of that age. Of the Miocene Tertiary beds of Diatoms, the Richmond and Maryland earths afford undoubted examples, while to the *Eocene* is referred by Ehrenberg the Aegina clay-marls and the *chalk-marls* of Mendon and Caltanisetta. Possibly some of our western beds may prove to be of similar origin. It is also worthy of note that fossil shields of diatomaceæ have been found by Dr. White of New Haven, in the hornstone of the Devonian and Silurian beds of New York, thus carrying back the existence of these frail but indestructible wonders to the very dawn of organic life.

I will conclude this letter, which has already reached an undue length, by giving a list of microscopic forms observed by my father in the vicinity of Waterville, which though not by any means complete may possibly be of service to those interested in such investigations.

List of Microscopic Forms observed in Emerson's Stream, Waterville, August, 1853.

PROTOZOA,	Melicerta ringens on leaves of Nuphar. Spongilla forming patches a foot or more in circumference. Arcella. Stentor polymorpha. Vorticella. Chaetonotus larus. Rotifer vulgaris.
CONFERVOID ALGAE,	Tetraspora gelatinosa. Vaucheria caespitosa. Rivularia, very abundant. Lemania fluviatilis, very abundant. Zygnema nitidum.
DESMIDIACEAE,	Euastrum verrucosum. Closterium lunula. Euastrum margaritaceum. Docidium. Micrasterias rotata. Arthrodesmus quadri-caudatus. Penium digitus.
DIATOMACEAE,	Eunotia gibba. Surierella splendida.

Along the coast may be found the usual variety of American marine forms, mostly parasitic upon the smaller Algæ. I would particularly mention that the beautiful "Isthmia nervosa" occurs in quantity upon the fronds of Polysiphonia at Portland.

Hoping that the above may be the information you desire,

I am, very sincerely yours,

L. W. BAILEY.

Fredericton, N. B., Nov. 13, 1862.

E. DESCRIPTIONS OF NEW FOSSILS.

Principal Dawson of McGill College, Montreal, C. E., visited the interesting Devonian deposits in Perry the past season, and kindly sent us an abstract of his observations, which is hereby presented. The letter has been previously published in the Proceedings of the Portland Society of Natural History, pages 99, 100, where may be found the drawings of the new species to which reference is made. Dr. Dawson has recently published what is really a monograph of the Later Devonian Flora of Northeast America, in the Quarterly Journal of the Geological Society, and gives the descriptions of many new species, with notes upon those already known, amounting in all to sixty-nine different species. The number is now increased to seventy-five by these additions from Maine, and will be increased still more in the future by material already in the hands of Mr. Hartt of St. John, N. B. Comparing the range of the same species of plants in Maine, New Brunswick and Canada with those in New York, it is clear that the more eastern deposits must lie between the Chemung and Hamilton groups of the Upper Devonian. It is very singular that simultaneously with the discovery by Dr. Dawson of a Dicotyledonous plant, the Syringoxylon in this Devonian series, Mr. Hartt should have found in New Brunswick the wings of insects, both remarkable and unexpected discoveries.

McGill College,
Montreal, Nov. 26, 1862.

Dear Sir:—I had the pleasure, in August last, of examining the locality of fossil plants at Perry, and with the aid of Jethro Brown, Esq., who kindly assisted me when there, and followed up the research after my departure, succeeded in obtaining several new plants and better specimens of some of the species previously known. With the exception of specimens of Cyclopteris Jacksoni and Psilophyton princeps found by Mr. Brown in red sandstone and shale on the Perry river, all our specimens were obtained from the original locality mentioned in the last report of the State Scientific Survey, which is a bed of grey sandstone about two feet in thickness, and apparently very limited in horizontal extent. It probably marks the spot where a stream flowing from the old Devonian land emptied into the waters in which the red sandstone and conglomerate were being deposited. The plants are all drifted, but they must have been derived from land at no great distance. The age of the deposit of red sandstone and conglomerate in which the bed occurs, I suppose to be that of the upper member of the St. John series, or the upper Devonian sandstone of Gaspe. Its total thickness appears to be about 1300 feet, and the lowest bed which I observed was a very coarse angular conglomerate or breccia. In respect to the geological relations of these beds to the underlying shales, &c., I have nothing to add to what is stated in your report of last year.

The new facts ascertained last summer in the fossil botany of Perry, may be summed up as follows:

1. Having carefully collected the fragments of fossil wood showing structure, I find among them in addition to Aporoxylon, and stems of ferns, portions of the wood of a conifer, of the genus Dadoxylon.

2. Fragments of a small Stigmaria resembling my S. exigua, from New York, but distinct, which I propose to name S. pusilla. Plate II. fig. 1. There are also some fragments of leaves which may be those of Sigillaria.

3. Specimens of Lepidostrobus Richardsoni, showing that these strobiles were attached in a row on one side of a slender stem bearing branchlets with short reflexed leaves; the whole showing that this plant was a new and beautiful species of Lycopodites.

4. Another Lycopodites with long slender leaves, which I propose to name L. comosus. Plate II. fig. 2.

5. A plant having the aspect of Calamites, but referable to the genus Anarthrocanna of Goeppert; I propose to name this, A. Perriana. Plate II. fig. 3.

6. A new Cordaites; or at least a lanceolate leaf, with broad base and uneven parallel nervation, and to be placed in this genus rather than any other. It may be named C. flexuosus. Plate II. fig. 4.

7. More perfect specimens of Cyclopteris Browniana,* showing that it was a large flabellate frond of most graceful aspect. I now suspect that it may be the same with the leaf from the Upper Devonian of Pennsylvania figured but not named by Rogers—Report on Pennsylvania, Vol. 2, Part 2, Plate 22. As suggested by Prof. Balfour in relation to the Pennsylvania plant, it nearly as much resembles the leaf of a conifer like Salisburia as a fern. Plate II. fig. 5.

---

* The drawing of Cyclopteris Browniana is taken from a specimen in the collection of Prof. Rogers of Boston, which is rather more perfect than any obtained by me.

J. W. D.

8. A new fern resembling Cyclopteris Jacksoni, but having a transversely pitted stem like that of C. Roemeriana, Goeppert; I propose to name this, C. Rogersi. Plate II. figs. 6 and 7.

9. A small but apparently new Sphenopteris. Plate II. figs. 8 and 8a.

10. A Trichomanites, which I believe to be new. Plate II. fig. 9.

11. Specimens of Leptophlœum rhombicum, showing that it bifurcated and bore long narrow one-nerved leaves, and rendering it probable that its fructification consisted of elongated strobiles with narrow pointed scales.

12. A new Hymenophyllites, too imperfect for description, but possibly identical with one found at St. John.

13. Two unknown objects which may be Carpolithes.

The whole of the above are now under examination, and I hope more fully to describe them in the course of the winter.

I am, with sincere regard, truly yours,

J. W. DAWSON.

Lake Sedgwick.

At the close of the field work for the season, after the appropriation for the survey had been exhausted, with the permission of the commissioners to whom we are responsible, the geologist of the survey at his own expense visited Lake Sedgwick (or Square lake) in the north part of Aroostook county, and collected some of the interesting Helderberg fossils found there for his own pleasure. The specimens collected were placed in the hands of Mr. Billings, Paleontologist of the Canada Survey, who was so kind as to describe them, although very much pressed by other duties. It was our intention to have presented these descriptions here; but as they have already been given to the world in the Proceedings of the Portland Society of Natural History, the crowded state of this report will render it unnecessary to repeat them. The thanks of all in Maine who are interested in geology are due to Mr. Billings for his aid in deciphering these relics of the past. The locality is an unusually interesting one, and we hope it will be visited by other collectors. Its richness can be judged of by the time occupied in collecting these fourteen new species. We spent only a single day at the ledge, and brought away several barrels full of the precious remains. The locality may be reached by boat, either by way of Portage lake or Fort Kent.

We will give, however, a list of the fossils from Lake Sedgwick. Nearly half the number were discovered last year by Mr. Packard, and were noticed on pages 240, 421, of the Preliminary Report:

CATALOGUE OF FOSSILS FROM SQUARE LAKE.

1. *Favosites gothlandica*, (Lamarck.)
2. *Zaphrentis*—allied to *Z. prolifica*, but probably distinct.
3. *Diphyphyllum*—several fragments three or four lines thick.
4. *Crinoid*—with moniliform column.
5. *Crinoid*—with a smooth round column.
6. *Fenestella.*
7. *Incrusting Bryozoon* on an *Orthoceras.*
8. *Strophomena rhomboidalis*, (Wahlenberg.)
9. *Strophomena punctulifera*, (Conrad.)
10. *Strophomena indenta*, (Conrad.)
11. *Strophomena perplana*, (Conrad.)
12. *Orthis*—like *O. discus*, (Hall.)
13. *Orthis*—a larger species of nearly the same form.
14. *Streptorhynchus ?* —one valve.
15. *Rhynchonella Mainensis*, (N. sp.)
16. *Rhynchonella nucleolata*, (Hall.)
17. *Rhynchonella Aspasia*, N. sp.)
18. *Rhynchonella*—like *R. bivalveata*, (Hall,) one imperfect specimen.
19. *Rensselaeria Portlandica*, (N. sp.)
20. *Eatonia medialis*, (Hall.)
21. *Leptocoelia ?* —one imperfect specimen of a N. sp.
22. *Retzia Maria*, (N. sp.)
23. *Retzia Hippolyte*, (N. sp.)
24. *Retzia dubia*, (N. sp.)
25. *Retzia Electra*, (N. sp.)
26. *Atrypa reticularis*, (Linne.)
27. *Athyris Blancha*, (N. sp.)
28. *Athyris Harpalyce*, (N. sp.)
29. *Spirifera macropleura*, (Conrad.)
30. *Spirifera varicosta*, (Conrad,) *S. Hesione* ? (Billings.)
31. *Platyceras*, allied to *P. subangulata.*
32. *Platyceras*, " "
33. *Loseonema Fitchi ?* (Hall.)
34. *Orthoceras rigidum*, (Hall.)
35. *Dalmanites Epicrates*, (N. sp.)
36. *Phacops Trajanus*, (N. sp.)
37. *Proetus macrobius*, (N. sp.)
38. *Proetus Junius*, (N. sp.)
39. *Bronteus Pompilius*, (N. sp.)
40. *Lichas ?* A fragment supposed to belong to a species of this genus.

Mr. Billings also examined a few fossils from Masardis, Telos Lake, Stair Falls and Moosehead Lake. All except the Masardis specimens were collected during the first year of the survey. For the details we would again refer to this valuable paper in the proceedings. There are some new species among them—both Lower Helderberg and Oriskany sandstone.

F. MINERALOGICAL NOTES.

Several new localities of minerals in Maine have been either mentioned to us or explored by us since the publication in the Preliminary Report of a Catalogue of the Minerals of Maine. We give herewith a list of all these localities, with a few corrections of the original list. Our obligations in this department are hereby acknowledged to Messrs. A. E. Verrill of Cambridge, Mass., Dr. N. T. True of Bethel, and E. Lewis Sturtevant of Winthrop:

ALBANY.—Oxide of Titanium in four-sided pyramids, brown and black tourmaline.

ANDOVER.—Magnetite.

BAILEYVILLE.—*Gold.*

BARING.—Gold.

BETHEL.—Rutile in lengthened prismatic crystals.

BOWDOIN, N. W. part.—*Rose quartz,* abundant.

BUCKFIELD.—Molybdenite, molybdine, *magnetite,* alum.

CALAIS.—*Pyrites.*

CARROLL.—Manganese wad.

COLUMBIA.—Gold.

CUTLER.—*Gold.*

DANVILLE.—Black tourmaline.

DENMARK.—Quartz crystals.

EAGLE LAKE PL.—Quartz crystals.

FREEPORT.—Feldspar in crystals, *rose quartz* near Hedgehog Mt., *garnet* (portions of one crystal found weighed fifteen pounds), scapolite, apatite, calcite (nail-head spar).

GREENWOOD.—*Beryl* (large), *mispickel* (not native arsenic), cassiterite or tin ore in small crystals, magnetite, bog-iron ore, molybdenite, zircon, *albite* in crystals, pyrochlore, *mica, rose quartz, garnet, fibrolite,* copperas, *corundum,* magnesite in a vein crossing the railroad.

HEBRON.—Cassiterite or tin ore, mispickel, idocrase, *lepidolite, amblygonite, rubellite, indicolite, green tourmaline, mica, beryl, apatite, albite.*

INDIAN TOWNSHIP.—Quartz crystals.

LITCHFIELD.—Spodumene, fibrolite, mica crystals, carbonate of copper on the east side of Oak Hill in a large boulder of amethystine quartz, pyrrhotine (boulder), labradorite (in boulders).

LUBEC.—*Copper ore* at West Quoddy Head.

MACHIAS.—Gold.

MINOT.—Beryl.

MONMOUTH.—Fibrous actinolite, apatite in six-sided yellow crystals, hornblende, beryl, oxide of titanium, elacolite, zircon, staurotide, andalusite, plumose mica, jasper containing crystals of silicate of iron, copperas, chlorite.

MOSCOW.—Gold.

NORWAY.—*Beryl, rose quartz, orthoclase* in crystals, cinnamon garnet, black tourmaline.

ORLAND.—Gold.

OXFORD.—*Garnet,* beryl, apatite, wad, mica crystals.

PARIS.—Amblygonite, yttrocerite, zircon, brookite, beryl, smoky quartz.

PHILLIPS.—Galena, pyrites, copper pyrites, plumbago.

POLAND.—Smoky quartz crystals.

PORTLAND.—Calcite (nail-head spar), *prehnite* massive and in small crystals.

POWNAL.—Rose quartz, feldspar crystals, scapolite, *black tourmaline,* garnet, actinolite, molybdenite, calcite, red ochre from a spring, manganese wad, specular oxide of iron, pyroxene, apatite, hornblende, mica crystals, pyrites.

PROSPECT.—Plumbago, galena.

RAYMOND.—Idocrase, yellow garnet, calcite, anorthite(?)

ROCKLAND and THOMASTON.—*White talc,* not kerolite.

SOMERSET COUNTY, No. 4, R. 18.—Chalybite.

UNION.—*Magnetite, bog-iron ore.*

WALES.—Axinite in boulder, alum, copperas.

WASHINGTON COUNTY, No. 7.—Native copper.

WEST BATH.—Copper ore, plumbago.

WOODSTOCK.—*Prehnite,* epidote, calcite.

YARMOUTH.—Molybdenite, black tourmaline, massive hornblende, feldspar crystals, chlorite, beryl, garnet.

Mount Mica in Paris is the most interesting locality of minerals known in the State; and we take pleasure in producing an interesting account of it, with notices of the minerals found, by Mr. A. E. Verrill of Cambridge, Mass. It was published originally in the *Oxford Democrat.* It seems that the elements Cæsium and Rubidium have not yet been found anywhere in North America except in Maine; and it is but very recently that they have been known in Europe.

Mt. Mica.

The well known locality, usually called Mt. Mica, and justly celebrated for the variety and beauty of the rare minerals it has offered, is situated on a low hill which forms one of that group of which Streaked Mountain is the crowning summit; or it may even be considered as a part of the northern flank of that mountain.

The excavation from which the minerals have been taken, is on the estate of Mr. Bowker, about two miles from Paris Hill, on the Buckfield road. It has been known for about forty years, and formerly afforded some minerals not now found, as well as finer specimens of the red and green tourmalines, than have been found for several years past. This is, as I believe, not because the locality has been exhausted, but the excavation has not been made in the proper direction to follow the centre of the vein. Yet within a few years past, there have been found here several interesting minerals, not known before. This mineral bearing vein consists chiefly of feldspar and albite, with mica and the various other minerals scattered through it in masses and crystals, passing through a coarse granite; and is of considerable width and of unknown extent, for it is concealed beneath the soil in either direction. It is worthy of remark that in Hebron, at a distance of about seven miles, there is another vein of similar character, and containing precisely the same minerals as this one, so that the same description will answer for both. The principal minerals that have been found here are the following.

Tin Ore. This ore, also called *Cassiterite,* was first noticed as a Paris mineral at a meeting of the Boston Society of Natural History, December 5, 1860, when I exhibited a fine specimen of the mineral, and gave a brief account of its mode of occurrence. This specimen I found in 1854, and it originally weighed at least five pounds. It was nearly pure, partly crystaline and partly massive, and occurred imbedded in albite. This specimen attracted considerable attention at the time, from the geologists and mineralogists present, on account of its size and purity. Since first finding this mineral I have found other smaller specimens, nearly every time that I have visited the locality; and a very beautiful specimen, imbedded in quartz, has also been found by Mr. S. R. Carter of Paris Hill. The specimens, as yet, have been found scattered through the vein, but it is possible that farther exploration might reveal larger quantities of it near the centre of the vein, which is now buried under the rubbish. This mineral has also been found in small quantities at the Hebron locality. The ore contains 78 per cent. of metallic tin, and if found in considerable quantity, would be very valuable; its color is dark brown or black, often with a rusty look, but when crystalized the sides of the crystals have a bright, but not metallic, lustre. It is most readily distinguished by its weight, its gravity being nearly as great as that of metallic iron. There is another locality of tin ore in Greenwood.

Amblygonite. This very rare and interesting mineral, not found before in America, has been discovered during the past summer, imbedded in the lepidolite, from both Paris and Hebron. Prof. G. J. Brush of Yale College, has given an account of its discovery and examination, in the September number of the American Journal of Science and Arts. This mineral is remarkable for containing, like lepidolite, a considerable amount of the rare metal Lithium, and probably, also, the newly discovered metals, Caesium and Rubidium. It is a translucent, feldspathic looking mineral, of a white or grayish color, cleaving perfectly in two directions, giving surfaces which have a bright vitreous lustre; and fuses readily in the flame of a candle, coloring the light with a beautiful crimson, owing to the lithia that it contains. It occurs in irregular masses, in the lepidolite.

Yttrocerite. This very singular and rare mineral has not before been noticed from Mt. Mica, but I have found a few good specimens. It is remarkable for containing the rare metals, Yttrium and Cerium, and has previously been found, in very small quantities only, in one or two localities in America. This occurs in small irregular masses, imbedded in Albite. Its color is dull violet, with a somewhat vitreous lustre; the hardness is less than that of feldspar, it being readily scratched by a knife.

Lepidolite. This beautiful mineral, which is generally considered rare, occurs abundantly at Mt. Mica, in masses of various sizes, some having been obtained weighing upwards of 100 pounds. These masses consist of an aggregation of small pearly scales, varying in color from pale rose to bright purple, giving to the mineral a very elegant appearance on the surfaces of fracture. Like *Amblygonite,* previously mentioned, it is interesting on account of containing a considerable amount of the metal Lithium, and, as has been recently demonstrated by Mr. O. D. Allen, in a very interesting paper published in Silliman's Journal for November, an unusually large amount of the newly discovered and rare metals, Rubidium and Cæsium. According to Mr. Allen, the proportion of Rubidium amounts to about 0.2 per cent., and of Cæsium to 0.3 per cent., which is a greater proportion than is contained in most other substances in which they have been found. The specimen analyzed was from the Hebron locality, but from the very close resemblance of all the minerals, there is no reason to suppose that there will be much difference in the composition of specimens from Mt. Mica.

Petalite. This mineral was formerly obtained here, in large and fine specimens, but none have been found, to my knowledge, for several years. Like Lepidolite, this also contains Lithium. The specimens that I have seen are small masses weighing one or two ounces; the color is gray or dull reddish with a glassy lustre, which becomes somewhat pearly on the cleavage surfaces. This is considered a rare mineral; in this country very few localities are known.

Tourmaline. Mt. Mica is perhaps more widely known as a mineral locality, by the rare and beautiful specimens of red and green tourmalines that have been obtained here, than on any other account. Some that were formerly found were an inch in diameter and transparent, approaching the ruby in beauty and value, but for several years none have been procured except more ordinary semi-transparent and brittle crystals of little value except as cabinet specimens. These are, however, of a great variety of colors and forms, and of considerable interest to mineralogists. The most common variety, here as well as elsewhere, is black, opaque and with but little lustre; this passes gradually into blue-black and blue varieties, which have been called *Indicolite.* These are generally opaque or nearly so, and are seldom found in well formed crystals. The green varieties, varying from very pale to deep grass green, are very abundant at this place, and though not often found in perfect crystals, are usually bright and nearly transparent. They are generally imbedded in mica or quartz, and frequently have a radiated structure consisting of thin, flat, or needle like crystals, diverging from a common centre. Sometimes crystals are green externally and red within, or the reverse, and sometimes one end is green while the other is red or blue. The red varieties are the most rare, and perhaps the most interesting. They occur of various shades, from pale pink to ruby red and deep crimson, and are generally transparent, and occasionally, are found in perfect crystals, when they are very beautiful. This variety has been called *Rubellite* from its color. In addition to these colors, pure white specimens are not uncommon, as well as various shades of brown, but these are not usually considered of much interest by collectors.

Beryl. Associated with the large crystals of black tourmaline and sometimes imbedded in them, very good specimens of beryl are sometimes met with. These are hexagonal prisms, generally somewhat irregular in form, and not often with the ends perfect, of a light green color, translucent, and vitreous in lustre. The hardness is greater than that of quartz.

Mica. Large quantities of this mineral, from which the locality has taken its name, may always be obtained. It occurs in imperfectly crystalized masses or sheets, often a foot or more in diameter, but seldom of sufficient purity or transparency to be of commercial value like that from Acworth and Grafton, N. H., which is used for the doors of stoves, lanterns, etc., and of late, to a considerable extent, for delicate photographic plates. The name *Muscovite* is also applied to this mineral.

Feldspar. The principal part of the vein in which all the minerals previously mentioned are found, consists of a grayish variety of feldspar, occurring in imperfectly crystalized masses, of but little interest or value, unless it be regarded in connection with the origin and nature of the vein and the relations of the different minerals, one to another, which are questions that cannot, with propriety, be discussed here.

Albite. The variety of albite which has been called by some mineralogists, *Cleavelandite,* occurs quite abundantly at Mt. Mica. It is generally in the form of masses consisting of flat flakes of about an inch in diameter, united together in various ways, and generally rather brittle or friable. The color is usually gray or yellowish, sometimes stained with brown; the lustre on the surfaces of the plates is pearly.

Quartz. Various varieties of quartz are met with here. The common transparent, crystalized form called *Rock crystal,* is often found filling cavities, with fine needle like crystals, pointing inward towards the centre from all sides. *Smoky quartz* is occasionally met with in good specimens, both massive and in perfect crystals, which are generally transparent or nearly so; the color is usually dark smoky brown. *Rose quartz* is found in masses of a light pink color, but not often transparent or sufficiently free from cracks to be of much beauty. Much better specimens of this variety have been obtained at another locality about a mile farther east. At this place a large mass was once obtained, as I have been told, for the purpose of making a large mortar for grinding the materials used in the manufacture of artificial teeth.

Blende. This ore, which consists of sulphur and zinc, has been found in small quantities. The only specimens that I have seen, were small masses or imperfect crystals, of a yellowish brown color.

Apatite. A massive variety of phosphate of lime, is not uncommon here. It is of a dull light green color, and rather soft.

Brookite. Small dark brown crystals having a brilliant lustre, are often found imbedded in albite and other minerals from this locality, which appear to be brookite or oxide of titanium, but I have not been able to procure a sufficient amount of the mineral to ascertain its nature with certainty.

Zircon or *Hyacinth.* Beautiful crystals of this rare mineral, well known as a gem, have been detected during the past year imbedded in the albite. They are square octahedrons of small size, with a brilliant lustre, semi-transparent, and bright red or brownish in color.

G. ECONOMICAL GEOLOGY.

A brief description was given in the first report upon the Geology of Maine, of all the mineral substances found within its limits that are of value in the Arts. That account will now be continued in the form of a sequel. We will notice the most important economical discoveries that have fallen under our notice during the second year of the Survey. We will speak of Iron, Gold, Tin, Antimony, Lead, Copper, Marble, Limestones, Gypsum, materials for the manufacture of Grindstones, Roofing Slate and Water Lime.

We have nothing of great consequence to add respecting the value and distribution of the Manganese, Arsenic, Zinc, Granite, materials for the manufacture of Glass, Flagging Stones and Clays, there described, although much information respecting their quantity remains to be acquired.

Iron.

Last year a very important ore of iron was described in Wade plantation, in Aroostook county. Late in the season one of us visited the locality, and immediately afterwards wrote the following letter to the Commissioners to whom we are responsible for the conduct of the Survey. The opinions expressed in it we still entertain.

Ashland, Me., Sept. 13, 1862.

To His Excellency Israel Washburn, Jr.,
and S. L. Goodale, *Esq.*

Honored Sirs:—According to the contract between myself as State Geologist, and the authorities, I am bound to make known to the State, discoveries of any valuable substances upon the public lands.

In accordance therewith, I am about to speak of the valuable deposit of iron ore in No. 13, R. 4, of Aroostook county, or "Wade plantation" of the maps. This deposit is in the south-east part of the township, upon the land occupied by Daniel Hickey. In my report of last year, it was noticed in two places—first on page 295, where Dr. Jackson's authority was quoted; and secondly, on page 435, in Assistant Packard's report. For details of position, amount, &c., I would respectfully refer you to those two places in my report.

The importance attached to this locality arises from the quality of the iron produced from it. For ordinary purposes I would not urge the matter upon your special attention; but at the present time, I regard the development of this iron ore a matter of national importance; and the results of its devolpment may form an era in our naval warfare, second only to the production of our iron gunboats. The ground of this startling assertion is the fact that the iron with which our national gunboats are built, will not withstand the force of improved ordnance; but such plates as can be manufactured from this ore, have stood every test that has been applied to them.

According to private experiments instituted by the English government, iron plates manufactured from scrap iron, (the same of which all our gunboats are constructed,) as well as from a variety of ores apparently the most unyielding, were shattered by a 230 lb. shot from an Armstrong gun. The plates which were manufactured from iron ore smelted in Woodstock, N. B., alone were left entire—the projectile having simply indented the plates in a slight degree. The trials were made six several times, with the same results. In consequence of these experiments, the British government uses chiefly the Woodstock iron for the manufacture of the plates—an ore which is obtained only five miles from Maine.

The Woodstock ore is a compact red hematite, rarely containing over 30 per cent. of metallic iron, but it differs in its general appearance from any ore I have seen west of Maine, and is easily recognized.

As respects geological position, the ores from New Brunswick and Maine are similarly situated; nor can there be any doubt that the ore from Maine will yield the same quality of iron as that from New Brunswick. So far as the situation is concerned, the ore in "Wade plantation" is admirably located for mining and smelting. The only misfortune attending its location is its great distance inland—it being ten miles west from Presque Isle; and to those so familiar as yourselves with the routes of communication in Aroostook county, and its connections with the seaboard, I need say nothing. Only the pig-iron, however, needs to be transported, or the iron in a state of readiness to be put into the furnace and rolled.

With these facts before me, I could no longer hesitate to believe that steps should be taken at once to erect the proper works for the smelting of this new ore, and for the manufacture of all the plates which are to be used in future in the construction of our gunboats.

Pardon me if my zeal for the integrity of our naval honor, or the desire to see the resources of the State developed, has led me to trespass upon your time. I could but think of the great confidence of our people in our gunboats, and the consternation which must ensue should our "Monitors" ever be pitted against such a craft as the "Black Prince" of England, a vessel plated with Woodstock iron, especially as we have in our hands the means of constructing more formidable boats than the "Black Prince"—more formidable, because with the same kind of iron, we have a better model.

Should the facts stated above appear worthy of consideration and inquiry, to your minds, we suppose the responsibility of further action will rest with the Government as to the mode of carrying on the manufacture of the iron, and of communicating thereon to the United States Government.

I need only say further, that all the members of the scientific corps of the State, will do all that is in their power to assist in the development of this ore, and that we always hold ourselves in readiness to perform any duty required of us by the State whether scientific or otherwise.

With great respect,

Your obedient servant,

C. H. HITCHCOCK,

State Geologist.

We requested Mr. George L. Goodale to report as fully as possible upon the iron ore from Wade plantation and upon some other ores recently discovered, particularly in Union. He has done so in a very satisfactory manner, and herewith we present his report. We feel much more sanguine respecting the prospects of the Union ore than his caution would allow him to express:

CHAS. H. HITCHCOCK, *M. A.*,

Geologist to the Scientific Survey of Maine.

SIR:—The following report upon the economical value of certain ores of iron, in this State, has been prepared at your request. Your careful study of the geological relations of these deposits has left me little to say in regard to their occurrence, and therefore my work is limited to the plain presentation of such facts as bear directly upon their importance to miners and dealers in iron.

It is not proposed to present accounts of all the ores of iron known to occur in Maine, but merely to notice those which promise to be of value as they are more fully explored and developed. The list is, consequently, small; including the following:

Limonite, at Wade plantation, near Houlton, and at Linneus—certain varieties of this species, at Katahdin Iron Works—Magnetite, at Union—Pyrites, at Jewell's island.

This list, in which has been placed the pyrites of Jewell's island, on account of its value in the manufacture of sulphuric acid and alum, of course excludes all the minor deposits, like those at Newfield and Shapleigh, Hodgdon and Clinton.

1. The ore found at Wade plantation, township 13 in range 4, is mainly limonite, with a lower per cent. of water than usual, occurring in such quantity as to deserve considerable attention. This bed was first described by Dr. C. T. Jackson, in 1837, as being "an ore of red hematite." His analysis, which I give below, indicates that under this name he included what we now call *limonite*, or hydrous per-oxide of iron. Limonite is often known by the name brown hematite.

The results of Dr. Jackson's were these:

"In one hundred grains the ore contains as follows—

Water,	6.00
Insoluble residue consisting of silex,	8.80
Peroxide of iron,	76.80
Oxide of manganese,	8.20
	99.80
Loss,	.20
	100.00

76.8 of peroxide of iron contains 53 of pure iron."

My analyses indicated that, in the specimen given to me, the quantity of water was somewhat greater than that found by Dr. Jackson. This and the fact that the amount of peroxide of iron detected in my analyses was smaller than that recorded above, lead me to entertain the opinion that the quality of the ore is, by no means, uniform. This ore is nearly identical with the one which is now extensively worked in Woodstock, N. B., and was considered by those who first examined the two beds, to be exactly similar.

Although it is my belief that the Woodstock ore yields no more pig iron than that in Wade plantation would produce with equal facility, I cannot consider that the two are precisely the same, mineralogically. The ore occurring at the former locality is more compact than the fissile rock found in Aroostook county. When the two specimens are seen on their edges they appear quite nearly alike, although the cleavage gives to one a slaty appearance.

It is now necessary to present some facts in regard to the facilities for reducing the Aroostook iron and bringing it to a market, prefacing what I have to say, by the remark that hematitic ores are most easily and advantageously reduced by charcoal. The pig iron thus obtained is changed into steel with great facility, and is readily affected by the Bessemer process, so called. This will be noticed further on.

In the first place, the ore is so bedded that it will afford natural drainage for at least 90 feet. This desideratum is often overlooked by those who enter hastily, and without forethought, upon the management of mines.

2dly. The supply of hard wood for charcoal is certainly sufficient to warrant one in saying that its cost would be trifling. Maple and beech wood are extensively used in charcoal-iron districts, and are here found in abundance.

3dly. Limestone is found quite near the deposit of iron, and would serve well as a flux in reduction.

4thly. Distance from navigable water by which the pig-iron can be brought to market. Here occurs the principal difficulty in the profitable management of mining operations in this vicinity. The whole distance, by the windings of the Aroostook river to the river St. John, can be estimated at 45 miles or thereabouts. Much of this way is obstructed by falls, and is rendered entirely impassable to rafts a portion of the year. This is a serious difficulty, but may be in a measure obviated by confining the transportation of the metal to such months as would allow of the safe and rapid conveyance of the iron down river by rafts. The portages at the falls would be comparatively trivial.

When the rafts carrying the metal had once reached the St. John, the iron could easily find its way, by water, to our eastern ports.

If it could be shown that ore similar to this was now being used to produce metal for purposes which demand this quality of ore and *no other*, it is plain that the demand would overcome the few obstacles which stand in the way of rendering the deposit accessible to the manufacturers. To be sure, it will be objected that a false value is thus given to a mine and its products, but it is obvious that what manufacturers need they will have even at a higher price. The difference between the cost of reducing metal from such an ore and that of obtaining iron from other ores, would be gladly paid by those who desired that particular variety of iron for a specific purpose which authorized the extra outlay. This is the case in the present instance. The Woodstock works which were commenced, upon a scale perhaps too large, went into operation a number of years ago. The quality of iron was deemed excellent and the products of the mine were readily disposed of in the Province of New Brunswick. Some unforeseen occurrence led to the abandonment of the furnace and mines by the first company, and they were at last obtained by the present firm which has the corporate name, "The Woodstock Charcoal Iron Company."

To one of the obliging partners of this firm, I am indebted for the following facts which he communicated to the Geologist of the Survey, in my presence:

The English Admiralty instituted experiments at Shoeburyness, England, in order to test the resistance which iron plates would offer to the heavy ordnance of Sir Wm. Armstrong. In that trial every plate was shattered except a triple plate made of Woodstock iron. This plate was indented by the shot but not pierced, and immediately attracted considerable attention. The fine results obtained by the Woodstock plate determined the use of the iron in mail-plating the ships in the English navy. An interesting account of the experiments testing these plates can be found in "The Artizan." The paper was prepared by Wm. Fairbairn, Esq., F. R. S., and gives us the following results—

Tensile strength in tons per square inch, . . 24.80

Scrap-iron plates were readily shattered by the shot.

For this purpose the Woodstock works are now busily engaged. The ore is brought two miles to the furnaces, and is now being rapidly reduced. Charcoal made on the company's grounds is employed in reduction. The limestone is brought a distance of a few miles.

The furnace is a blast of old style and will soon be replaced by one having greater capacity and being much more economical. The fan-engine is also old-fashioned. The boiler is ingeniously heated by escape-air from the furnace itself.

The following are the proportions of iron-ore, fuel and lime:

3 barrows of ore, 450 pounds each, 1,350 pounds.
20 bushels charcoal.
70 pounds limestone.

The metal is cast into pigs of 90 pounds each.

The ore is obtained from two localities, in one of which a portion of the mineral has a bright red streak, and goes by the name of red hematite at the mine. It is undoubtedly, compact red hematite. In another locality the ore is plainly hydrous peroxide of iron. 200 pounds of one of these ores is usually reduced in a charge with 1,150 of the other.

The metal which is reduced from the ore is a fibrous silver-grey iron which has a thready fracture indicating great tenacity.

From what has now been said, it is plain to see that, relying upon the statements of the proprietors of the Woodstock mines, this ore is now being used extensively in the English navy. It is employed because it is their best iron.

This description has been given in order that it may be understood that in Maine we have an ore which will yield an iron equal in every respect to that which is sent to Liverpool from the river St. John.

It is not for me to decide whether private individuals should embark in the enterprise of developing the resources of the mine on the Aroostook, because it is my opinion that mining operations in New England should be very cautiously undertaken. The value of a mine does not wholly depend upon the per cent. of metal which it will yield, but also, largely, perhaps principally, upon the cost of *labor*, *reduction* and *transportation.*

But it certainly appears reasonable that a matter of such importance as this, of plating our ships of war with metal as good or better than those of our now friendly neighbors, (may we long be friendly,) should receive the careful attention of our government. It would be well to learn, before it is too late, whether our plates made of scrap iron can be shattered as the scrap-iron plates were broken at Shoeburyness.

A variety usually referred to limonite (Beudant) occurs in the immediate vicinity of Houlton. It appears to be, by no means, uniform in quality. The geologist of the survey has already referred to this in his report, and it only remains for me to say that the ore contains too much manganese and too little iron to be of much use in the economical manufacture of the latter metal. A small proportion of oxide of manganese is not considered injurious to the reduced metal or to the steel manufactured from an ore containtaining it.

But it is my opinion that a diligent metallurgical search under proper supervision, and with facilities greater than those placed in the hands of members of the survey corps, will be able to discover on our side of the boundary, near Houlton, a bed of ore continuous with that in Woodstock. The discovery of a bed of ore as good as that in Woodstock, or in Wade plantation, would have such an important bearing upon the growth of eastern Aroostook county as to make this a matter of much importance.

That the grounds upon which the opinion is based may be fully understood, the following considerations are presented:

1st. The general direction of the rocks in and around the Woodstock mines would cause us to search for the deposit of ore, if continuous, in the immediate vicinity of Houlton, and not far from the manganesian ore referred to.

2d. The same ore has been seen, by the superintendent of the Woodstock works, in Richmond, a township lying between Houlton and the mines.

3d. The occurrence of the manganesian variety of hematite in nearly the line of strike.

It is for the above reasons that the owners of the fields in which the poorer ores are found, are advised to examine their portion of the town very carefully. The large bed of ore described by Dr. Holmes in Linneus is undoubtedly similar to that in Woodstock; though it is somewhat magnetic.

Bog-iron ore, which is an hydrous per-oxide of iron, is found in many parts of our State, and, in some localities, in quantities which warrant active mining operations.

The most important one is at a point south-west of the Ebeeme mountains, called Katahdin Iron Works. The whole territory belonging to the company, which erected furnaces, has been specially explored by Dr. Jackson of Boston. He was engaged by the company, I have been informed, to make a thorough survey of the township. As a result of this examination, furnaces were soon at work, and excellent iron was produced. The quality and cheapness of the iron produced, and not the percentage of metal detected in the laboratory is the true test of the value of a mine. In this case, the reduced metal was carried by mules to Bangor, and of course the distance of transportation caused the profits to be materially lessened. The property is now in litigation and the work is entirely suspended.

An excellent account of the situation and extent of these works was published by John C. Houghton, B. A., in the Report of the Scientific Survey for 1861.

2. *Magnetite.* Specimens of a superior ore from Union were placed in my hands, in June last, for analysis. The ore is one of remarkable purity, yielding according to a gentleman who had had the ore analyzed by a Boston chemist, a percentage of pure iron as large as 70. I was not able to obtain as large a proportion of iron as this, my highest result being 64 of iron. But this is a percentage so large as to warrant the erection of a furnace, *provided*,

1st, There is enough of the ore to keep the furnace well supplied.

2d, Fuel can be cheaply provided.

3d, The metal can be easily transported to tide water.

Perhaps I may be blamed for thus having insisted strongly upon the many elements which must enter into the question of the practicability of erection of furnaces. But there have been so many lamentable failures in New England mining because these points were not appreciated, that I feel justified in keeping them plainly before those interested in mining operations. And the more so in a report upon the economical value of iron ores, a metal which the coal regions of our Middle States furnishes so cheaply. To refer again to the iron at Union, let me observe, that there is abundance of lime, to be used as a flux, very near the bed ; that there is a possibility that the old canal can be reopened for the transportation, and that the ore is of a superior quality.

3. *Pyrites.* Of late years English manufacturers have employed sulphur prepared from Iron Pyrites, which is a bi-sulphide of iron, in preference to crude commercial sulphur. The method of obtaining sulphur from the pyrites was described in the chemical report for 1861. Since writing that report, I have visited Jewell's island in Casco bay, where there is a large deposit of iron pyrites, where an alum factory was erected in 1836 and afterwards abandoned. After a thorough examination of the locality, assisted much by the amiable and intelligent owner of the island, Capt. Chase, I was convinced that the alum and copperas works were erected in a part

of the island poorly adapted to obtain the best material for manufacturing. The best deposit is at the other end of the island and appears sufficiently rich in pyrites to authorize the establishment of sulphur works. The outlay would be comparatively slight, and under present circumstances would yield a fair return.

This concludes what I have to say concerning the economical value of the iron ores of Maine. Compelling myself to write plainly, fairly and briefly, the report may appear to present many discouragements to those who may have intended to embark in iron mining in this State. But the value of the work of a survey is not enhanced by giving exaggerated estimates of the mineral wealth of a State. If I have succeeded in advising that speculators and proprietors use more caution than heretofore in commencing mining operations in any place where a few handfuls of metal are found, I shall be truly gratified. Knowing that you entertain the same opinions as these which come from the Laboratory, I have, sir, presented the matter with the more freedom.

In conclusion, let me offer the thanks of the Survey to Dr. H. T. Cummings of Portland, and to Prof. P. A. Chadbourne of Brunswick, for their many kind attentions and the liberal use of their laboratories.

With high regard,

I am, sir,

Yours respectfully,

GEO. L. GOODALE.

Gold.

Gold has been for a long time known to exist in the valley of Sandy river. An examination of the country last September shows that the rocks there contain the peculiar veins of quartz in which metallic gold is disseminated. We could not feel, however, that the precious element is very abundant in this portion of the State.

The opinion expressed concerning the auriferous character of the country upon the upper river St. John, is still entertained after another inspection of the country. The limits of this region have been enlarged by our observations. It must extend nearly to Moosehead lake on the Penobscot, and perhaps along the dividing ridge between Maine and Canada, even to the New Hampshire corner. Where the Canada road crosses this belt in Sandy bay, the veins look exceedingly promising, and the banks of the streams are full of the "black sand" so generally accompanying gold. The veins are unusually abundant here. It is only a short distance over the line to where gold is washed out of the Chaudiere river, and the rocks are continuous across the line.

Newspaper reports have expatiated largely upon the gold of Orland, east of S. B. Swasey & Co.'s mills. This spot we visited, and were not favorably impressed by the indications. The rock is granite, not containing many quartz veins, but the bed of the brook contains many boulders from an auriferous region. Our opinion is that gold occurs here, but that it has been derived exclusively from the transported materials. The question to be asked next, is where were these auriferous materials broken off? We must examine the tables of drift striae in the vicinity to learn. Upon reference we find that the boulders must have come from the north and west of north. For many miles in that direction the rock is schistose with some veins of quartz. It is the great mica schist belt of central Maine, extending from the eastern Schoodic lakes to Portland. If this is auriferous, then gold may yet be found exceedingly plentiful in the State. And we would recommend to any persons living in this district who may wish to find gold, to search for quartz veins, and then to test the value both of the veins and of the soil near the ledges. In Orland there is an abundance of very bright yellow scales of mica in the stream, which an unpractised eye would certainly mistake for the precious metal.

The finest auriferous belt brought to light the past season crosses the St. Croix river above Calais. The rock is a mica schist full of quartz veins and beds. An examination of these veins near the railroad bridge in Baileyville showed us several pieces of bright flake gold. The best locality is on the west side of the river upon some ledges through which a passage for the railway has been excavated. There is a considerable pyrites in the schist, so much so that the action of the air decomposes it and gives the whole ledge a rusty appearance. After our departure a mining company—we believe the same that works at Lubec—effected a lease of the property, and have sunk a shaft, for the purpose of experimenting upon the value of the quartz. Alluvial washings on the river have not promised so well. Nor does the rock at the railroad bridge (Sprague's falls) afford as many quartz veins.

Across the river in New Brunswick, upon land of Mr. Bolton of St. Stephens, is another locality where gold has been found. Its distance from Sprague's falls cannot be very great, as it is about nine miles north-west from the Calais bridge. The exact locality of the gold is in a plumbaginous slate, very black and greasy. Near it is a large boss of quartz, with sub veins of quartz running through it; and there are also near by veins of quartz containing pyrites. These two localities are the most promising of anything seen by us in the St. Croix country.

Mr. Esty of Calais, who is interested in the gold mines of Nova Scotia, showed us a large mass of quartz on Bog brook in Hardscrabble in Calais, where he had found a few specimens of gold. The rock is syenite and the boss of quartz is largely of a carnelian character, not good enough, however, to be dug out for an ornamental stone.

An excursion to the east part of Tallmadge revealed the existence of a great number of quartz veins in the schist. Whether they contain gold in very minute quantities, such as would be developed by a crusher, we could not determine. We found no specks in them visible to the naked eye.

We were informed that in Cutler there are auriferous veins; and that Mr. Steadman of that place sent a ton of the rock to be crushed, and the yield was one hundred dollars. Quartz that pays less than twenty dollars to the ton is not worth crushing; but every dollar above twenty is a net profit. In Columbia, also, according to the papers, gold has been found and we have no reason to doubt the truth of the statement; as all these localities are in the same schistose rock that enters the State from New Brunswick on the St. Croix river.

From the accounts received concerning the gold rocks of Nova Scotia, we have every reason to believe that this new gold field in Maine is very similar to the Nova Scotian one. Both are somewhat different from those in California and along the Apallachian region of the United States and Canada. The great peculiarity of the Nova Scotia gold consists in its dissemination through the quartz in such fine particles that it is rarely visible. A ton of Nova Scotia quartz, in which not a particle of gold can be seen, will yield richly to the crusher and amalgamator. If a few preliminary experiments upon the St. Croix quartz yield good results, then it will be for the interest of the proprietors to erect works for crushing and amalgamating upon the St. Croix river in Calais or Baring.

A mass of pyrites has been handed to us for examination from

Machias. It appears auriferous, but we have not been able yet to have it tested. Another large mass of beautiful pyrites occurs in Calais, upon the river, which we would recommend to those interested to have assayed for gold. Oftentimes the auriferous pyrites is more valuable than the quartz.

Tin.

No additional discoveries of tin ore have been made the past season. A visit was made to the vein upon Mt. Mica, but no more masses of the ore have been found. Small crystals of the oxide sometimes are picked up. We suggested the examination of the gangue of the vein for tin. Mr. Goodale has assayed some of the specimens, in which he found a very small per cent. of tin, but not enough to pay for working. Quite a large excavation has been made in the rock, but this has been done entirely by mineralogists, who find here many beautiful minerals.

The agent of a mining company has since visited the locality, and has leased the property for ninety-nine years. Doubtless it will now be thoroughly explored, and the value of the gangue for ore soon be made known to the public.

Dr. Holmes informs me that he obtained a piece of what he now supposes to be tin ore from Paris, forty years ago, when the locality was first discovered. The catalogue of minerals will show one or two other localities of tin ore in Oxford county, which we hope will also be carefully examined by all who are interested in them.

Antimony.

A very remarkable vein of the sulphuret of antimony has recently been opened in Prince William, New Brunswick. It is not a great distance from the Maine boundary; and the rock is the same mica schist which contains the gold. Hence we should not be surprised to learn of the discovery of similar veins in Maine.

But we learn from Surgeon General Hamlin, U. S. A., that in the eastern part of the State there is a valuable bed of this ore known to him. We hope its quality and quantity may soon be made known to the public.

Lead.

A new lead mine has been opened the past season (1862) on Denbo Point, Lubec. We give here a letter relating to it from

Prof. Forrest Shepherd, whose skill and energy in developing metallic veins is so well known to the public, and who has of late been much interested in mining lands in Maine:

EASTPORT, November 14, 1862.

Prof. C. H. Hitchcock:

DEAR SIR:—The late discoveries of sulphuret of lead by the "Maine Mining and Manufacturing Company" at what is known as the Old Comstock, or Lubec Lead Mine, are truly wonderful—far surpassing the most sanguine expectations and calculations of the highly esteemed and much lamented Prof. Manross. The new veins discovered by, and the former ones more judiciously opened by General J. N. Palmer, expose to view on the face of the cliff, I may safely say, hundreds of tons of galena in vertical veins with very little foreign admixture. Collectively they will probably quite equal if not surpass the extraordinary mine recently opened eighty miles from New York, on the New York and Erie Railroad, near Port Jervis. Gen. P. has introduced very simple machinery for cleaning the ore effectually, so that he can smelt it with the greatest ease. A steam engine is now being erected at Denbo Point, four miles from the Lubec Mine upon a vein which I am informed promises a rich return of Silver Lead.

P. S. Gen. Palmer found by trial that the machinery adopted by Mr. Collum at the Lead Mine would not answer, and therefore rejected it and substituted the simple jig, Dolly tub and buddle, for the ore after it had passed twice through the rollers, having first passed through Blake's crusher.

CALAIS, Nov. 14, P. M.—I have arrived here and am at once presented with a surface specimen of surface gold taken, or said to have been taken from a gravel bed in Baileyville on the railroad. It is embedded in quartz having slate almost black, like that in Nova Scotia, on one side. I am inclined to believe it genuine, and I have engaged a miner to open one or more of the quartz veins for a trial.

I have recently found additional specimens of Ox. Tin at Mt. Mica, Paris, and secured said mount and adjacent grounds for further exploration.

Believe me very sincerely yours,

FORREST SHEPHERD.

Copper.

Several valuable veins of copper ore have recently been discovered at West Quoddy Head, in Lubec, by Prof. Shepherd. We insert his letter respecting them. The south part of Washington county seems to be a very rich metalliferous region, and deserves further exploration:

EASTPORT, July 22, 1862.

It is with great pleasure that I have recently observed six well defined spar veins from one to two feet in diameter, four of which show copper ore on the surface. These veins are situated on the land of Benj. Fowler, Esq., at West Quoddy, near the Carrying Place.

One vein appears rich in the yellow sulphuret, and another yields specimens of the purple or horse-flesh ore, and all give promise of future mineral wealth. Still another vein on the same property, although somewhat subdivided on the surface, yet carries yellow copper ore, accompanied with magnetic iron pyrites. Were this property on Lake Superior it would probably be taken in hand at once.

FORREST SHEPHERD.

Upon page 307 of the Preliminary Report, mention was made of the discovery of native copper in No. 7, near Carroll. The specimens were shown us subsequently, and are the genuine mineral. We made an effort to find the vein, in company with Mr. Levi Bailey, who procured them at first. He was unable to find the exact spot. The rock is a coarse granite, and is not promising for such ores. The true locality must be further south-west. The rocks on Nickatou's lake have been described to us in such a way as to make us anxious to explore them for copper.

We have already spoken of the copper mine in Woodstock, and of the peculiar syenitic rock containing the veins. The occurrence of a similar rock on the Maine side of the line should be examined carefully for metalliferous veins.

The probability of finding copper ores in the north-west part of the State, on the river St. John, is confirmed by the wonderful extent and richness of the copper ores in Lower Canada in the same formation. It is but recently that the Canadian rocks were supposed to be as destitute of copper ores as the corresponding rocks in Maine are now esteemed. The region in Maine being still a wilderness, the prospect of finding valuable ores for some time yet, is lessened.

Marble and Limestones.

The marbles to be found in Maine will occur chiefly upon the belt of Helderberg limestone running from Matagamon (East Branch Penobscot) river north-easterly. All that we have learned additional respecting them the past year, is given in Dr. Holmes' report. An excellent way to ascertain the limits of this formation would be to send a party on foot through the woods to examine and describe every outcrop. Without doubt the statuary marble variety would be discovered in this way. Valuable limestones for the manufacture of lime would, at all events, thus be discovered.

Other localities of good limestones have been visited the past season. That at Carroll surpassed anticipation; and similar beds can be found in the vicinity, and in adjoining towns. It would be a great desideratum to find limestone near Moosehead lake.

Dr. Holmes thinks there is a good water lime in the south part of Aroostook county. The specimens have not yet been tested.

In our map of the country bordering upon Penobscot bay, the distribution of the limestone bands is given more correctly than ever before. And in the accompanying text, descriptions are given of several beds not known at the time of writing our first report.

It is stated in an early portion of the present Report that new openings of good limestone could probably be opened in Thomaston. Those considerations we venture again to call attention to. To be sure the quantity of surface now quarried is so great that it will take long to exhaust it; still it is well to provide for the future, especially wherever it is possible for land owners to realize something by the enhanced value of their property.

Mr. Robinson of Thomaston, has discovered a few new outcrops of limestone, since the printing of that portion of the report relating to limestones. He writes as follows respecting them: "The localities that I mentioned in my letter, are in the range between the Cochrane quarry in South Thomaston and the most southern appearance of the limestone on the bay at Thomaston. This latter locality and the intermediate openings, all agree in the character of the limestone with the Cochrane quarry."

Gypsum.

Upon page 418 of the first report, it is suggested that perhaps a rock containing gypsum enters Maine from New Brunswick. Certain considerations make this suggestion still stronger. 1. The wide-spread distribution of conglomerate boulders over the eastern part of the State—already alluded to. 2. The discovery by Prof. Shepherd of a poor quality of gypsum between Grand and Big western Schoodic lakes. 3. The discovery of this conglomerate rock in ledges on the route of the proposed turnpike between Princeton and Milford west of Wawbawsoos lake, by W. W. Sawyer of Calais, as communicated to us privately. And it is Mr. Sawyer's belief that he has found the gypsum itself in connection with the conglomerate. These considerations are sufficient to make an examination of this region with reference to this object very desirable.

We have not taken the pains to ascertain whether gypsum or plaster of Paris could be obtained more cheaply from this new locality, should our anticipations be realized, than from New Brunswick and Nova Scotia by water, but cannot doubt that its discovery in Maine would stimulate its use very much by farmers, especially upon those farms which are now suffering for the want of it.

Grindstones.

I am informed that hornstone or flint rock makes a most excellent material for grindstones. If so, Maine need never again go outside of her limits for these essential articles, for in every portion of the State there are mountains of it. We have spoken of this rock often under the name of metamorphic slate and silicious slate. The most prolific localities are in Oxford and York counties, Moosehead lake, Portage lake, and the southern part of Washington county. In fact no rock is so uniformly distributed over the State as this.

Roofing Slate.

Dr. Holmes has described opportunities for quarrying roofing slate near Patten. Our own scientific researches have led us to define more closely the limits of the roofing slate belt, upon which the best quarries are located, from Patten to Pleasant Ridge on the Kennebec river. We found an excellent place for a quarry on Moses P. Townshend's farm in Pleasant Ridge; and others might be specified in Foxcroft, Barnard, Sebec, Kingsbury, Howard and Bowerbank. And there must be others in the vicinity.

The conclusion which we have derived from a second year's exploration of the State, is that when her mineral resources shall have become fully known, every one will be astonished at their immense extent and value.

INDEX.

INDEX.

www.ingramcontent.com/pod-product-compliance
Lightning Source LLC
LaVergne TN
LVHW090545110826
845146LV00001B/31

* 9 7 8 1 4 2 5 5 4 9 7 3 2 *